S0-CAS-502

The Nature of the
Surface Chemical Bond

The Nature of the Surface Chemical Bond

Editors

T.N. Rhodin

School of Applied and Engineering Physics,
Cornell University, Ithaca, New York

G. Ertl

Institut für Physikalische Chemie der
Universität München,
Munich, F.R. Germany

1979

NORTH-HOLLAND PUBLISHING COMPANY
AMSTERDAM · NEW YORK · OXFORD

ISBN: 0 444 85053 8

Published by:

North-Holland Publishing Company – Amsterdam, New York, Oxford

Sole distributors for the U.S.A. and Canada:

Elsevier North-Holland, Inc.
52 Vanderbilt Avenue
New York, N.Y. 10017

Library of Congress Cataloging in Publication Data

Rhodin, Thor N
The nature of the surface chemical bond.

Includes indexes.
1. Chemical bonds. 2. Surface chemistry.
I. Ertl, Gerhard, joint editor. II. Title.
QD461.R46 541'.224 78-25676
ISBN 0-444-85053-8

PRINTED IN THE NETHERLANDS

This book is dedicated to the work of H.S. Taylor and G.M. Schwab who, in both teaching and research, contributed substantially to the concepts of surface structure and chemical reactivity, the microscopic details of which are now in a major state of study and clarification.

PREFACE

The principles which define the structure of and interactions among atoms and molecules at interfaces in solids is basic to understanding the behavior of matter. These phenomena are succinctly embodied in the concept of chemical bonding at surfaces. Pauling in his famous contribution, "The Nature of the Chemical Bond", fully realized the importance of this for molecular phenomena. In a sense, his famous publication anticipated a new stage in the understanding of physical chamical behavior in terms of microscopic concepts. The usefulness of this approach, however, is not limited to the repetoire of the chemist but identifies with many of the concepts of physics in the realm of fundamental interactions among particles. Expansion in the understanding of the nature of bonding between atoms in molecules was founded strongly on the cornerstone of quantum chemistry and flowered in the unifying principles it provided in rationalizing a great body of thermodynamic and kinetic information. This in turn resulted from the expansion of precise experimental microscopic measurements of the physical chemistry of molecules, which also started about that time.

A corresponding growth of factual information on the microscopic properties of atoms and molecules associated with surfaces of liquids and solids was initiated in the 1950's and has grown rapidly since then. A critical feature of this body of information is that it represents a reliable and systematic body of facts on the microscopic behavior of a fascinating class of materials, e.g. atoms and molecules arranged in localized clusters and long range two and three dimensional arrays both at free surfaces and at interfaces. It is not surprising that the powerful spectroscopic instrumentation of this period has been extensively applied to the study of properties of atoms and molecules at external surfaces. This class of materials can be effectively prepared and characterized and is at the same time most accessible to this type of measurement. The flowering of surface science which started about the middle of this century has depended, to a great extent, on the previous body of quantum mechanical understanding of free molecules. It benefited, in addition, from the subsequent happy confluence of certain additional experimental, theoretical, and practical developments. The experimental stage began with the ultra-high vacuum physical techniques initiated by the brilliant work of Langmuir, Dushman, Alpert and others. This subsequently was reinforced by the development of effective methods for the preparation and characterization of a broad class of single crystal surfaces of metals, semiconductors and insulators.

A major shift forward in this progression was associated with a transformation from the study of surface processes involving *macroscopic* parameters, such as volume, mass, and energy change per mole, to those which probed *microscopic* parameters such as energy and structure associated with partial monolayers of adsorbed gases. Two of the several important experimental techniques which heralded the advent of the new look in surface science, were thermal desorption spectroscopy and low-energy electron-diffraction. It is too easy perhaps at this time to be over aware of the uncertainties associated with these approaches and to forget the tremendous impact they had in the dramatic shifting of gears from a macroscopic to a microscopic methodology in surface physics and chemistry. The development of other spectroscopy techniques involving the interaction of electrons and photons with solid surfaces, nucleating from these beginnings, has in many ways dwarfed them in complexity and scope. A most natural outcome of this flood of high quality microscopic experimental facts has been an expansion in the areas of theoretical solid state surface physics and quantum chemistry. An independent but additional important factor in the expansion of surface science was the rather belated recognition by government and industry of the enormous practical benefits to be gathered from its application to pressing ecological and community challenges relating to a wide variety of problems including energy conservation and distribution, nutritional quality, industrial catalytic processes, and automotive emission control.

Thus the science and application of surface phenomena is rapidly coming of age, as is dramatically indicated by a mushrooming expansion in the publication of journals, monographs and books on the subject including many excellent proceedings, reviews and treatises. Availability of texts integrated about a central theme in surface science at the graduate student or research scientist level, however, has not kept up with the press of current research progress. The objective of this writing is to provide such a monograph centered on the theme of the nature of the surface chemical bond. It is our intent to provide an effective perspective upon which to form a valid basis for the interpretation of characteristic physical configurations and chemical states of electrons, atoms, and molecules in, near, or at the boundary layer between a solid and a gas.

An approach to the subject of surface science from the viewpoint of the nature of chemical bonding provides an effective basis upon which to establish a useful set of general principles relating to electron and atomic structure and energetics of solid surfaces. Such is the primary objective of this volume. Although this approach can provide a useful point of departure for establishing an informed perspective on the subject, it is clear that many gaps still exist in our knowledge. A text dealing with the field of surface chemical bonding at this time is, in one sense, more an account of potential achievement rather than a documentation of past accomplishments. Nevertheless, development of a general interpretative approach conducive to providing a critical perspective is our main objective. It is also intended to respond to an important need to render this material in a well integrated, indexed and referenced manner.

An outline of the treatment follows. Grimley starts out with theoretical considerations of molecular orbital schemes needed to describe the electronic features of clean and chemisorbed metal surfaces. This provides the general formulation rigorously derived from principles in solid state physics. Of the several proposed molecular orbital schemes, the density functional formalism due to Hohenberg and Kohn, and Kohn and Sham, has provided a remarkably effective basis for useful calculations of chemisorption binding energies on simple metals. The long used and probably best-known molecular orbital scheme for describing the electron structure of clean metals in both the bulk and surface configuration, namely the Hartree–Fock theory, is subsequently discussed in detail in terms of both the dynamic Hartree–Fock and the multi-configurational Hartree–Fock formulations. The implications of these and other molecular orbital schemes in connection with chemisorption are presented with reference to concepts of band structure and chemical bonding in terms of the tight-binding, muffin-tin and jellium models for metallic solids. This approach is then extended to applications involving embedded clusters and the versatility of using the Hamiltonian description as originally suggested by Anderson. The molecular orbital description is finally concluded with a consideration of its relevancy to the important effect of long range interactions among atoms and molecules in overlayers.

One is forced to bring together the theoretical techniques of solid state physics with those of quantum chemistry if one is to address properly the problem of the electronic structure of clean metal surfaces and adsorbate behavior. Messmer subsequently considers the usefulness of the so-called "independent particle" approach in a theoretical treatment which attempts to emphasize more strongly the rôle of the adsorbate.

This is done by combination of the physical and chemical concepts, within certain restricted aspects, of electron structure theory. Applications of this viewpoint are therefore discussed, first with reference to the usefulness of metal complexes as surface models, and then in applications of the cluster model to the definition of the electron structure of atomic and molecular chemisorption for several appropriate gas–metal adsorption systems.

The electron spectroscopic methods have been among the most useful of the multitude of experimental methods which have become widely used in studies of chemisorptive bonding on solids. It is useful in making a transition in our presentation of chemical bonding concepts from a theoretical to an experimental approach to consider the theoretical framework upon which a description of some of these spectroscopies is based and their subsequent application to the interpretation of electronic features of clean and chemisorbed metal surfaces. This is done in the third chapter by Rhodin and Gadzuk through a presentation of the theoretical formulation and interpretation in some detail of relevant principles involving three of the most effective among the electron spectroscopies, namely; photoemission, Auger emission and the electron energy loss methods. This is followed with an extensive consideration of the applications of these methods in combination with appropriate additional data on low-energy electron-diffraction and adsorption energetics to a variety of experimental applications, from clean metal surfaces, to both atomic and molecular adsorption and to more complex adsorption systems relating to chemisorption and reaction of hydrocarbons on transition metals. An emerging clarity in the interpretation of chemisorptive bonding on surfaces of both simple and transition metals resulting from the use of a combination of measured parameters from different experimental sources for selected cases provides encouragement that some such broad-based set of interpretative principles will eventually be established. The development of a monolithic mass of published data on surface behavior makes it apparent that interpretation based on generalized principles must be developed to cope with these advances in new information. This challanging requirement can only be assuaged in part by the increasing availability of critical reviews and monographs. The importance of developing unifying principles derived from bulk physics and chemistry to the ordering of our understanding of the surface chemical bond is clearly indicated.

Systemization of detailed experimental data in terms of unifying principles is carried one step further in the treatment presented in the fourth chapter by Van Hove, of chemical bonding at surfaces through recent advances in our knowledge of surface crystallography. No experimental method has provided such a wealth of microscopic data on long range ordering on surfaces as that derived from low-energy electron-diffraction. This was anticipated by Farnsworth, Germer and Lander in their historic studies of ordered surface chemical overlayers. It continues to be valid today inspite of some recognized experimental shortcomings and theoretical limitations intrinsic in the present state-of-the art. Based on the assumption that the body of structural data developed from low-energy electron-diffraction work will prove to be essentially substantiated by newer complementary data, important general principles on bond length, bond angle and charge transfer can be inferred. Some of the conclusions in this regard, although tentative in some cases due to the limited data base, pose the encouraging observation that studies of surface crystallography are advancing into a new stage of interpretative validity with reference to the nature of chemical bonding at surfaces.

An additional challanging aspect of the nature of the surface chemical bond deals with the microscopic features of the energetics of chemisorption. Important critical questions raised by Ertl in his treatment of this subject in the last chapter keynote some of the basic factors to be considered. It is important that the connection between the energetics of adsorption, the effects of surface chemical interactions and the nature of surface chemical bonding all be defined in microscopic detail. Particular attention is focussed on the analogy between chemical bonding in chemical complexes and metal molecular clusters and that of extended surfaces to clarify the importance of local coordination and bond formation. These considerations are extended to the interpretation of the "ensemble" and "ligand" effects for chemisorption on alloy surfaces. Classification of information on the effects of surface orientation and surface step structures is included. The important factor of variation of the adsorption energy across the surface is subsequently considered in terms of predicted calculations and observations taken from infrared spectroscopy. Finally, this interpretative approach is applied to a discussion of simple dissociative reactions on surfaces, the properties of ordered adsorbed phases and thermodynamic properties.

It is clear that this monograph is a beginning in the continuing effort to document an important subject. It is likely to experience much refinement in the foreseeable future. Although most of the examples are taken from gas–metal systems, many of the principles are equally applicable to gas–semiconductor and gas–insulator systems. A great deal more systematic data on the microscopic parameters associated with surface chemical bonding are required before the subject can be completely documented. An important additional extension of these concepts in the future for example, will inevitably be related to chemical bonding at *internal surfaces*; a study area of great academic and practical importance.

Sincere appreciation is extended to our fellow authors as well as to our colleagues and students for the benefit of many fruitful discussions with them and for their making research results available to us prior to publication. We are also sincerely grateful for the patient support of our secretaries and drafting people in the preparation of the manuscript. One of us (TNR) is particularly thankful for the support of the National Science Foundation and of the Cornell Materials Science Center, while the other (GE) gratefully acknowledges support by the Deutsche Forschungsgemeinschaft (Sonderforschungsbereich 128).

Thor Rhodin
Ithaca, New York 14853
September 7, 1978

Gerhard Ertl
Münich, Germany
September 7, 1978

CONTENTS

CHAPTER 1

METAL SURFACE BONDING

T.B. GRIMLEY

Donnan Laboratories, University of Liverpool, Liverpool, England

Contents

The nature of the surface chemical bond
Edited by Th.N. Rhodin and G. Ertl

1. Introduction

The traditional attitude of quantum chemists is that most chemical bonds can be *understood* at the level of an orbital approximation, although quantitative agreement with experiment, and the development of predictive ability will usually require a configuration interaction (CI) calculation. The molecular orbital (MO) scheme based on the Hartree–Fock (HF) approximation is still the most familiar orbital approximation, and it is therefore very interesting to note that it does not contain in any obvious way the bonds and bond properties so central to chemistry. This is a famous problem in chemistry. The HF MO's are usually obtained as symmetry orbitals (i.e. they belong to particular irreducible representations of the point group of the molecule). But for closed-shell molecules, a unitary transformation of the *occupied* MO's leaves all measurable quantities unaltered, and therefore, by suitably choosing this transformation, the localized bond properties can be brought into prominence (Lennard-Jones and Pople 1950, Hall 1950, see also Coulson 1970). Methane is a classic case. The eight valence electrons go into four symmetry MO's, $1a_1$, and the three degenerate orbitals, $1t_{1x}$, $1t_{1y}$ and $1t_{1z}$, but a unitary transformation can take these four MO's into four equivalent orbitals, in each of which the electron density is localized to a large extent along a particular C–H bond. This new description is entirely equivalent to the old one as far as observable quantities are concerned, but we emphasize that orbital energies are not observable, and consequently to identify ionization energies with the negative orbital energies (Koopmans' theorem) is usually a better approximation for symmetry MO's than for localized ones.

The metals, and the electron-deficient boron hydrides are well-known examples of the failure of ball-and-stick models of chemical bonds, and this failure is reflected in the fact that the unitary transformation to give localized orbitals does not work. In both the metals, and the boron hydrides, we find atoms with coordination number much larger than their normal valency. Boron, normal valency three, has coordination number 6 in $B_{10}H_{14}$, and sodium, normal valency one, has coordination number eight in the metal. Of course these systems present no difficulties when we use symmetry MO's to describe their electronic structures since these MO's allow quite naturally the multi-centre bonds which is the key concept for these systems. But as already mentioned, no unitary transformation of the occupied MO's can take them into two-centre bond orbitals; there are not enough occupied MO's to provide the required bonds. In B_2H_6 for example, we need nine bond orbitals, but we have only eight occupied MO's, and in Na metal we need four bond orbitals per atom, but have only one occupied MO per atom.

For a metal we are committed to multi-centre bonds once we decide to construct Bloch MO's. These are eigenfunctions of the translation operators defining the lattice, and they describe electrons distributed equally between every unit cell. We note however that this description fails as the lattice parameter is increased. With Na metal for example, we know that, as the lattice is uniformly dilated, the ground state eventually goes over to a collection of Na atoms with the valency electrons in localized 3s atomic orbitals, not doubly occupying the lowest de-localized Bloch MO's. However, it is not difficult to ensure the correct dissociation products (Na atoms) using Bloch MO's. What we have to do is to drop the implicit assumption that the Bloch orbitals are occupied by electrons with $\uparrow$ and $\downarrow$ spins, but to allow instead the Bloch orbitals for $\uparrow$ and $\downarrow$ spins to be different. The ground state for a large lattice parameter can then be correctly obtained as the anti-ferromagnetic ordering of spins between the two simple cubic sub-lattices.

At this point we consider briefly a question which has been asked since the earliest days of quantum chemistry. Are the (known) electronic structures of the atoms convenient starting points for the discussion of the electronic structures of molecules and solids? With obvious exceptions such as the solid rare gases, and in spite of the fact that the first successful theory of the covalent bond in H_2 by Heitler and London (1927) took H atoms as its starting point, the answer to this question is generally "No" if we are interested in the situation near the equilibrium bond lengths, because then the system is, in a general sense, on the "metal" side of the Mott (1958) transition, where the MO description is the natural one. This is true even for rare-gas compounds. We conclude therefore that, for studying the electronic structures of metal surfaces in the first place, and secondly for the chemisorption bond, that an MO scheme is an appropriate starting point. We remark further, that, for studying systems near their dissociation limits (transition states in catalysis perhaps), where the Heitler–London scheme would certainly be appropriate, a spin polarized MO scheme works perfectly satisfactorily (Gunnarsson and Johansson 1975).

In this chapter we shall be concerned entirely with MO schemes, and although the Hartree–Fock theory, first in its single configuration form, and then in its multi-configuration, and therefore exact form, is still probably the best-known MO scheme, there are two others which are better suited to the discussion of the large systems one has to deal with in studying metal surfaces and chemisorption. One is the density functional (DF) formalism which has been the basis for some good calculations of chemisorption binding energies on simple metals (see subsect. 6.4), the other is the dynamic Hartree–Fock (DHF) theory which has distinct advantages in chemisorption theory over the multi-configuration Hartree–Fock (MCHF) approach. Both these schemes, like MCHF, are in principle exact.

2. The density functional formalism

This formalism, which is due to Hohenberg and Kohn (1964) and Kohn and Sham (1965), reduces the N-electron problem to the solution of a one-particle Schrödinger equation with a potential energy which would have to be calculated self-consistently from the solutions of the one-particle equation. The potential energy is a functional of the electron density, but this functional is unknown because all the complications of the N-electron problem are contained in it. However, there are very useful approximations to it (local density approximations, see below) which enable good calculations to be made of the electronic structure and bonding at the surfaces of simple metals (subsects. 5.4 and 6.4).

Hohenberg and Kohn (1965) and Kohn and Sham (1965) considered a system of electrons in its ground state (assumed non-degenerate) in a static external potential $w(\boldsymbol{r})$ (usually that of the nuclei), and showed that the ground-state energy, excluding the nuclear repulsion, is a universal functional $E_0[n]$ of the electron density $n(\boldsymbol{r})$;

$$E_0[n] = \int \mathrm{d}\boldsymbol{r}\, w(\boldsymbol{r})n(\boldsymbol{r}) + T[n] + V[n]$$

where $T[n]$ and $V[n]$ are the expectation values of the kinetic and potential energies of the electrons. It is usual to separate the Coulomb energy and write

$$V[n] = \tfrac{1}{2}\int \mathrm{d}\boldsymbol{r}\,\mathrm{d}\boldsymbol{r}'\frac{n(\boldsymbol{r})n(\boldsymbol{r}')}{|\boldsymbol{r}\quad\boldsymbol{r}'|} + E_{\mathrm{xc}}[n]$$

so that E_{xc} describes all other contributions (the so-called exchange and correlation contributions) to the potential energy. The variation principle implies that the functional derivative

$$\frac{\delta}{\delta n(\boldsymbol{r})}(T[n]+V[n])=\mu-w(\boldsymbol{r}) \tag{1.1}$$

where μ is a Lagrange multiplier associated with the constancy of the electron number N. In a large system, μ is the chemical potential of the electrons. Kohn and Sham (1965) used eq. (1.1) to obtain a one-body form of the many-electron problem. They define a one-body potential

$$\begin{aligned} v_{\text{eff}}[n;\boldsymbol{r}] &= v_{\text{H}}(\boldsymbol{r})+\frac{\delta E_{\text{xc}}[n]}{\delta n(\boldsymbol{r})} \\ v_{\text{H}}(\boldsymbol{r}) &= w(\boldsymbol{r})+\int \mathrm{d}\boldsymbol{r}'\frac{n(\boldsymbol{r}')}{|\boldsymbol{r}-\boldsymbol{r}'|} \end{aligned} \tag{1.2}$$

and then show that by solving self-consistently the equations*

$$\begin{aligned} (-\tfrac{1}{2}\Delta^2+v_{\text{eff}}[n;\boldsymbol{r}])\psi_\alpha(\boldsymbol{r}) &= \varepsilon_\alpha\psi_\alpha(\boldsymbol{r}) \\ n(\boldsymbol{r}) &= \sum_{\alpha=1}^{N}|\psi_\alpha(\boldsymbol{r})|^2 \end{aligned} \tag{1.3}$$

the sum on α being over the lowest N states, the correct electron density $n(\boldsymbol{r})$ is obtained. The correct ground-state energy is also obtained as

$$E_0=\sum_{\alpha=1}^{N}\varepsilon_\alpha-\tfrac{1}{2}\int \mathrm{d}\boldsymbol{r}\,\mathrm{d}\boldsymbol{r}'\frac{n(\boldsymbol{r})n(\boldsymbol{r}')}{|\boldsymbol{r}-\boldsymbol{r}'|}-\int \mathrm{d}\boldsymbol{r}\,n(\boldsymbol{r})v_{\text{xc}}(\boldsymbol{r})+E_{\text{xc}}[\text{n}] \tag{1.4}$$

with

$$v_{\text{xc}}(\boldsymbol{r})=\delta E_{\text{xc}}[n]/\delta n(\boldsymbol{r}). \tag{1.5}$$

It remains therefore to find the universal function $E_{xc}[n]$, but this is of course a very difficult task.

It has often been emphasized (see Hedin 1974 for example) that the similarity between eqs. (1.3) and the HF model can be a source of confusion. There is no theoretical reason to expect the one-particle wavefunctions and eigvenvalues of eqs. (1.3) to have the meanings that have come to be associated with those of the HF model. In particular, the eigenvalues ε_α should not be regarded as approximations to the true ionization and affinity levels of the N-electron system as they are in the

* Atomic units are used, however energies will be given in Rydbergs or eV.

HF model. This is important. The true ionization and affinity levels are the eigenvalues of another one-particle equation (the DHF equation with an energy-dependent electron self-energy operator), and it is because the familiar HF equation, with its energy-independent self-energy operator, is a useful approximation to the DHF equation, that its eigenvalues can approximate the true ionization and affinity levels (see sect. 3). Approximate versions of the DF one-particle equations (1.3) may have eigenvalues close to those of the HF equation because the large term in the potential, the Coulomb or Hartree term v_H, is present in both equations, but there is no theoretical reason why the eigenvalues of the DF and DHF equations should coincide.

The most commonly used approximate version of the DF formalism is the local density approximation defined by

$$E_{xc}[n] \simeq \int d\boldsymbol{r}\, \varepsilon_{xc}\{n(\boldsymbol{r})\}n(\boldsymbol{r}) \tag{1.6}$$

where $\varepsilon_{xc}(n)$ is the exchange–correlation energy per electron of an electron gas of uniform density n. From eqs. (1.2) and (1.6)

$$v_{eff}[n;\boldsymbol{r}] \simeq v_H(r) + \frac{d\varepsilon_{xc}(n)n}{dn}. \tag{1.7}$$

Separating the exchange energy we write

$$\varepsilon_{xc}(n) = -\tfrac{3}{4}(3n/\pi)^{1/3} + \varepsilon_c(n) \tag{1.8}$$

where $\varepsilon_c(n)$, the correlation energy per electron of a uniform electron gas of density n, is usually taken as some interpolation between the low-density formula of Wigner (1938), and the high-density one of Gell-Mann and Brueckner (1957), for example

$$\varepsilon_c(n) = -0.056n^{4/3}/(0.079 + n^{1/3}). \tag{1.9}$$

Equations (1.7)–(1.9) lead to

$$v_{eff}[n;r] = v_H(r) - \left(\frac{3n}{\pi}\right)^{1/3} \frac{0.056n^{2/3} + 0.0059n^{1/3}}{(0.079 + n^{1/3})^2} \tag{1.10}$$

which we can compare at once with the popular $X\alpha$ potential

$$v_{eff}^{X\alpha}[n;\boldsymbol{r}] = v_H(\boldsymbol{r}) - 6\alpha(3n/\pi)^{1/3}. \tag{1.11}$$

The local density approximation certainly provides a good description of the electronic structures of simple (i.e. non-transition metal) metal surfaces, and of their interactions with first row atoms (Lang 1976). For dealing with weak chemisorption, and with chemical reactions on metal surfaces where original bonds are being broken and new ones made, it will generally be necessary to allow the electron density $n(\boldsymbol{r})$ to be different for different spins, and to introduce therefore an index σ, either $\uparrow$ or $\downarrow$, and thus to go over to the spin density functional (SDF) formalism (Kohn and Sham 1965, Stoddart and March 1971, Gunnarsson et al. 1972, Gunnarsson 1975).

3. The dynamic Hartree–Fock theory

Roothaan's (1960) equations for the HF MO's expressed in terms of a basis set $\{\phi_i\}$ of non-orthogonal atomic orbitals are familiar. In matrix form

$$\boldsymbol{\psi} = \boldsymbol{\phi A} \tag{1.12}$$

and

$$(\varepsilon \boldsymbol{S} - \boldsymbol{F})\boldsymbol{A} = \boldsymbol{0} \tag{1.13}$$

where $\boldsymbol{S}$ is the overlap matrix for the basis set, and $\boldsymbol{F}$ is the HF matrix. Also there is a HF Green function matrix $\tilde{\boldsymbol{G}}$ (the tilde is to remind us that the basis set is non-orthogonal) which satisfies the equation

$$(\varepsilon \boldsymbol{S} - \boldsymbol{F})\tilde{\boldsymbol{G}} = \boldsymbol{1} \tag{1.14}$$

and whose poles, which are the roots of the secular equation

$$\det|\varepsilon \boldsymbol{S} - \boldsymbol{F}| = 0 \tag{1.15}$$

are the familiar orbital energies. Through Koopmans' (1933) theorem, these orbital energies were long ago regarded as approximations to the ionization and affinity levels, but we now know that the poles of the exact Green function matrix would be the exact ionization and affinity levels (i.e. the single hole and single particle excitations), and the poles of the HF Green function matrix are naturally approximations to these quantities. However, the fact that the poles of the HF Green function matrix are the Koopmans ionization and affinity levels can be confusing.

In eq. (1.14), the Green function matrix is closely connected with the resolvent operator of differential equation theory. But an equation formally identical with (1.14) defines a much more complicated entity, namely the matrix of the Fourier time transforms of Zubarev's (1960) double time Green function defined on the same non-orthogonal basis set $\{\phi_i\}$ (Grimley 1974, see also Bonch-Bruevich and Tyablikov 1962). But $\boldsymbol{F}$ is now a very complicated operator which is energy dependent because it contains the self-energy $\boldsymbol{M}(\varepsilon)$;

$$\boldsymbol{F}(\varepsilon)=\boldsymbol{H}_0+\boldsymbol{M}(\varepsilon) \tag{1.16}$$

where $\boldsymbol{H}_0$ is the matrix of the Schrödinger operator for a single electron moving in the field of the nuclei. The real space operator corresponding to the matrix $\boldsymbol{F}(\varepsilon)$ is written $\hat{F}(\boldsymbol{r};\varepsilon)$, and has been called (Hedin 1974) the dynamic HF operator, but an equally appropriate name is the Dyson operator. Its right and left eigenfunctions $\{u_\alpha(\boldsymbol{r};\varepsilon)\}$ and $\{u^+_\alpha(\boldsymbol{r};\varepsilon)\}$ form a bi-orthogonal set;

$$\begin{aligned}\hat{F}(\boldsymbol{r};\varepsilon)u_\alpha(\boldsymbol{r};\varepsilon)&=E_\alpha(\varepsilon)u_\alpha(\boldsymbol{r};\varepsilon)\\ u^+_\alpha(\boldsymbol{r};\varepsilon)\hat{F}(\boldsymbol{r};\varepsilon)&=E_\alpha(\varepsilon)u^+_\alpha(\boldsymbol{r};\varepsilon),\end{aligned} \tag{1.17}$$

$$\int \mathrm{d}\boldsymbol{r}\, u^+_\alpha(\boldsymbol{r};\varepsilon)u_\beta(\boldsymbol{r};\varepsilon)=\delta_{\alpha\beta}. \tag{1.18}$$

In terms of these functions, the one-electron Green function is given by

$$G(\boldsymbol{r},\boldsymbol{r}';\varepsilon)=\sum_\alpha \frac{u_\alpha(\boldsymbol{r};\varepsilon)u^+_\alpha(\boldsymbol{r}';\varepsilon)}{\varepsilon-E_\alpha(\varepsilon)} \tag{1.19}$$

and it is therefore diagonal in this representation;

$$G(\alpha,\beta;\varepsilon)=\delta_{\alpha\beta}/[\varepsilon-E_\alpha(\varepsilon)]. \tag{1.20}$$

If the equation

$$\varepsilon_\alpha=E_\alpha(\varepsilon_\alpha) \tag{1.21}$$

is satisfied for some energy ε_α, then G has a pole at ε_α with residue η_α say, and from the spectral resolution of G we know that these poles are the exact ionization and affinity levels of the N-electron system. The functions $u_\alpha(\boldsymbol{r};\varepsilon_\alpha)$ which satisfy eqs. (1.17) and (1.21) are known in the quantum chemistry literature (see for example Nerbrant 1975 and references therein) as Dyson orbitals. In terms of these orbitals, the electron density is expressed as

$$n(\boldsymbol{r})=\sum_{\alpha=1}^{N}\eta_\alpha u_\alpha(\boldsymbol{r};\varepsilon_\alpha)u^+_\alpha(\boldsymbol{r};\varepsilon_\alpha) \tag{1.22}$$

the sum on the α being over the N lowest eigenvalues ε_α. Finally, the formula for the ground-state energy is

$$E_0=\frac{1}{2\pi\mathrm{i}}\int_C \mathrm{d}\varepsilon(\varepsilon\,\mathrm{Tr}\,\boldsymbol{S}\tilde{\boldsymbol{G}}+\mathrm{Tr}\,\boldsymbol{H}_0\tilde{\boldsymbol{G}}) \tag{1.23}$$

where the contour C consists of the real axis, and an infinite semicircle in the upper half-plane. We note that eq. (1.23) assumes its simplest form in the representation afforded by the functions u because then

$$\boldsymbol{S} = \mathbf{1}, \qquad \tilde{G}(\alpha, \beta; \varepsilon) = G(\alpha, \beta; \varepsilon) = \delta_{\alpha\beta}/[\varepsilon - E_\alpha(\varepsilon)]$$

so that

$$E_0 = \sum_{\alpha < \varepsilon_F} \eta_\alpha[\varepsilon_\alpha + H_0(\alpha, \alpha)]. \tag{1.24}$$

The HF model provides a simple version of the above general theory where the operator $\boldsymbol{M}$ in eq. (1.16) is energy independent, and consequently the residues η_α are all unity. We note that these residues are in fact the amplitudes of the lines in the single hole excitation spectrum of the system.

There is a considerable literature on better-than-HF self-energy operators (see for example Devreese et al. 1974, Nerbrant 1975), but so far no experience has been gained with the dynamic HF (DHF) method in chemisorption theory. An important reason for introducing the DHF formalism into chemisorption theory is the following. If chemisorption is a localized phenomenon, a sensible way to treat it is with the embedded cluster model where a refined model is used over the cluster, and a relatively crude one elsewhere (Grimley and Pisani 1974, Grimley 1974, 1976). Now even if we start out with an energy-independent self-energy $\boldsymbol{M}$ over the cluster, embedding the cluster into the rest of the substrate introduces an energy-dependent complex addition to $\boldsymbol{M}$ (embedding is in effect described by an optical potential over the cluster). Thus there are no essentially new computational problems if we use an energy-dependent self-energy over the cluster from the start. A better-than-HF model over the cluster can therefore be handled in a natural way using a self-energy $\boldsymbol{M}(\varepsilon)$. By comparison it seems much more difficult to embed a cluster in which one goes beyond HF by a CI circulation.

4. The multi-configuration Hartree–Fock theory

The solution of eqs. (1.12) and (1.13) leads to an orthogonal set of HF MO's. For a closed-shell system the HF ground state Ψ_0 is obtained by doubly occupying the lowest MO's. But of course states of the same symmetry as Ψ_0 exist among the excited configurations where there are vacant MO's below the highest occupied one. If we call these states Ψ_1, $\Psi_2, \ldots$, then the expansion

$$\Psi = a_0\Psi_0 + a_1\Psi_1 + \cdots \tag{1.25}$$

with coefficients a_0, a_1, ... determined variationally, will give an exact manifold of N-electron states. Unfortunately the series (1.25) converges too slowly for practical use because the Ψ_i, being orthogonal, interact very little. But instead of determining the MO's variationally to give the best Ψ_0 [this is what we do when we solve eqs. (1.12) and (1.13)] and then constructing the series (1.25), we can use the variation principle to determine simultaneously the best MO's to use in the series (1.25), and the corresponding CI coefficients. The convergence is then much improved because the variation principle now ensures that the Ψ_i will overlap as much as possible (Veillard 1966, Veillard and Clementi 1967). This is the multi-configuration HF theory widely used in the quantum theory of small molecules (see Wahl and Das 1970 for a review), and already applied to a 4-atom cluster model of H_2 chemisorption by copper and nickel (Melius et al. 1976). Of course for practical calculations the series (1.25) still has to be truncated, but even so, there must be grave doubts as to whether clusters large enough to model chemisorption accurately (see ch. 2) can be handled by this method with realistic computing times.

5. Electronic structure of metal surfaces

Although it may be self-evident that we cannot expect to treat chemisorption by metals unless we have a theoretical model good enough to treat also the electronic structure of the clean metal, this does not mean that we actually have to calculate the clean surface electronic structure. We would have to do so if we were interested say in the chemisorption binding energy for all surface bond lengths up to infinity, but not perhaps if we require properties near the equilibrium surface bond length (this bond length itself, the chemisorption binding energy, the vibration frequency of the surface bond etc.). Thus when chemisorption at the equilibrium bond length is a localized phenomenon, a cluster model, or an embedded cluster model will provide a good description, although the metal cluster, lacking the translational symmetry of the clean surface clearly cannot describe all aspects of its electronic structure, and the embedded cluster will do so only in a relatively low approximation. But in general it is useful to consider the clean surface problem first, not only because in the case of overlayers the chemisorption system retains most of the features of a (possibly reconstructed) clean surface problem (subsect. 6.2), but also because, just as in the theory of molecules, some knowledge of the electronic structures of the free partners in a bond is useful in considering how bonds can form between them. We might, for example, ask the question: What is the rôle of surface states in chemisorption?

For a semi-infinite crystal, the fact that half the crystal planes are missing means that the remaining ones are not translationally equivalent so there is only a two-dimensional translation group with irreducible representations labelled by the component $\boldsymbol{k}_{\parallel}$ of the usual crystal momentum parallel to the surface. The MO's therefore satisfy Bloch's theorem in the parallel direction only;

$$\psi(\boldsymbol{k}_{\parallel}, \boldsymbol{r}+\boldsymbol{a}) = \exp(\mathrm{i}\boldsymbol{k}_{\parallel}\cdot\boldsymbol{a})\psi(\boldsymbol{k}_{\parallel}, \boldsymbol{r}) \qquad (1.26)$$

where $\boldsymbol{a}$ joins equivalent atoms in the surface plane. But Bloch's theorem is not required from plane to plane going into the crystal, and consequently MO's for which

$$|\psi(\boldsymbol{k}_{\parallel}, \boldsymbol{r}+n\boldsymbol{c})|^2/|\psi(\boldsymbol{k}_{\parallel}, \boldsymbol{r})|^2 \to 0 \qquad \text{as } n\to\infty \qquad (1.27)$$

can exist. Here $\boldsymbol{c}$ joins equivalent atoms in consecutive planes parallel to the surface. MO's for which eq. (1.27) is true are surface states; we do not distinguish surface and sub-surface states as some authors do. Of course, most MO's do not satisfy eq. (1.27) but instead the ratio on the left in this equation oscillates as we increase n. These are the volume states. If we normalize over the first N layers of the semi-infinite crystal, the ratio of the squared modulus of a surface state wavefunction in the surface region to that of a volume state is of the order $1:N$. But we expect there to be of the order N volume states for one surface state, and therefore the contributions of surface and volume states to the electron density, and in general to the expectation value of any one-electron operator, in the surface region are expected to be similar. One cannot therefore reach a general conclusion about the rôle of surface states in chemisorption; the situation will be different for different systems.

5.1. Elementary methods

Even simple models of semi-infinite metals are not amenable to analytic treatment. A completely solvable model is a tight binding solid with one atomic orbital per atom, and although this model was useful in early investigations (Goodwin 1939), it does not by any means exhibit the full range of phenomena to be expected in real metals. This should not surprise us; the chemistry of the first row atoms depends on their 2s and 2p valency electron structure, and for the second row we need to consider the 3s, 3p and 3d orbitals (even though the latter are unoccupied in the ground state) to understand all aspects of their chemistry.

To illustrate some important features of the electronic structure of a metal surface, the easiest thing to do now is to make Extended Hückel calculations on a chain of Al atoms using Quantum Chemistry Program Exchange (QCPE) program number 95. Such calculations will show that, even though the 3s and 3p levels are far apart in the free atom ($\varepsilon_{3s} = -10.74$ eV, $\varepsilon_{3p} = -5.73$ eV), at the nearest neighbour distance in the metal ($d_{Al-Al} = 0.286$ nm), there is strong overlap of the 3s and $3p_z$ orbitals so that the 3s and $3p_z$ bands are strongly hybridized with a hybridization gap between them, and that there are two degenerate surface states (one for each end of the chain) in this hybridization gap. The results will not be shown here, instead we show the results of a tight-binding (i.e. ordinary Hückel) calculation for a chain of 15 Al atoms with parameters exactly as in the Extended Hückel calculation;

$$\alpha_{3s} = -10.74 \text{ eV}, \qquad \alpha_{3p} = -5.73 \text{ eV}, \qquad \zeta = 1.364$$
$$H_{ij} = \tfrac{1}{2} S_{ij}(\alpha_i + \alpha_j) \tag{1.28}$$

where S_{ij} is the overlap integral. The computing time is less than 0.1 second for an ICL 1906S. In fig. 1.1 we exhibit the energy bands, the surface states, and by means of vertical lines, the AO coefficients for the even surface state. The lines on the atomic 3s level refer to the 3s AO's, those on the atomic 3p level to the $3p_z$ AO's. All MO's, including those for the surface states, are either even or odd on inversion in the centre of the chain, and for a long enough chain, the energies of the even and odd states coincide. It is clear from fig. 1.1 that the 3s and $3p_z$ orbitals are participating almost equally in the surface state MO's, which is what we would expect in a hybridization gap. We note also that the degree of s–p_z hybridization on each atom is not the same, a fact realized long ago (see for example Koutecký 1965).

There is an extensive literature on surface states beginning with Tamm (1932), and although a review exists (Davidson and Levine 1970), modern theoretical work, and practical schemes of calculation have not been adequately covered (but see Forstmann 1977).

The tight binding approximation (TBA) has been extensively used both for bulk band structures, and for surface properties of transition metals (for a recent review see Allan 1976). Most of the work has concentrated on the d bands with hybridization with the s–p bands being held over for later consideration. As we have just seen, this also means that consideration of some important surface state structure (that asso-

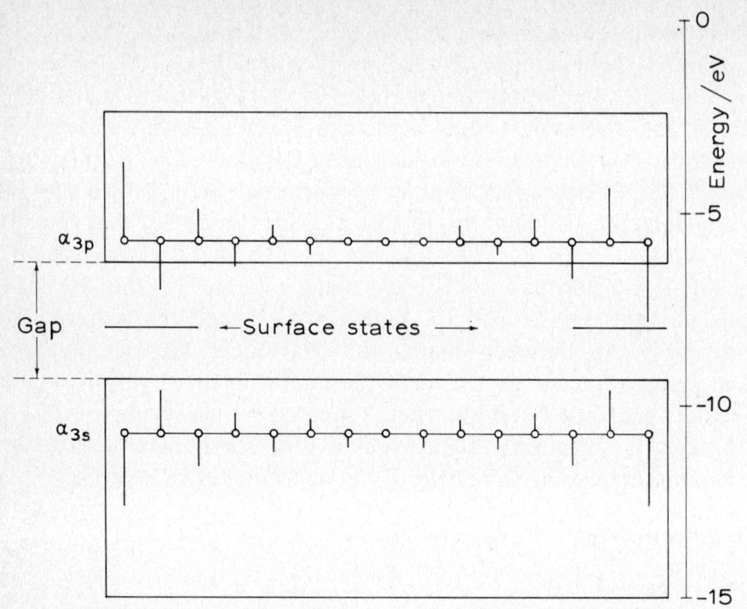

Fig. 1.1. Tight binding electronic structure of a chain of fifteen aluminium atoms. There are two bands separated by a hybridization gap, and surface states in this gap. The vertical lines represent the atom orbital coefficients in the molecular orbital of the even surface state. α_{3s} and α_{3p} are the atomic 3s and 3p levels.

ciated with hybridization gaps) is also left till later. The mathematical simplicity of the TBA equations has stimulated work on methods of looking at the electronic structures of solids from a local point of view, whereby one tries to obtain successively better and better approximations to the local density of states by including successively more and more atoms in a local model, and then terminating the series in some way. An important procedure here is the recursion method of Haydock et al. (1972, 1975).

To introduce the basic idea, consider a semi-infinite chain of atoms with one AO per atom (fig. 1.2). In the TBA, $\boldsymbol{S} = \boldsymbol{1}$ in eqs. (1.13) and (1.14), and we define the Green function matrix $\boldsymbol{G}$ in the AO representation by

$$(\boldsymbol{1}\zeta - \boldsymbol{F})\boldsymbol{G} = \boldsymbol{1}, \qquad \zeta = \varepsilon + \mathrm{i}0. \tag{1.29}$$

It has the property that ρ_i, the local density of states on the orbital ϕ_i is given by

$$\rho_i(\varepsilon) = -\frac{1}{\pi}\,\mathrm{Im}\,G_{ii}(\zeta). \tag{1.30}$$

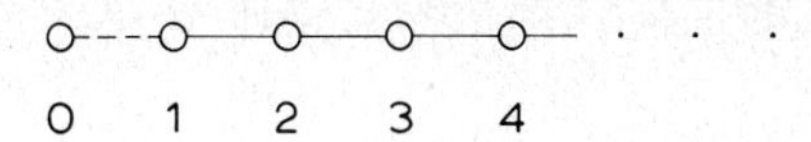

Fig. 1.2. A chain of atoms to illustrate the recursion method. Having broken the 0–1 bond, atom 1 becomes the end atom of the chain.

To calculate G_{00}, we break the $0-1$ bond to give a system with Green function matrix $\boldsymbol{G}^0$, and with atom 1 now the end atom of a semi-infinite chain. Dyson's equation relates $\boldsymbol{G}$ to $\boldsymbol{G}^0$,

$$\boldsymbol{G} = \boldsymbol{G}^0 + \boldsymbol{G}^0 \boldsymbol{V} \boldsymbol{G} \tag{1.31}$$

where $\boldsymbol{V}$ represents the $0-1$ coupling. In our case, eq. (1.31) gives a simple algebraic equation for G_{00},

$$G_{00} = G^0_{00}/(1 - G^0_{00}\beta^2 G^0_{11}) \tag{1.32}$$

where $\beta = V_{01}$ is the hopping matrix element between orbitals 0 and 1. If we take our energy zero at the atomic level $G^0_{00} = 1/\zeta$

$$G_{00} = 1/(\zeta - \beta^2 G^0_{11}). \tag{1.33}$$

But having broken the $0-1$ bond, atom 1 is in exactly the same situation as atom 0 was originally. Consequently $G^0_{11} = G_{00}$, and there is a continued fraction expansion for G_{00},

$$G_{00} = \cfrac{1}{\zeta - \cfrac{\beta^2}{\zeta - \cfrac{\beta^2}{\zeta - \dots}}} \tag{1.34}$$

but also of course the closed formula

$$G_{00} = 1/(\zeta - \beta^2 G_{00}). \tag{1.35}$$

Solving eq. (1.35) for G_{00}, and choosing the sign of the square root to make $\mathrm{Im}\, G_{00} \leq 0$ so as to agree with eq. (1.30) we have

$$G_{00}(\varepsilon) = [\varepsilon - \mathrm{i}(4\beta^2 - \varepsilon^2)^{1/2}]/2\beta^2, \qquad |\varepsilon| \leq 2\beta \tag{1.36}$$

so that

$$\begin{aligned} \rho_0(\varepsilon) &= (4\beta^2 - \varepsilon^2)^{1/2}/2\pi\beta^2, \qquad && |\varepsilon| \leq 2\beta \\ &= 0, && |\varepsilon| > 2\beta \end{aligned} \tag{1.37}$$

which is a well-known result.

To exploit this simple method in more general situations, the essential step is to obtain the basis orbitals (linear combinations of the original

AO's) in which the matrix $\boldsymbol{F}$ is tri-diagonal as it is in the AO basis for the linear chain. This basis is generated systematically starting with the AO ϕ_0 on which the local density of states $\rho_0(\varepsilon)$ is required. But the situation is more complicated than that for the linear chain, and in the new basis, the diagonal elements of $\boldsymbol{F}$ are all different, and so are the non-diagonal elements. If we denote the former by a_n, $n = 1, 2, \ldots$, and the latter by b_{n-1}, $n = 2, 3, \ldots$, then in place of eq. (1.35) we find the continued fraction expansion

$$G_{00} = \frac{1}{\zeta - a_1 - b_1^2 t_1(\zeta)}$$

$$t_1(\zeta) = \frac{1}{\zeta - a_2 - b_2^2 t_2(\zeta)} \tag{1.38}$$

$$t_2(\zeta) = \ldots.$$

In practice the matrix elements a_n and b_n converge quickly to constant values a and b, so the series can be terminated at the Nth stage ($N \sim 20$) by putting $a_n = a$, $b_n = b$ for $n \geq N$. This gives

$$t(\zeta) = \frac{1}{\zeta - a - b^2 t(\zeta)} \tag{1.39}$$

which is solved for $t(\zeta)$. Then, working back up the system of equations (1.38) we obtain $G_{00}(\zeta)$.

Another procedure for obtaining information on the local density of states directly is the moment method (Desjonquères and Cyrot-Lackmann 1975, and references therein). The rth moment μ_r of the local density of states $\rho_0(\varepsilon)$ is

$$\mu_r = \int \mathrm{d}\varepsilon\, \varepsilon^r \rho_0(\varepsilon) \tag{1.40}$$

so if we expand $(\mathbf{1}\zeta - \boldsymbol{F})^{-1}$

$$(\mathbf{1}\zeta - \boldsymbol{F})^{-1} = \sum_{n=0}^{\infty} \boldsymbol{F}^n / \zeta^{n+1}$$

we have from eqs. (1.29), (1.30) and (1.40)

$$\mu_r = (F^r)_{00} = \sum_{i,j,\ldots} F_{0i} F_{ij} \ldots F_{l0}. \tag{1.41}$$

In eq. (1.41), $0ij \ldots l0$ is a closed walk on the lattice with step length the nearest neighbour distance. The problem is now to construct $\rho_0(\varepsilon)$ from a knowledge of its first N moments, and here it seems that the best way to proceed is to relate the moments μ_r calculated from eq. (1.41) to the members of the set $\{a_n,b_n\}$ in eqs. (1.38), and so to get G_{00} as a continued fraction again. But this way of proceeding has the advantage that the higher moments μ_r are easier to calculate than the higher members of the set $\{a_n,b_n\}$.

5.2. The muffin-tin model

The TBA does not provide us with an ab initio method for calculating electronic structures, the elements of the matrix $\boldsymbol{F}$ in the AO representation are obtained by fitting TBA band structures to those obtained by accurate calculations. To make an accurate calculation we need to calculate the actual potentail seen by a valency electron in a metal. At the present time, the quickest way to proceed to obtain believable information on the surface, and bulk electronic structure of a transition metal is to substitute the actual one-electron potential by its muffin-tin version, write down the general solution to Schrödinger's equation, for given parallel momentum $\boldsymbol{k}_{\parallel}$, inside and outside the metal, and join the two solutions together at the metal/vacuum interface.

The large amount of work done on LEED theory using the muffin-tin model has led to a generally available suite of programs (Pendry 1974) which can also be used to calculate bulk band structures, and with some minor additions, surface state energies too. It should however be remembered that muffin-tin potentials are less satisfactory represen-

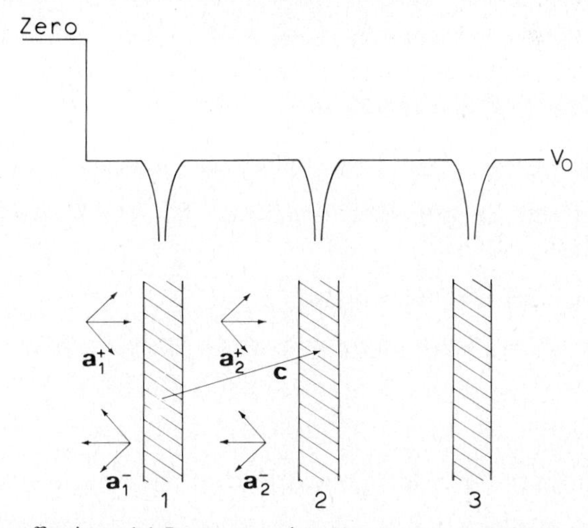

Fig. 1.3. The muffin-tin model. Beam expansions between layers, and between the surface barrier, and the first muffin tin layer are illustrated.

tations of the actual potential at surfaces than in the interior, that the surface barrier is a problem, and that there is usually no self-consistency. We now review the method in its simplest form.

The one-electron potential is that of layers of identical muffin tins parallel to the surface which are represented by a potential step of height $|V_0|$ placed at one half the layer spacing outside the first muffin-tin layer (fig. 1.3). The vector **c** joins equivalent atoms in adjacent layers, and a reciprocal vector of the surface net is denoted $\boldsymbol{g}$. All solutions of Schrödinger's equation for the semi-infinite metal have definite values of $\boldsymbol{k}_\parallel$, the parallel component of the crystal momentum. In the region of constant potential V_0 between the $(n-1)$th and nth muffin-tin layers the wavefunction with energy ε is expressed in the beam expansion

$$\sum_{g}\{a_{ng}^{+}\exp[\mathrm{i}\boldsymbol{K}_{g}^{+}\cdot(\boldsymbol{r}-n\boldsymbol{c})]+a_{ng}^{-}\exp[\mathrm{i}\boldsymbol{K}_{g}^{-}\cdot(\boldsymbol{r}-n\boldsymbol{c})]\} \tag{1.42}$$

with

$$\begin{aligned}\boldsymbol{K}_{g}^{\pm}&=[\boldsymbol{k}_\parallel+\boldsymbol{g},\ \pm(2\varepsilon-2V_0-|\boldsymbol{k}_\parallel+\boldsymbol{g}|^2)^{1/2}]\\&=(\boldsymbol{k}_\parallel+\boldsymbol{g},\ K_{gz}^{\pm}).\end{aligned} \tag{1.43}$$

Therefore, inside the crystal between the surface barrier and the first muffin-tin layer, the wavefunction, referred to an origin in the barrier is

$$\psi^{\text{in}}(\varepsilon,\boldsymbol{k}_\parallel,\boldsymbol{r})=\sum_{g}\{a_{1g}^{+}\exp(\mathrm{i}\boldsymbol{K}_{g}^{+}\cdot\boldsymbol{r})+a_{1g}^{-}\exp(\mathrm{i}\boldsymbol{K}_{g}^{-}\cdot\boldsymbol{r})\} \tag{1.44}$$

and outside the crystal where the potential is zero

$$\begin{aligned}&\psi^{\text{out}}(\varepsilon,\boldsymbol{k}_\parallel,\boldsymbol{r})=\sum_{g}a_{0g}\exp(\mathrm{i}\boldsymbol{K}_{0g}\cdot\boldsymbol{r})\\&\boldsymbol{K}_{0g}=[\boldsymbol{k}_\parallel+\boldsymbol{g},-\mathrm{i}(-2\varepsilon+|\boldsymbol{k}_\parallel+\boldsymbol{g}|^2)^{1/2}]=(\boldsymbol{k}_\parallel+\boldsymbol{g},-\mathrm{i}\lambda_{g})\end{aligned} \tag{1.45}$$

for bound states ($\varepsilon<0$). The condition that (1.44) and (1.45) join smoothly at the barrier is

$$a_{1g}^{+}=[(\lambda_g-\mathrm{i}\,\boldsymbol{K}_{gz}^{-})/(\mathrm{i}\,\boldsymbol{K}_{gz}^{+}-\lambda_g)]a_{1g}^{-} \tag{1.46}$$

and using matrix notation ($\boldsymbol{a}_1^{+}$ a column matrix with elements a_{1g}^{+}, and so on) we write

$$\boldsymbol{a}_1^{+}=\boldsymbol{J}\boldsymbol{a}_1^{-}. \tag{1.47}$$

Next we introduce the reflection matrix $\boldsymbol{R}$ of the crystal with no surface barrier

$$\boldsymbol{a}_1^{-}=\boldsymbol{R}\boldsymbol{a}_1^{+} \tag{1.48}$$

so from eqs. (1.47) and (1.48), the eigenvalue condition for the one-electron energies of the semi-infinite metal is

$$\det|\mathbf{1} - \boldsymbol{RJ}| = 0. \tag{1.49}$$

We note that $\boldsymbol{R}(\varepsilon,\boldsymbol{k}_\parallel)$ is expressed (Pendry 1974) in terms of four matrices $\boldsymbol{Q}^{\mathrm{I}} - \boldsymbol{Q}^{\mathrm{IV}}$ which embody the scattering properties of a single layer;

$$a_n^+ = \boldsymbol{Q}^{\mathrm{I}}(n)\boldsymbol{a}_{n-1}^+ + \boldsymbol{Q}^{\mathrm{II}}(n)\boldsymbol{a}_n^-$$
$$\boldsymbol{a}_{n-1}^- = \boldsymbol{Q}^{\mathrm{III}}(n)a_{n-1}^+ + \boldsymbol{Q}^{\mathrm{IV}}(n)\boldsymbol{a}_n^-$$

which we solve for $\boldsymbol{R}(n-1)$

$$\boldsymbol{R}(n-1) = \boldsymbol{Q}^{\mathrm{III}}(n) + \boldsymbol{Q}^{\mathrm{II}}(n)\boldsymbol{R}(n)[\mathbf{1} - \boldsymbol{Q}^{\mathrm{II}}(n)\boldsymbol{R}(n)]^{-1}\boldsymbol{Q}^{\mathrm{I}}(n). \tag{1.50}$$

In our simple example all layers are identical, so the dependences on the layer index n can be dropped, $\boldsymbol{R}(n-1) = \boldsymbol{R}(n)$. The reflection matrix $\boldsymbol{R}$ is calculated in LEED theory (computer programs are published, Pendry 1974), $\boldsymbol{J}(\varepsilon,\boldsymbol{k}_\parallel)$ the reflection matrix of the surface barrier, is a diagonal matrix with diagonal elements given in eq. (1.46), and the energies ε for which eq. (1.49) is satisfied are the eigenvalues of electrons in the semi-infinite crystal.

If we exhibit the bulk band structure for given $\boldsymbol{k}_\parallel$ we have a one-dimensional band system as k_z, the perpendicular component of $\boldsymbol{k}$ is varied. This is the $\boldsymbol{k}_\parallel$ sub-band system with allowed ranges of energy,

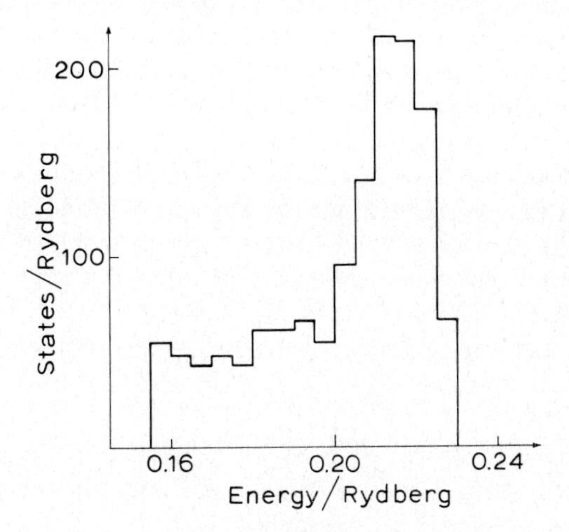

Fig. 1.4. Density of states in the surface state band on (100) Cu calculated by Gurman (1976).

and forbidden gaps. If eq. (1.49) can be satisfied for an energy ε in such a gap, then this corresponds to a surface state.

As an example of the results obtained with this model, but not using exactly the above formalism, we show in fig. 1.4 Gurman's (1976) results for the surface states on (100) Cu using nine g-values in eq. (1.42). The surface barrier height is 0.47 Hartree, and surface states, associated with an s–d hybridization gap in the $\boldsymbol{k}_{\|}$ sub-bands, are found in a narrow band 0.04 Hartree wide some 0.18 Hartree below the Fermi level. There is a band of surface states because the solution of eq. (1.49) varies with $\boldsymbol{k}_{\|}$. However, it is not necessary for a solution to be found in the hybridization gap for all values of $\boldsymbol{k}_{\|}$, and in fact such a solution exists, giving a surface state, for roughly three out of four values of $\boldsymbol{k}_{\|}$.

5.3. The chemical pseudo-potential equation

The only computational schemes from quantum chemistry which can compete with the muffin-tin model for both speed and storage requirements, are the semi-empirical Extended Hückel, and CNDO-type schemes. But these schemes are not generally regarded as having predictive ability in chemistry, and therefore, although the programming work necessary to organize them for layer-by-layer calculations has been started (Fassaert and van der Avoird 1976, Dovesi et al. 1976) (these programs are not generally available, so that only cluster calculations are generally reported), it is hard to believe that they could be as useful as the muffin-tin model in surface studies. However, we mentioned in subsect. 5.2 that the mathematical simplicity of the TBA equations has stimulated useful work on calculating local densities of states by more direct methods than by first solving a (large) secular equation and then constructing a local density of states from the eigenvalues and eigenvectors, and if these methods are to have a really secure foundation we would like to be in a position to calculate the localized orbitals for which the TBA equations would describe exactly the one-electron eigenvalue problem. This possibility was first studied by Anderson (1969).

We wish to solve the one-electron eigenvalue problem

$$\hat{F}\psi_\mu = \varepsilon_\mu \psi_\mu \tag{1.51}$$

and make the simplifying, though unnecessary, assumption that the self-consistent potential in $\hat{F}$ is the sum of atomic potentials $V_n(r)$ as it is in the muffin-tin model for example:

$$\hat{F} = -\tfrac{1}{2}\nabla^2 + \sum_n V_n(r). \tag{1.52}$$

On each atom n, localized orbitals ϕ_n are defined by a pseudo-Hamiltonian $\hat{F}^{(n)}$

$$\hat{F}^{(n)}|n\rangle = (-\tfrac{1}{2}\nabla^2 + V_n)|n\rangle + \sum_{m\neq n}[V_m|n\rangle - |m\rangle\langle m|V^{(n,m)}|n\rangle]$$
$$= E_n|n\rangle \tag{1.53}$$

E_n is the atomic eigenvalue, and $V^{(n,m)}$ is arbitrary at present.

The orbitals $\{\phi_n\}$ are assumed to be a complete set for the eigenvalue problem (1.51), so we expand

$$\psi_\mu = \sum_n \phi_n c_{n\mu}$$

so that

$$\hat{F}\psi_\mu = \sum_n c_{n\mu}E_n\phi_n + \sum_p c_{p\mu}\sum_m |\phi_m\rangle\langle\phi_m|V^{(p,m)}|\phi_p\rangle \tag{1.54}$$

if we make $V^{(n,n)} = 0$. We require the right hand side of eq. (1.54) to be equal to

$$\varepsilon_\mu \sum_n c_{n\mu}\phi_n$$

which is true if

$$E_n c_{n\mu} + \sum_p \langle\phi_n|V^{(p,n)}|\phi_p\rangle c_{p\mu} = \varepsilon_\mu c_{n\mu} \tag{1.55}$$

provided the orbitals ϕ_n are linearly independent.

The system of linear equations (1.55) has a non-trivial solution if

$$\det|(E_n - \varepsilon_\mu)\delta_{nm} + \langle\phi_n|V^{(m,n)}|\phi_m\rangle| = 0. \tag{1.56}$$

This equation is exact for all choices of $V^{(m,n)}$. If we choose $V^{(m,n)} = 0$, the ϕ_n are the MO's ψ_μ themselves; if we choose $V^{(m,n)} = \hat{F}$, the ϕ_n are Wannier functions. A useful choice is

$$V^{(m,n)} = V_n, \qquad m \neq n \tag{1.57}$$

which in eq. (1.53) leads to the so-called chemical pseudo-potential (CPS) equation

$$(-\tfrac{1}{2}\nabla^2 + V_n)|\phi_n\rangle + V^n_{\text{ps}}|\phi_n\rangle = E_n|\phi_n\rangle \tag{1.58}$$

where

$$V^n_{\text{ps}} = \sum_{m\neq n}(V_m - |\phi_m\rangle\langle\phi_m|V_m). \tag{1.59}$$

Consequently, if we write the secular equation (1.56) in the form

$$\det|\varepsilon\delta_{nm} - A_{nm}| = 0 \tag{1.60}$$

then the eigenvalues of eq. (1.58), the atomic eigenvalues in fact, are the diagonal elements of $\boldsymbol{A}$, and from the eigenfunctions ϕ_n we calculate the non-diagonal elements:

$$A_{nn} = E_n = \langle n| -\tfrac{1}{2}\nabla^2 + V|n\rangle - \sum_{m\neq n} \langle n|m\rangle\langle m|V_m|n\rangle$$

$$A_{nm} = \langle n|V_n|m\rangle. \tag{1.61}$$

Thus a tight binding secular equation (1.60) with hopping matrix elements between all atoms, but with no overlap matrix of rigorously justified. Of course we get a secular equation of the same form as (1.60) in the Wannier representation, but the usefulness of the CPS equation approach is that $V^n_{\rm ps}|\phi_n\rangle$ is small, both because $V^n_{\rm ps}$ has only contributions from atoms other than n, and because the two contributions in eq. (1.59) roughly cancel (Anderson 1969). Therefore the atomic orbitals ϕ^0_n satisfying

$$(-\tfrac{1}{2}\nabla^2 + V_n)\phi^0_n = E_n\phi^0_n \tag{1.62}$$

are approximate solutions of the CPS equation, and if we construct the matrix elements (1.61) using these orbitals, and solve the secular equation (1.60) we expect to have a useful approximation to the true eigenvalues of eq. (1.51). So far, little experience has been gained with this scheme for metal surfaces, and reference should be made to two recent papers (Bullett and Cohen 1977, Bullett 1977).

At this point we consider what advantages, if any, the CPS scheme has over the familiar LCAO–MO approach to solving eqs. (1.51) and (1.52). The latter approach leads to the eigenvalue problem

$$(\varepsilon\boldsymbol{S} - \boldsymbol{F})\boldsymbol{c} = \boldsymbol{0} \tag{1.63}$$

with real symmetric matrices $\boldsymbol{S}$ and $\boldsymbol{F}$. This eigenvalue problem is solved routinely in quantum chemistry, and there is no advantage in having $\boldsymbol{S} = \boldsymbol{1}$ if this requires us to solve the non-Hermitian eigenvalue problem posed by the CPS equation (1.58) before we can construct the matrix $\boldsymbol{A}$ to use in eq. (1.60). Indeed having to solve the CPS equation is a distinct disadvantage. Nevertheless, having done this, the matrix $\boldsymbol{A}$ is relatively easy to construct compared with $\boldsymbol{F}$ because F_{mn} involves 3-centre integrals $\langle\phi_m|V_p|\phi_n\rangle$ whereas there are no 3-centre integrals in $\boldsymbol{A}$. This is a great computational simplification to be set against the fact that the

CPS basis set is not known until we have solved a non-Hermitian eigenvalue problem, whereas the LCAO–MO basis set is given.

The familiar Extended Hückel scheme simply avoids the 3-centre integrals by using the approximation

$$F_{mn} = \tfrac{1}{2}CS_{mn}(F_{mm} + F_{nn}) \tag{1.64}$$

with $C \sim 1$, and work so far with the CPS scheme (Bullett and Cohen 1977, Bullett 1977) avoids solving the CPS equation by using the atomic orbitals ϕ_n^0 instead. Finally this is a suitable place to mention that Batra and his co-workers (see Ciraci and Batra 1977 for example, and references therein) have implemented the LCAO–MO scheme with a resolution of the multi-centre integral problem by the method first proposed by Lafon and Lin (1966). This is evidently a useful development but so far no applications to metal surfaces have been reported. Practical calculations by both methods considered in this subsection are made on thin slabs of material, the size of the secular equation for each $\boldsymbol{k}_\parallel$ being of course determined by the number of layers in the slab.

For didactic work, the simple tight binding scheme, organized for calculations at given $\boldsymbol{k}_\parallel$ on slabs of material, is most convenient. For (100)Al, a calculation on a 9-layer slab giving the energies and wave-functions of the occupied states for 210 values of $\boldsymbol{k}_\parallel$ in the surface Brillouin zone executes in $1\frac{1}{2}$ minutes on an ICL 1906S (divide by 8 for a CDC 7600), and gives energy bands, and surface states qualitatively similar to those found by Caruthers et al. (1973) who used a very carefully constructed potential for an aluminium slab.

5.4. Jelliums

In the jellium model of a semi-infinite metal, the positive ions are smeared out into a uniform charge distribution n_+ terminating abruptly at the surface, $z = 0$;

$$\begin{aligned} n_+(z) &= n_0, \qquad z < 0 \\ n_+(z) &= 0, \qquad z > 0 \end{aligned} \tag{1.65}$$

and then one tries to solve the DF one-particle equations (1.3) for interacting electrons in the field of this charge distribution. Thus, it is assumed that electron interaction effects, and self-consistency, are more important than band structure effects, the latter coming of course from the non-uniform part of the electron–ion potential, neglected in eq. (1.65). This is a useful starting point for the study of the surface properties of the simple metals where, band gaps being small, there is not too great a dependence of many surface properties on the crystal face – the jellium model of course having no crystallinity.

Most results have been obtained with the local density approximation explained in sect. 2, and as an example we show in fig. 1.5 the electron density $n(z)$ for the jellium model of aluminium. A jellium is characterized by a sphere, radius r_s, defining the volume per valency electron,

$$\tfrac{4}{3}\pi r_s^3 = 1/n_0. \tag{1.66}$$

r_s is between 2 and 6 A.U. for simple metals ($r_s = 2.07$ for aluminium).

The early importance of the jellium model was due to the fact that, with a properly self-consistent $v_{eff}[n;\boldsymbol{r}]$ in eq. (1.3), a work function calculation could be contemplated. In this connection it is important to note that, although the solutions of eq. (1.3) do not represent quasi-particle states, nevertheless, the work function φ *is* connected with the energy ε_F of the highest occupied state in eq. (1.3) exactly as one would expect if they were (Schulte 1974), namely $\varphi = -\varepsilon_F$.

Crystal structure effects can be introduced into the jellium model by perturbation theory. $\Delta V(\boldsymbol{r})$ being the difference between the electron–ion (periodic) pseudo-potential, and the potential of the uniform positive background of the jellium model, and $\Delta n(\boldsymbol{r})$ the change in the electron density when one electron is removed from the jellium,

$$\Delta\varphi = \int \mathrm{d}\boldsymbol{r}\Delta V(\boldsymbol{r})\Delta n(\boldsymbol{r}) \tag{1.67}$$

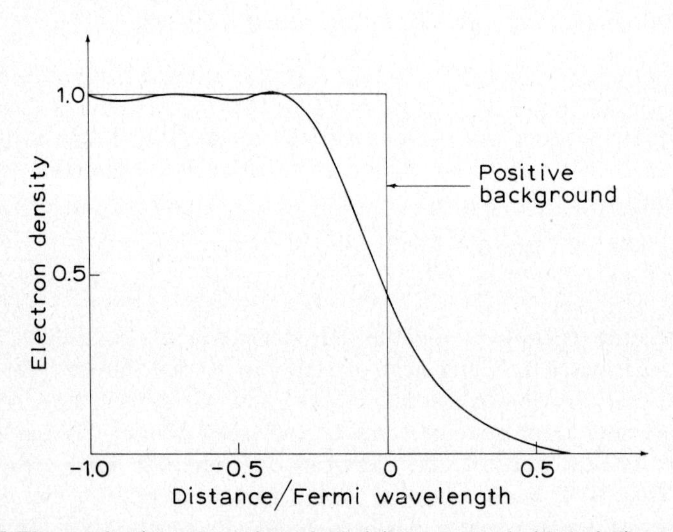

Fig. 1.5. Electron density at the surface of aluminium calculated with the jellium model by Lang and Kohn (1970). One Fermi wavelength = $2\pi/k_F = 0.35$ nm for $r_s = 2$.

is the change in the work function induced by $\Delta V(\boldsymbol{r})$. Since $\Delta n(\boldsymbol{r})$ is only finite near the surface ($z = 0$), the behaviour of ΔV near the surface determines $\Delta\varphi$ which therefore comes to depend on the crystal face exposed to $z = 0$.

6. Chemisorption

Much of the early thinking on chemisorption by metals brought the molecular orbital/valence bond dichotomy into surface studies. It was accepted that the electronic structure of the metal substrate was to be described by an itinerant electron model, yet much of the experimental data from chemisorption and catalysis could be interpreted most easily in terms of the existence of surface compounds, complexes, or molecules, with properties similar to gas-phase counterparts (see for example Sachtler 1969), and therefore, by inference with bonds similar to those of small molecules. The problem then was to show how a *local* picture of chemisorption could become valid when the substrate itself did not have localized bonds. The suggestion that only surface (and therefore localized) states of the substrate were utilized was not convincing because surface states were not thought to be sufficiently widespread phenomena, and the more general observation that the efficient screening of an electron disturbance in a metal necessarily localizes that disturbance seemed to make any localization of electrons in the chemisorption bond depend on the self-consistent response of an interacting system of itinerant electrons to the presence of the adsorbate. Yet all the tools necessary to discuss the localization of electrons in surface bonds were available in 1954 (Lifshitz 1948, 1964, Baldock 1953, Koster and Slater 1954), the key concepts being provided once more by the LCAO–MO scheme. Chemisorption can clearly be a localized phenomenon if there are molecular orbitals localized on the adsorbate and only a few metal atoms near it in the metal surface. But this is not a necessary condition. Chemisorption is a spatially localized phenomenon if the adsorbate-induced change in the local density of states on a metal AO is significant only for a few metal atoms near the adsorbate. This condition is clearly met if the important adsorbate-induced electronic structure consists of localized MO's, but the fact that it can be met when no such MO's are present is the reason why localized chemisorption is a widespread phenomenon.

6.1. Bonds and bands

The principles, but not the realistic details, of chemisorption are easily illustrated by TBA computer calculations on chains of atoms. In fig. 1.6 we show some results for the states arising from the s orbitals of a 10-atom chain. The Hückel parameters are:

$$\alpha = 0, \qquad \beta = 1$$

and the energy is in units of β. The band width is 3.838 instead of 4 as it would be for an infinite chain. With finite numbers of atoms, the band width, and the sublimation energy are always under-estimated.

Next we add chemisorbed atoms at each end of the chain. The hopping integral (β') to these atoms is 1.5, and we exhibit the twelve energy levels for two values of the adatom parameter α_A;

$$\alpha_A = 0.5, \qquad -1.0. \tag{1.68}$$

For $\alpha_A = -1.0$ there are two split-off states below the bottom of the band with energies -2.3075 and -2.3015 and with wave functions localized on the adatoms and the chain ends. We call these chemisorption states. The states are either even or odd on inversion in the centre of symmetry at the middle of the chain, and for an infinitely long chain these two chemisorption states are degenerate. In our 10-atom chain, an electron in a split-off state is shared 27 : 33 between the adatom

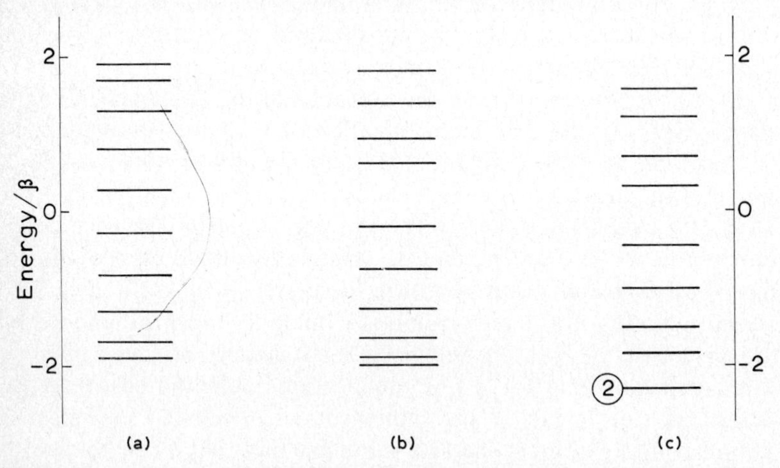

Fig. 1.6. Energy levels of Hückel chains of one-electron atoms. (a) 10-atom substrate chain, (b) substrate with adatoms at both ends, $\alpha_A = 0.5$, (c) substrate with adatoms at both ends, $\alpha_A = -1.0$. The two lowest energy states here are the nearly degenerate chemisorption states. Energies in units of β.

and the substrate, and a localized chemisorption bond is formed. But such localized MO's are not essential for chemisorption. If we calculate the adatom binding energy D from the change in energy of the occupied orbitals in the two cases (1.68) we find

$$D = 2.207, \qquad 2.354$$

so the chemisorption binding energy does not depend strongly on the existence or otherwise of localized MO's.

A final point we illustrate with this simple model is that the chemisorption is a spatially localized phenomenon. To do this we exhibit in table 1.1 the changes in the atomic orbital occupation numbers Δn_i, $i = \mathrm{A}, 1, 2, \ldots, 5$, brought about by chemisorption.

Even for this simple system, the disturbance in the charge density does not propagate significantly into the chain, and we note particularly that for $\alpha_A = 0.5$ where there are no chemisorption states, the disturbance is better localized than for $\alpha_A = -1.0$ where there are.

Next we take a chain with s and p_z orbitals on each atom (aluminium) as in subsect. 5.1 with parameters (1.28), and add a hydrogen atom at each end of the chain, with parameters

$$\alpha_A = -6.802 \text{ eV} \qquad (\text{i.e.} \quad \zeta = 0.5). \tag{1.69}$$

The orbital expansion represented by (1.69) is to simulate roughly the Coulomb repulsion effect of the partial occupancy of the hydrogen ls orbital by ↑ and ↓ spin electrons such as we expect in the adatom. First we take the H–Al coupling calculated from eq. (1.28) for an H–Al distance 1.5 A.U. This is a strong chemisorption situation. The surface

TABLE 1.1
Changes in the AO occupancies per spin for chemisorption on 10-atom tight binding chains

i	Δn	
	$\alpha_A = -1.0$	$\alpha_A = 0.5$
A	0.190	−0.101
1	−0.120	0.065
2	0.018	−0.010
3	−0.047	0.025
4	−0.013	0.007
5	−0.028	0.015

states at -7.93 eV on the Al_{15} substrate are removed by hydrogen chemisorption (as a result of adsorbate-induced s–p_z hybridization), and chemisorption states are formed below and above the band system at -15.07 and -1.55 eV respectively. In fig. 1.7 we show the contribution $\Delta\rho_\sigma(\varepsilon)$, $\sigma = \uparrow$ or $\downarrow$, which the two chemisorbed hydrogen atoms make to the density of states (the adsorbate-induced density of states) constructed from the 30 levels before, and 32 after chemisorption, by giving each level a width of 1 eV. The negative δ-functions at -7.93 eV denote the removal of the original surface states, and $\Delta\rho_\sigma$ is normalized

$$\int_{-\infty}^{+\infty} d\varepsilon \Delta\rho_\sigma(\varepsilon) = 2 \tag{1.70}$$

because two hydrogen 1s levels are added to the Al_{15} substrate. As expected, the only prominent features of $\Delta\rho_\sigma$ are the two chemisorption states, and the negative δ-functions where the original surface states have been removed.

If we add a zero width pπ band to our simple system at -5.73 eV, the Fermi level is fixed at this value, and with the H atoms adsorbed we find the lowest 18 levels of the s–p_z band system occupied giving an occupancy per spin, $n_{A\sigma}$, of the hydrogen 1s orbital of $n_{A\sigma} = 0.415$ and so justifying roughly our choice of eq. (1.70).

Figure 1.7 suggests that a reasonably good description of chemisorption in this case would be to say that the hydrogen atom couples to the surface state at -7.93 eV to form bonding and anti-bonding levels at -15.07 and -1.55 eV respectively. But such a description certainly overlooks

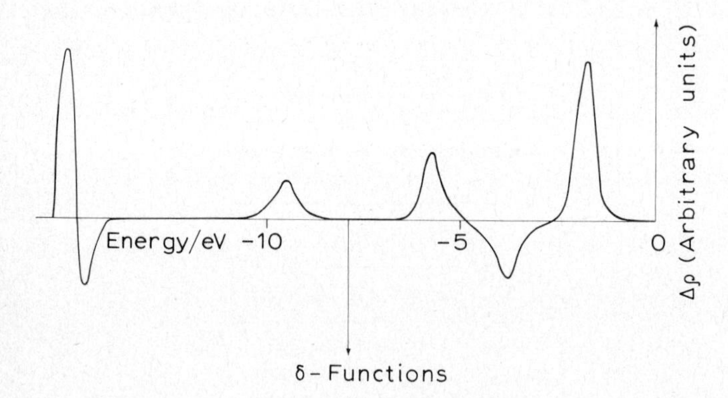

Fig. 1.7. Adsorbate-induced density of states for hydrogen atoms adsorbed at both ends of a tight-binding chain of fifteen aluminium atoms. The energies are in eV, and there are two negative δ-functions at -7.93 eV where the surface states have been removed by chemisorption.

structure in the density of states which would be important in a binding energy calculation. Thus, if we calculate the one-electron energy contribution ΔE_σ defined as*

$$\Delta E_\sigma = \int_{-\infty}^{\varepsilon_F} d\varepsilon \, \Delta\rho_\sigma(\varepsilon)\varepsilon \tag{1.71}$$

for the above simple description, we find -10.00 eV per hydrogen atom whereas the correct value is -7.55 eV. This warns us that chemisorption binding energies can depend on relatively insignificant features in $\Delta\rho_\sigma$.

The important feature of the model chemisorption system we are using here is that it shows surface states on the clean substrate being removed by chemisorption. This is a well-known phenomenon in surface electron spectroscopy (see ch. 4). Nevertheless, we should not assume that surface states are essential for chemisorption, and to examine this point we set to zero the matrix elements of the s–p_z hybridization which are responsible in our model for the existence of the surface states, and recalculate the one-electron energy contribution (1.71). The result is -7.48 eV, i.e., only 0.07 eV above the correct value, so this calculation demonstrates how a chemisorption binding energy can be quite insensitive to much of the detailed band structure of the metal. Of course the general criterion for this insensitivity here is that the s–p interaction should be small compared to the adsorbate–metal interaction. In the calculations reported above, the matrix elements describing these interactions in the Hamilton matrix were in the ratio 1:2 approximately. This is quite large, and yet a theoretical model which ignores the s–p interaction, the associated band gap, and the surface states will succeed in a chemisorption binding energy calculation. However, our satisfaction with this state of affairs is tempered by the realization that the adsorbate-induced density of states, which is the object of much experimental and theoretical study at the present time, may show prominent features whose existence is actually of no consequence in determining the strength of the chemisorption bond. The removal of surface states by chemisorption is perhaps the most important of these.

Finally we note that, from a formal point of view, the adsorbate-induced change in the density of states is obtained from the Green function matrices $\boldsymbol{G}^f$ and $\boldsymbol{G}$ of the clean substrate and of the substrate-adsorbate system,

*Of course we calculated ΔE_σ from *sums* over discrete levels; an integration is appropriate for an infinitely large system.

$$\Delta\rho_\sigma = -\frac{1}{\pi}\mathrm{Im}\,\mathrm{Tr}(\tilde{\boldsymbol{G}}\boldsymbol{S} - \boldsymbol{G}^f) \tag{1.72}$$

if we use an orthogonal basis set for the clean substrate problem, but not for the substrate–adsorbate system where the overlap matrix is $\boldsymbol{S}$, and consequently there is a tilde on the Green function matrix. In eq. (1.72)

$$\tilde{\boldsymbol{G}} = (\zeta\boldsymbol{S} - \boldsymbol{F})^{-1}, \quad \boldsymbol{G}^f = (\zeta\boldsymbol{1} - \boldsymbol{F}^f)^{-1}, \quad \zeta = \varepsilon + \mathrm{i}0. \tag{1.73}$$

If we use as basis set for the clean substrate, the substrate orbitals themselves, ψ_k (including any surface states), and add to these the adatom orbital ϕ_A, or two such orbitals if the substrate has two surfaces like our Al_{15} chain, then it is easy to show (see for example Grimley 1970, 1975) that, for the type of model considered here

$$\Delta\rho_\sigma = -\frac{1}{\pi}\mathrm{Im}\frac{1 - \partial q_A(\varepsilon)/\partial\varepsilon}{\zeta - \varepsilon_A - q_A} \tag{1.74}$$

where ε_A is the diagonal element F_{AA} of $\boldsymbol{F}$ (the orbital energy of ϕ_A in the system), and $q_A(\varepsilon)$ is the chemisorption function defined by

$$q_A = \sum_k \frac{|V_{Ak} - \varepsilon S_{Ak}|^2}{\zeta - \varepsilon_k} \tag{1.75}$$

involving the non-diagonal elements F_{Ak} and S_{Ak} of $\boldsymbol{F}$ and $\boldsymbol{S}$, and the MO energies ε_k of the clean substrate. If we write

$$q_A(\varepsilon) = \alpha(\varepsilon) - \mathrm{i}\Gamma(\varepsilon) \tag{1.76}$$

then

$$\begin{aligned} w(\varepsilon,\varepsilon') &= \pi\sum_k |V_{Ak} - \varepsilon S_{Ak}|^2\delta(\varepsilon' - \varepsilon_k) \\ \Gamma(\varepsilon) &= w(\varepsilon,\varepsilon) \end{aligned} \tag{1.77}$$

and α is the Hilbert transform of $w(\varepsilon,\varepsilon')$

$$\alpha(\varepsilon) = \frac{1}{\pi}\int_{-\infty}^{+\infty} \mathrm{d}\varepsilon' \frac{w(\varepsilon - \varepsilon')}{\varepsilon - \varepsilon'} \tag{1.78}$$

In a semi-infinite substrate, Γ is a continuous function of ε (not just a discrete set of δ-functions), and so therefore is $\Delta\rho_\sigma$, except for the possible existence of a few poles at ε – values for which

$$\varepsilon - \varepsilon_A - \alpha = 0, \qquad \Gamma = 0. \tag{1.79}$$

Such poles denote chemisorption states with energies in band gaps, or completely outside the band system. There is however an interesting point to be made* in connection with eqs. (1.74)–(1.78) when we pass from a finite to a semi-infinite substrate using a theoretical model, like those considered in this section, which do not contain the electron–electron interaction in an adequate form. It is that, whereas for finite clusters of substrate atoms, electrons are generally conserved when the chemisorption bond is formed without any change in the position of the Fermi level, when we pass to the semi-infinite substrate with Γ a continuous function of ε, they are not. Now from actual calculations on clusters we know (Grimley and Mola 1976) that, for the clean substrate, the highest occupied levels defining ε_F are usually degenerate, that there are more levels at ε_F than electrons to fill them, and that most of these levels do not participate in the chemisorption bond. Thus chemisorption proceeds without changing ε_F. As an example we quote from the results for the on-site chemisorption of hydrogen by a 124-atom tight-binding cluster of one-electron atoms (cubium) built up from one substrate atom and its first shells of nearest neighbours. For the clean substrate there are 16 levels at ε_F occupied by only 8 electrons, but none of these levels have the right symmetry to participate in the bonding of the hydrogen atom.

On the other hand, when we use eqs. (1.74)–(1.78) for a semi-infinite substrate, and fix the Fermi level at its substrate value, we find that, for a one-electron atom being chemisorbed, the number of electrons brought to the system, calculated as

$$\Delta n = \sum_{\sigma} \int_{-\infty}^{\varepsilon_F} d\varepsilon \, \Delta\rho_\sigma \tag{1.80}$$

is *not* in general unity because in a continuous spectrum the concept of degeneracy has no meaning. Consequently electrons cannot be added without changing ε_F by an infinitesimal amount $\Delta\varepsilon_F$ such that

$$2\rho_\sigma^f(\varepsilon_F)\Delta\varepsilon_F = 1 - \Delta n \tag{1.81}$$

where $\rho_\sigma^f(\varepsilon)$ is the density of states per spin at ε_F in the clean substrate.

The resolution of this problem seems to raise some fundamental issues (private communication from Dr. Doyen) which we shall not discuss here. Instead we simply remark that actual calculations (Grimley and Mola 1976) confirm that results obtained with clusters converge onto those obtained with a semi-infinite substrate with $\Delta\varepsilon_F$ determined from eqs. (1.80) and (1.81).

* Dr. G. Doyen of the University of Munich raised this question.

6.2. Tight-binding solids

Some of the earliest studies of the chemisorption of single atoms by semi-infinite solids (Koutecký 1956, Grimley 1958, see Koutecký 1965 for a review) were made with the TBA, partly because of its mathematical simplicity, and partly because it was believed to be a useful approximation for transition metal d bands. As mentioned in the last section, with this model one does not automatically conserve electrons when the chemisorption bond is formed, and we discussed there one way to handle this problem. There is another way (Allan and Lenglart 1970), which is certainly interesting in didactic work, in which electrons are conserved by making suitable adjustments to one or two diagonal elements of the tight binding Hamiltonian matrix.

Consider a semi-infinite chain A123 . . . , consisting of an adatom A, and substrate atoms 1,2,3, . . . Let α_A, α_1 and $\alpha_n = \alpha(n \geq 2)$ be the diagonal elements of H, γ the A–1 coupling, and let all other nearest neighbour couplings be β. The TBA equations for the atomic orbital coefficients $c_n(\varepsilon)$ in the MO of energy ε are

$$\begin{aligned} (\varepsilon - \alpha_A)c_A &= \gamma c_1 \\ (\varepsilon - \alpha_1)c_1 &= \gamma c_A + \beta c_2 \\ (\varepsilon - \alpha)c_n &= \beta(c_{n-1} + c_{n+1}) \qquad n \geq 2. \end{aligned} \tag{1.82}$$

In a real metal, the electron–electron interaction leads to efficient screening of the disturbance in the electron density caused by changing the tight-binding parameters on atoms A and 1 from the clean substrate values α and β. Therefore, let us require that this be true also in our model, although there are no explicit electron–electron interaction terms.

We set

$$c_n = \exp(-in\theta) + r\exp(in\theta), \qquad n \geq 2 \tag{1.83}$$

with θ in the interval $(0,\pi)$, and r a complex munber of modulus unity. Eq. (1.83) describes an incoming Bloch wave travelling in the perfect region of the substrate, and being reflected from the imperfect (surface) region with phase shift determined by r. θ determines the energy according to the familiar equation

$$\varepsilon = \alpha + 2\beta\cos\theta. \tag{1.84}$$

Using the first four members of eq. (1.82), eqs. (1.83) and (1.84), we can now calculate for any energy ε in the band (1.84), the AO coefficients c_A and c_1. They depend on α_A and α_1 of course, but if there is no change in the number of electrons on the substrate atoms outside the surface region (i.e., for $n \geq 2$), we conserve electrons in chemisorption if

$$P \sum_{\varepsilon \text{ occ.}} |c_A|^2 = \tfrac{1}{2} + \Delta n$$
$$P \sum_{\varepsilon \text{ occ.}} |c_1|^2 = \tfrac{1}{2} - \Delta n. \tag{1.85}$$

Here P is a normalizing constant depending on how many occupied energy levels in the band (1.84) we sample in the summation;

$$P \sum_{\varepsilon \text{ occ.}} (|c_A|^2 + |c_1|^2) = 1. \tag{1.86}$$

$2\Delta n$ is the charge on atom A, $-2\Delta n$ that on 1. From eq. (1.85) we determine values of α_A and α_1 which are self-consistent in the sense that they lead to AO coefficients c_A and c_n ($m = 1,2,\ldots$) which conserve electrons and localize the charge disturbance in chemisorption to the diatomic molecule A–1. But having determined α_A and α_1 this way, it will be essential to verify that for these values, there are no occupied localized chemisorption states split-off the band [eq. (1.84)]. Eq. (1.83) does not allow for the existence of such states, and if they do exist, the whole procedure outlined above fails because such states having tails in the perfect region of the substrate contribute to the charge density there. In this case the way to proceed is to enlarge the surface region until the occupied chemisorption states have negligible amplitude outside it. Appelbaum and Hamann (1975, 1976) do this in their work on silicon using, however, a much more elaborate model than the one considered here. Thus, if we take atom 2 into the surface region (but with unchanged parameter α of course), the conditions analogous to eqs. (1.86) and (1.87) are

$$P \sum_{\varepsilon \text{ occ.}} |c_i|^2 = \tfrac{1}{2} + \Delta n_i, \qquad (i = A,1,2,)$$
$$P \sum_{\varepsilon \text{ occ.}} \sum_i |c_i|^2 = 3/2 \tag{1.87}$$

and these conditions will determine α_A and α_1 self consistently with occupied chemisorption states provided the penetration of these states beyond atom 2 can be neglected. The further generalization is obvious.

For tight binding solids, the most convenient way to formulate the whole of the above procedure is to use Green functions. Thus eq. (1.87) is written

$$-\frac{1}{\pi}\mathrm{Im}\int_{-\infty}^{\varepsilon_F} \mathrm{d}\varepsilon G_{ii}(\varepsilon) = \tfrac{1}{2} + \Delta n_i \qquad (i = \mathrm{A},1,2)$$
$$\sum_i \Delta n_i = 0 \tag{1.88}$$

and Dyson's equation is used to calculate $\boldsymbol{G}$ from the Green function matrices of the clean substrate, and the free atom (Allan 1970, see Grimley 1975 for a review).

The question as to the importance of the conduction band in chemisorption by transition metals is not yet settled. Cluster calculations include the "conduction band" (as 4s and 4p atomic orbitals in the basis for the first transition series when LCAO schemes are used, and as s and p partial waves in the MS–Xα scheme), and there is no doubt that it ought to be included in calculations on semi-infinite metals too. But as we have seen in subsect. 6.2, it will not be necessary to treat the conduction band – d band hybridization if chemisorption itself provides a stronger mixing than that in the clean substrate. Thus, a model with tight binding d bands and a conduction band described by plane waves orthogonalized to the d orbitals, but with no conduction band – band interaction otherwise, should be a useful starting point.

6.3. The muffin-tin model

It is a relatively simple matter to add an *overlayer* of chemisorbed atoms (but not a single chemisorbed atom) to the model of the clean metal discussed in sect. 5 (see Pendry 1974). The simplest case to treat is that of an overlayer with the same unit mesh as the clean metal. Then, referring to fig. 1.3, layer 1 is the overlayer, and layers 2,3,... are the substrate. Layer 1 therefore has different scattering properties from the rest, but otherwise we can proceed as in subsect. 5.2. Thus we introduce the reflection matrix $\boldsymbol{R}'$ of the crystal with an overlayer, but with no surface barrier;

$$\boldsymbol{a}_1^- = \boldsymbol{R}'\boldsymbol{a}_1^+ \tag{1.89}$$

and the reflection matrix $\boldsymbol{J}$ of the surface barrier from eqs. (1.46) and (1.47) to arrive at the determinantal equation

$$\det|\mathbf{1} - \boldsymbol{R}'\boldsymbol{J}| = 0 \tag{1.90}$$

for the eigenvalue spectrum of the system at every $\boldsymbol{k}_\parallel$. Comparing this with the spectrum for the clean substrate calculated as in subsect. 5.2

will lead to the overlayer-induced density of states as a sum of contributions from every $\boldsymbol{k}_{\|}$.

The importance of the muffin-tin model is that it enables us to make calculations on the eigenvalue spectra and wave functions for overlayers on transition metals. Its main drawback is that it does not provide an expression for the total energy so that questions about stability, bond lengths, and structure cannot be answered. We must also remember that in its simplest form the model has no self-consistency, either in the sense of eq. (1.80) or in the Hartree–Fock sense of the wavefunctions reproducing the assumed potential from which they were determined. To attend to both these points, the phase shifts which are needed to determine the scattering properties of a layer of muffin tins must be calculated from the wavefunctions in a cycle of iterations to self-consistency. Finally we note that the muffin-tin zero in the overlayer (chosen here to be the same as in the bulk for simplicity), and the position of the surface barrier, are parameters in the model.

6.4. Jellium

The problem of a single atom interacting with a semi-infinite jellium has been solved in the local density approximation of sect. 2 (Lang and Williams 1976a, Gunnarsson et al 1976). The ground-state potential energy curve and the adsorbate-induced density of states are among the important quantities calculated (see table 1.2 and fig. 1.8). The binding

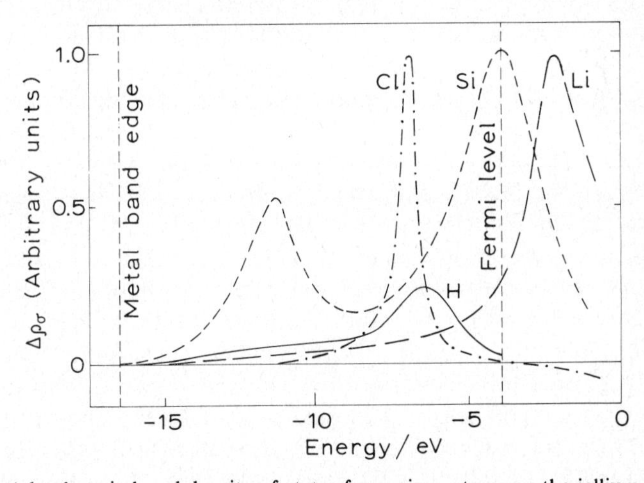

Fig. 1.8. Adsorbate-induced density of states for various atoms on the jellium model of aluminium ($r_s = 2$). Only the in-band contribution to $\Delta\rho_\sigma$ is shown. (Constructed from Lang and Williams 1976a, b).

TABLE 1.2
Binding energies in eV of atoms on the jellium model of aluminium ($r_s = 2$)

Atom	H[a,b]	Li[a]	O[a]	Si[c]	Cl[c]
Binding energy	1.5	1.3	5.4	3.0	3.6

[a]Lang and Williams 1976a, [b]Gunnarsson et al. 1976. [c]Lang and Williams 1976b.

energies are clearly reasonable, showing that the jellium model describes the important contributions to the chemisorption bond on simple metals in an essentially quantitative way, although we would naturally not expect a perfect description.

Regarding the densities of states several comments need to be made. Firstly, the core levels are not shown, although these too are calculated as split-off states far below the bottom of the jellium band, at $-52\,\text{eV}$ for Li 1s for example. Secondly, the local density approximation gives very efficient screening, so the calculations are made assuming that

$$\Delta n = \sum_{\sigma} \int_{\text{in-band}}^{\varepsilon_F} d\varepsilon \Delta\rho_\sigma(\varepsilon) \tag{1.91}$$

the integration being from the bottom of the jellium band, is automatically 1, 1, 4, 4, 5 for H, Li, O, Si, and Cl respectively. Thirdly, there is no indication in fig. 1.8 of the bonding/anti-bonding structure in $\Delta\rho_\sigma$ familiar in quantum chemistry with calculations which do not include the continuous spectrum in the basis set. Here the Kohn–Sham one-particle equations (1.3) admit solutions going up continuously above ε_F, and through the vacuum level, so, whilst there must be, in the case of hydrogen for example, a further contribution to $\Delta\rho_\sigma$ above ε_F, there is no reason to suppose that this part of $\Delta\rho_\sigma$ is anything other than a very broad featureless peak. Finally, we remind ourselves that the density of Kohn–Sham one-particle states is not the same as the density of ionization and affinity levels, although the two densities should be similar for de-localized states with only small electron correlation effects.

For the model used in the calculations just described there are no crystal structure effects, and no possibility therefore to discuss either different crystal surfaces or adsorption at different sites on a given surface. These are serious deficiencies because, in a strict sense, the whole of surface chemistry is left untouched; in surface chemistry (and catalysis) we are intimately concerned with the substrate geometry, the geometry of the adsorption sites and so on. However, crystal structure

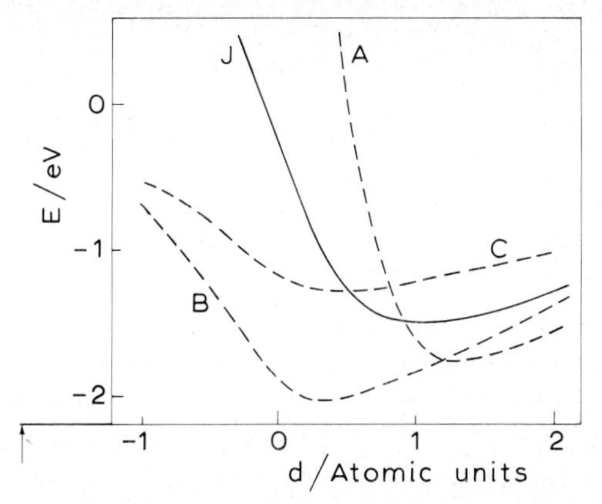

Fig. 1.9. Potential energy curves for hydrogen adsorbed on (100)Al calculated using jellium model of aluminium ($r_s = 2$) with crystal structure effects included through eq. (1.92). (A) on site, (B) bridged, (C) centred, (J) jellium result. d is the distance to the jellium edge, and the arrow gives the position of the outermost layer of Al ions 0.101 nm from the jellium edge. 1 atomic unit ≡ 0.0529 nm. (Calculations by Gunnarsson et al. 1976).

effects can be introduced into the jellium model by perturbation theory as mentioned already in subsect. 5.4. If $\Delta V(\boldsymbol{r})$ is the quantity used in eq. (1.67), and $\Delta n(\boldsymbol{r})$ the adsorbate-induced electron density calculated for the jellium model, then in lowest order,

$$\Delta D = \int \mathrm{d}\boldsymbol{r} \Delta V(\boldsymbol{r}) \Delta n(\boldsymbol{r}) \tag{1.92}$$

is the change in the chemisorption binding energy induced by $\Delta V(\boldsymbol{r})$. Results of calculations by Gunnarsson et al. (1977) showing how this term introduces the adsorption geometry into the jellium model are shown in fig. 1.9 for hydrogen on (100) Al. From these results it seems that crystal structure effects are rather too large to be given accurately by a lowest-order perturbation theory calculation.

6.5. Embedded clusters

For small molecules elaborate calculations going beyond the HF limit are possible, and chemical accuracy in bond dissociation energies can be achieved. Such calculations have generally been made using the CI technique. We have already seen that chemisorption by metals is often a spatially well-localized phenomenon, so the possibility of treating a small cluster of atoms – the adsorbate, and its nearest neighbour shell of metal atoms for example – accurately, and the rest of the metal, and its coupling to the cluster in a lower approximation suggests itself (Grimley 1974). Such a cluster embedding scheme has been tested in model calculations of hydrogen chemisorption by simple solids using the HF approximation over the cluster and a tight binding model elsewhere (Grimley and Pisani 1974). In these calculations the cluster was reduced to its minimum size, the adatom and one metal atom. The formal equations describing different embedding schemes have been derived (Grimley 1976a, b, see also Gunnarsson and Hjelmberg 1975) but extending the actual calculations to larger clusters embedded either in simple metals or in transition metals presents some computational problems. Since the foundations of the HF approximation are very well understood, embedding a HF cluster in a tight binding metal is a technical problem on which it is not appropriate to dwell here. Instead we make a remark about the reason for embedding a cluster calculation in this fashion. For a large enough cluster, embedding it makes no difference of course; the cluster is so large that the disturbance caused by forming the chemisorption bond does not penetrate to its boundary with the rest of the metal, and neither does the disturbance at its boundary, caused by embedding, propagate back to the chemisorption bond. But for a small cluster, both these disturbances overlap, and it is not immediately obvious that embedding does not do more harm than good. More computations are needed to explore these matters, but it is clear that embedding a good cluster calculation in a tight binding model of the metal must be better than simply calculating everything in TBA, and it would be very surprising if the embedded cluster calculation did not give a better representation of the true state of affairs than a straightforward cluster calculation. After all, the recursion and moment methods for calculating local densities of states, discussed briefly at the end of subsect. 5.1, can be regarded as schemes for embedding a single atom with a few discrete states in a tight binding substrate, and we could hardly claim that the final density of states so obtained is a worse representation of the state of affairs at the solid surface than that provided by the free atom.

6.6. Anderson's hamiltonian

This theoretical model, originally proposed by Anderson (1961) in another context, has been extremely useful in forming concepts in chemisorption theory because it contains the essential features of the chemisorption problem, the formation of a bond between a semi-infinite metal with itinerant electron states, and an atom with localized ones, in the clearest possible way. The HF version of the model is an MO theory using as basis set, not localized AO's, but a mixture of localized and non-localized orbitals $\{\phi_A, \phi_k\}$ consisting of the adatom orbital ϕ_A, and a set of states $\{\phi_k\}$ describing electrons in the semi-infinite metal such as we used to write eq. (1.74). If we limit the set $\{\phi_k\}$ to a band of states deemed to be important for chemisorption (the d bands of the transition metals in early work) the basis, though non-orthogonal at the equilibrium adatom–metal distance, is not complete, and the theory can proceed along lines already familiar in quantum chemistry [in the early work (Grimley 1967a, b, Edwards and Newns 1967) the non-orthogonality of the basis was ignored], except that Green function techniques are conveniently used to handle the essentially continuous spectrum of states in the metal band. But it is really by use of these techniques that the characteristic features of the chemisorption bond to metals is brought out, and we have here a good example of how progress and understanding in chemistry and physics can depend on the use of an appropriate mathematical technique.

We have already set down in eqs. (1.72)–(1.79) the equations for the adsorbate-induced density of states for the HF version of this model, and we remark here, firstly that the formalism treats the in-band and localized chemisorption state contributions to $\Delta\rho_\sigma$ in a unified way, secondly that it shows how, by changing the parameters, chemisorption states appear and disappear in a continuous way, thus indicating that their existence is not a prerequisite for chemisorption, and thirdly that it concentrates our attention on the chemisorption function of eq. (1.75) and in particular on its imaginary part, which here determines the nature of the chemisorption bond. This imaginary part, $\Gamma(\varepsilon)$, is the link between chemisorption theory and the bonding in an ordinary diatomic molecule. If $\Gamma(\varepsilon)$ is small everywhere except in a narrow region round some energy ε_0, the surface bond is in many ways similar to the bond in a complex A–M_n where the n metal atoms in M_n have the geometry of the adsorption site, and provide a single molecular orbital with energy ε_0, and the right symmetry to bond to A. To see this we suppose that in eq. (1.77)

$$w(\varepsilon,\varepsilon') = \pi|\varepsilon S - V|^2\delta(\varepsilon' - \varepsilon_0) \tag{1.93}$$

so that

$$\alpha(\varepsilon) = |\varepsilon S - V|^2/(\varepsilon - \varepsilon_0) \tag{1.94}$$

and

$$\Gamma(\varepsilon) = \pi|\varepsilon S - V|^2\delta(\varepsilon - \varepsilon_0). \tag{1.95}$$

Then eq. (1.79) becomes

$$(\varepsilon - \varepsilon_A)(\varepsilon - \varepsilon_0) = (\varepsilon S - V)^2 \tag{1.96}$$

which is nothing more than the usual secular equation of MO theory for the energies $E1$ and $E2$ say, of the bonding and anti-bonding levels formed from the adatom orbital ϕ_A at ε_A, and another orbital on M_n with energy ε_0. Correspondingly, the adsorbate-induced density of states $\Delta\rho_\sigma$ is obtained from eq. (1.74) in terms of three δ-functions;

$$\Delta\rho_\sigma = \delta(\varepsilon - E1) + \delta(\varepsilon - E2) - \delta(\varepsilon - \varepsilon_0) \tag{1.97}$$

which is a structure similar in some respects to that in fig. 1.7.

To understand the significance of the form (1.95) we consider (Grimley 1975) the surface of a tight binding d band solid with a hydrogen atom in the surface plane in the centred position (fig. 1.10). The metal states are linear combinations of all the d orbitals on all the atoms, but up to its nearest neighbours, the hydrogen 1s orbital couples to the group orbital (belonging to the unit representation A_1 of C_{4V})

$$\psi_d = \frac{1}{\sqrt{8}} \sum_{n=1}^{4} d_{3z^2-r^2}(n) - 2\, d_{3z^2-r^2}(5) \tag{1.98}$$

so that

$$V_{Ak} = \sqrt{8}\, V \langle \psi_d | \phi_k \rangle, \qquad S_{Ak} = \sqrt{8}\, S \langle \psi_d | \phi_k \rangle$$

where V is the hopping matric element, and S the overlap between the hydrogen 1s orbital and $d_{3z^2-r^2}(1)$. Using these results in eq. (1.77)

$$w(\varepsilon, \varepsilon') = (\lambda\varepsilon + 1)^2 \Gamma(\varepsilon'), \qquad \lambda = -S/V$$

$$\Gamma(\varepsilon) = 8\pi |V|^2 \sum_k |\langle \psi_d | \phi_k \rangle|^2 \delta(\varepsilon - \varepsilon_k)$$

$$= 8\pi |V|^2 \rho_d(\varepsilon) \tag{1.99}$$

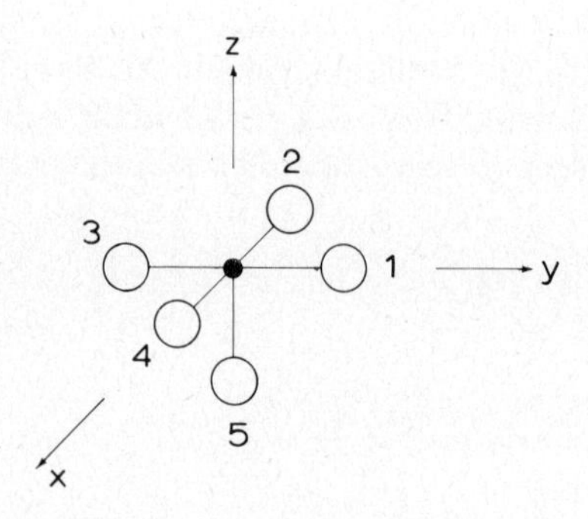

Fig. 1.10. Hydrogen in a 4-fold hole on the (100) face of a face-centred cubic solid.

$$\rho_d(\varepsilon) = \sum_k |\langle\psi_d|\phi_k\rangle|^2 \delta(\varepsilon - \varepsilon_k). \tag{1.100}$$

$\rho_d(\varepsilon)$ is the spectral density of the group orbital (1.98) in the d bands of the clean substrate, i.e. it is the d band density of states on the orbital ψ_d. If ρ_d is small everywhere, except in a narrow region round ε_0 so that

$$\rho_d(\varepsilon) \simeq \delta(\varepsilon - \varepsilon_0) \tag{1.101}$$

we recover eqs. (1.93)–(1.95). Calculations by the recursion method (Kelly 1974) show that eq. (1.101) is not in fact a particularly good approximation to the spectral density (1.100) of the group orbital (1.98), and consequently the concept of a surface molecule A–M_n is rather rough and ready. But even so, the importance of Anderson's model is that it provides a theoretical framework for examining this widespread concept, and giving it a precise meaning.

If we try to use Anderson's model to treat an atom interacting with a simple metal, we at once encounter a new mathematical problem, namely, the basis set $\{\phi_A, \phi_k\}$ becomes overcomplete. This happens because there are free-electron-like states ϕ_k with energies going up continuously from the bottom of the conduction band, and up through the vacuum level. These states $\{\phi_k\}$ are themselves now a complete set, and the adatom orbital ϕ_A is not needed in the basis. This means firstly, that the overlap matrix with elements S_{Ak} is singular, and secondly, that if we try to expand the system MO's ψ_μ in terms of the basis set $\{\phi_A, \phi_k\}$ as in eq. (1.12), then the expansion is not unique unless a subsidiary condition is imposed (Grimley 1974, 1976a). A useful subsidiary condition is

$$\sum_k S_{Ak} A_{k\mu} = 0, \qquad \text{all } \mu \tag{1.102}$$

which ensures that the orbital ϕ_A is used as much as possible in the expansion (1.12). One consequence of the condition (1.102) is that it makes the traces of $\tilde{G}S$ and $\tilde{G}$ equal so the latter can be used in eq. (1.72) to calculate $\Delta\rho_\sigma$.

The importance of Anderson's model for chemisorption by the simple metal lies in its ability to reproduce quite accurately the first-principles results like those in fig. 1.8 for an adatom on a jellium (Muscat and Newns 1977). These authors have shown that a model consisting of a semi-infinite free-electron gas bounded at its surface by a vertical potential step, and an adatom potential chosen to give an appropriate energy ε_A in eq. (1.74) for a given r_s-value of the jellium, can reproduce very well the adsorbate induced density of states obtained from a first-principles calculation. This is particularly remarkable when we reflect that the simple model used has no screening, and no self-consistency except that introduced by the parameterization of the adatom potential, so that eq. (1.80) with $\Delta n = 1$ for hydrogen is not automatically satisfied for arbitrary ε_A. Therefore, one thing we learn from this work is that, if the adatom level ε_A is adjusted to give the correct Δn in eq. (1.80), we are likely to have a better approximation to the correct $\Delta\rho_\sigma$ than if ε_A is determined by HF self-consistency. This way of proceeding was of course used by Allan and Lenglart (1970) for transition-metal substrates (see subsect. 6.2), but we are not concerned with tight-binding solids here.

The importance of Anderson's model in reproducing the exact atom–jellium results is not that it tells us anything new about this system – the exact results provide all the information – but that it may open the way to including the conduction band in chemisorption by transition metals in a simple but now proven way.

To amplify some of the above remarks, and to see the Anderson model in relation to the exact Hamiltonian, we set down some basic equations. To use the formalism of second quantization with the overcomplete basis set $\{\phi_A, \phi_k\}$, we introduce operators $b^+_{i\sigma}$ and $a_{i\sigma}$ which create and destroy electrons with spin σ in the basis orbitals ϕ_i ($i = A,k$). Because of overcompleteness, the creation operator is not $a^+_{i\sigma}$ the Hermitian adjoint of $a_{i\sigma}$, but the operator $b^+_{i\sigma}$ defined by

$$b^+_{i\sigma} = \sum_j a^+_{j\sigma} S_{ji} \tag{1.103}$$

in terms of the elements of the (singular) matrix S. The a and b^+ operators satisfy the usual anti-commutation rules of destruction and creation Fermion operators; the a and a^+ operators do not (Grimley 1970). The exact Hamiltonian for the electron system can be written in terms of the a and a^+ operators, and matrix elements formed with the overcomplete basis,

$$H = \sum_{\sigma ij} \langle i|\hat{h}|j\rangle a^+_{i\sigma} a_{j\sigma} + \tfrac{1}{2} \sum_{\sigma\sigma'} \sum_{iji'j'} \langle ij|1/r_{12}|i'j'\rangle a^+_{i\sigma} a^+_{j\sigma'} a_{j'\sigma'} a_{i'\sigma} \tag{1.104}$$

where $\hat{h}$ is the Schrödinger operator for an electron moving in the field of the nuclei, and in the second term we have the (2-electron) matrix elements of the Coulomb interaction $1/r_{12}$ of a pair of electrons. For our present purpose we reduce eq. (1.104) to

$$H = \sum_\sigma E_A \tilde{n}_A + J_A \tilde{n}_{A\uparrow} \tilde{n}_{A\downarrow} + \sum_{k\sigma} \varepsilon_k \tilde{n}_{k\sigma} + \sum_{k\sigma} (V_{Ak} a^+_{A\sigma} a_{k\sigma} + V_{kA} a^+_{k\sigma} a_{A\sigma}) \tag{1.105}$$

which might be referred to as Anderson's Hamiltonian in an overcomplete basis, although, since Anderson (1961) did not treat the overcompleteness problem, this description is arbitrary. In eq. (1.105) $\tilde{n}_{i\sigma} = a^{+}_{i\sigma}a_{i\sigma}$, and the only Coulomb term retained from eq. (1.104) is that for two electrons in the localized orbital ϕ_A. As a matter of fact eq. (1.105) with $J_A = 0$ does not describe exactly the model used by Muscat and Newns (1977), but the difference is unimportant here. For the formula $\langle \tilde{n}_{A\sigma} \rangle$, the expectation value of $\tilde{n}_{A\sigma}$ is

$$\langle \tilde{n}_{A\sigma} \rangle = -\frac{1}{\pi}\mathrm{Im}\int_{-\infty}^{\varepsilon_F} \mathrm{d}\varepsilon \tilde{G}_{AA}(\varepsilon + \mathrm{i}0) \tag{1.106}$$

involving the exact Green function $\tilde{G}_{AA}$, but in the HF approximation this becomes

$$\langle \tilde{n}_{A\sigma} \rangle = \frac{1}{\pi}\mathrm{Im}\int_{-\infty}^{\varepsilon_F} \frac{\mathrm{d}\varepsilon}{\zeta - \varepsilon_A - q_A}, \qquad \zeta = \varepsilon + \mathrm{i}0 \tag{1.107}$$

with

$$\varepsilon_A = E_A + J_A \langle \tilde{n}_{A\sigma} \rangle \tag{1.108}$$

and

$$q_A = \sum_k V_{Ak}(V_{kA} - \varepsilon S_{kA})/(\zeta - \varepsilon_k) \tag{1.109}$$

the chemisorption function q_A being changed from that in eq. (1.75) for an incomplete basis by using the subsidiary condition (1.102). The point we wish to demonstrate is that ε_A and $\langle \tilde{n}_{A\sigma} \rangle$ have to be determined self-consistently for every distance between the adatom and the metal. This is the HF self-consistency referred to above. But having determined ε_A in this way, we know that electrons will not be automatically conserved according to eq. (1.80) (because there are no Coulomb terms in eq. (1.105) for electrons in the states ϕ_k), unless we are willing to adjust E_A. Now E_A varies with the adatom–metal distance because the adatom potential in $\hat{h}$ overlaps the metal potential, and what Muscat and Newns (1977) have shown is that we can ignore the HF self-consistency problem because there is an entirely reasonable choice of the adatom potential, which determines E_A as a function of the adatom–metal distance such that we can put $J_A = 0$ in eq. (1.108), and determine through eq. (1.74) a very good approximation to the correct atom–jellium $\Delta\rho_\sigma$, and of course very nearly satisfy eq. (1.80) with $\Delta n = 1$ for hydrogen as a consequence.

Anderson's Hamiltonian is important in another connection; it is simple enough to be treated in approximations higher than HF so that the influence of electron correlations in the adatom on chemisorption phenomena can be studied (Brenig and Schönhammer 1974). The important question here is the correlation structure in $\Delta\rho_\sigma$. In any system, the spectral function

$$\rho_\sigma(\varepsilon) = -\frac{1}{\pi}\mathrm{Im}G(\varepsilon + \mathrm{i}0)$$

has peaks at the poles of the Green function. The poles of the HF Green function occur at the MO orbital energies ε_μ of eq. (1.15) so we are led to a Koopmans (1933) spectrum for $\Delta\rho_\sigma$. But the poles of the exact Green function (1.20) occur at the exact ionization and affinity levels of the system, and there are more of these levels than there are HF orbital energies, so we find correlation structure in $\Delta\rho_\sigma$. A simple example will make this clear. Consider the surface molecule discussed above in connection with the HF $\Delta\rho_\sigma$ of eq. (1.97). There are two orbital energies $E1$ and $E2$, but four ionization and affinity levels because an electron can be added or removed with simultaneous excitation of the system, a process not allowed for in the HF Green function. We note however that such a process is allowed for in a delta–HF calculation, although such a calculation does not proceed via a Green function. In any case, either by an exact calculation or by delta–HF we are led to expect four positive δ-functions* not two in eq. (1.97). The Anderson model, though simple, is not simple enough to be handled exactly, but we know (Brenig and Schönhammer 1974) that the actual appearance of $\Delta\rho_\sigma$ depends on the parameters in the Hamiltonian. Doyen and Grimley (1977) have discussed in some detail the exact $\Delta\rho_\sigma$ for the model with a zero-width band of states ϕ_k, in connection with adsorbate-induced photoemission, and whenever new electronic processes in adsorbates are under discussion the Anderson model is likely to be used to elucidate the important concepts.

6.7. Interactions in overlayers

A characteristic feature of chemisorption by metals is the existence of a through-bond (indirect) interaction between adsorbates such that ordered overlayers are expected. This interaction belongs to a wide class of phenomena among which are to be found the indirect interaction between localized magnetic moments in metals (see Kittel 1963) and the

* The strengths of these δ-functions, being connected with the residues at the poles of G, are less than unity, not equal to unity as in the Koopmans spectrum.

retarded (indirect) van der Waals interaction between neutral atoms (Casimir and Polder 1948). The former is a close relative of the adatom interaction to be considered here. In the latter each atom interacts with the same boson field (the vacuum electromagnetic field) which therefore transmits an influence from one atom to the other. Perturbation theory has to be taken to fourth order to obtain the van der Waals interaction correctly. For two adatoms on a metal surface each one interacts with the same fermion field (that describing electrons in the clean metal), which therefore transmits the influence of one to the other. Again, the interaction is found in fourth-order perturbation although the details are different from the case of van der Waals forces because, amongst other things, the field operators obey different commutation rules. The practical importance of this through-bond interaction between adatoms is that it provides a long range ordering force in overlayers, and its study should therefore help us to understand the stabilities of LEED structures. Theoretical investigations using simple models reveal the richness of the phenomenon. The interaction between two adatoms on the same crystal plane is not necessarily isotropic, and it is an oscillatory function of their separation (i.e. sometimes attractive, sometimes repulsive), but with a periodicity which does not coincide with that of the surface net because it is determined in part by the electronic structure of the metal in such a way that, with a rigid-band model, changing the degree of band filling changes completely the through-bond interaction. These features are all to be found in the early papers on the subject (Grimley 1967a, b, Grimley and Walker 1969) as well as in the later ones (Einstein and Schrieffer 1973, Burke 1976). From the sign of the through-bond interaction between a pair of adatoms on nearest or next-nearest neighbour sites, one can argue (Einstein and Schrieffer 1973) about the stability of LEED structures. There are slight dangers in this approach to LEED structures because the through-bond interaction in an overlayer has a many-adatom character, i.e. the total interaction energy is not simply the sum of pair interactions; there are terms depending on the relative positions of three, four, etc. adatoms (Grimley and Walker 1969). For this reason, and also because experimental observations are usually made on ordered overlayers not on pairs of atoms (but see Bassett 1975), it is useful to approach this problem by comparing the binding energy per adatom of an ordered overlayer on a semi-infinite substrate with that of a single atom on the same substrate. In this way all many-adatom effects are included from the beginning.

As a simple example of this approach we consider hydrogen-like atoms on the (001) surface of a simple cubic tight binding solid with chemisorption of a single atom described by Anderson's Hamiltonian (1.105) with $J_A = 0$, and the basis set $\{\phi_A, \phi_k\}$ assumed orthogonal. This is the same model as that used by Einstein and Schrieffer (1973) to investigate the pair interaction. We consider a p(1 × 1) overlayer and use the Hamiltonian

$$H = E_A \sum_{\alpha\sigma} n_{\alpha\sigma} + \sum_{k\sigma} \varepsilon_k n_{k\sigma} + \sum_{k\alpha\sigma} (V_{\alpha k} a^+_{\alpha\sigma} a_{k\sigma} + V_{k\alpha} a^+_{k\sigma} a_{\alpha\sigma}). \tag{1.110}$$

Here α goes over all adatoms in the overlayer, and there are no tildes because the basis is orthogonal. We impose periodic boundary conditions over M surface atoms in the x and y directions, and with the z axis perpendicular to the surface we label the (001) planes of the substrate by an index $m = 0,1,2,\ldots$ On each plane we form the 2-dimensional Bloch sums

$$\psi_m(\boldsymbol{k}_\parallel) = M^{-1} \sum_{\boldsymbol{m}} \phi_{m\boldsymbol{m}} \exp(\mathrm{i}\boldsymbol{k}_\parallel \cdot \boldsymbol{m}), \qquad m = 0,1,2,\ldots \tag{1.111}$$

where $\boldsymbol{m}$ is a 2-component vector specifying the position of an atom in the (001) plane so that m and $\boldsymbol{m}$ together specify a particular atom, and $\phi_{m\boldsymbol{m}}$ is the atomic orbital on this atom. $\boldsymbol{k}_\parallel$ is the parallel wave vector. Similarly, from the overlayer orbitals $\phi_{A\boldsymbol{m}}$ we form

$$\psi_A(\boldsymbol{k}_\parallel) = M^{-1} \sum_{\boldsymbol{m}} \phi_{A\boldsymbol{m}} \exp(\mathrm{i}\boldsymbol{k}_\parallel \cdot \boldsymbol{m}). \tag{1.112}$$

If we use these layer orbitals (1.111) and (1.112) as basis, the Hamiltonian (1.110) is diagonal with respect to $\boldsymbol{k}_\parallel$, and within each $\boldsymbol{k}_\parallel$ manifold there is a 1-dimensional chemisorption problem (like those considered in subsects. 6.1 and 6.2) to solve. Thus the chemisorption function (1.75) and the adsorbate-induced density of states (1.74) have to be calculated at all $\boldsymbol{k}_\parallel$ in the surface Brillouin zone, as does the one-electron energy contribution

$$\Delta E_\sigma(\boldsymbol{k}_\parallel) = \int_{-\infty}^{\varepsilon_F} \mathrm{d}\varepsilon (\varepsilon - \varepsilon_F) \Delta\rho_\sigma(\boldsymbol{k}_\parallel, \varepsilon).$$

Summing ΔE_σ over $\boldsymbol{k}_\parallel$ leads to a formula for the binding energy $D_{p(1\times1)}$ per atom in the overlayer from which we obtain

$$\Delta D = D_{p(1\times1)} - D_{single} \tag{1.113}$$

the difference between the binding energy per atom in the overlayer and that of a single adatom on the surface. Some results (Grimley and Rosales 1976) obtained in this way are shown in fig. 1.11. The energy

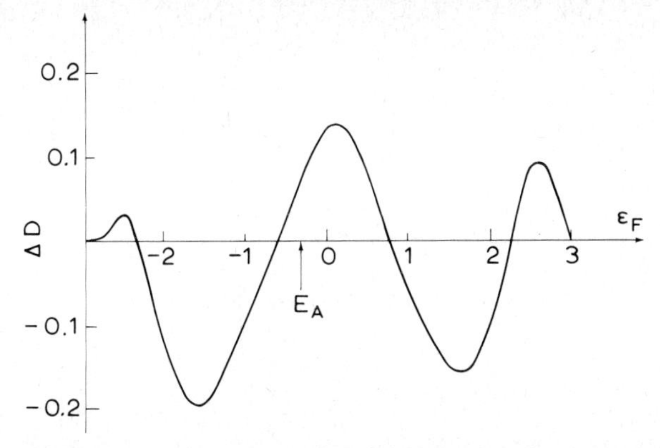

Fig. 1.11. The difference between the binding energy per atom in a p(1 × 1) overlayer on the (001) surface of a simple cubic tight-binding solid and the binding energy of a single atom on the same surface as the Fermi level sweeps through the band. Adatom level $E_A = -0.3$, and the adatom–substrate hopping matrix element is 2.0. Energy in units of 2β where β is the substrate hopping matrix element.

unit is twice the tight-binding substrate hopping matrix element, the mean band energy is taken as zero, the adatom level $E_A = -0.3$, the adatom–metal hopping matrix element is 2.0, and we show how ΔD varies as the Fermi level ε_F sweeps through the band from -3 to $+3$. Where ΔD is positive, islands of ordered structure will tend to form even at low surface coverage, where ΔD is negative, these calculations only show that islands of p(1 × 1) structure will not form at low coverage; the c(2 × 2) structure could be stable in these regions.

We note that the parameters chosen for this calculation describe a fairly strong adatom–substrate interaction, though not strong enough to give a split-off chemisorption state for a single adatom. However, for many values of $\boldsymbol{k}_\parallel$ there is a split-off chemisorption state for the overlayer. Such overlayer chemisorption states are more easily split-off than single adatom chemisorption states because the $\boldsymbol{k}_\parallel$ sub-bands are narrower than the bulk band; in the simple model treated here they are only one third as wide as the bulk band.

The most urgently needed work with overlayers on metals (as with the pair potential) is to use more realistic substrates and a model with built-in screening and self-consistency so that the through-space as well as the through-bond interaction is properly included. In this connection we note that whilst Appelbaum and Hammann (1975) have used such a model to study hydrogen overlayers on (111)Si they do not calculate the overlayer binding energy. Si is in any case a simpler solid than a transition metal.

References

Allan, G., 1970, Ann. Phys. (Paris) **5**, 169.
Allan, G., 1976, Electronic structure of transition metal surfaces, in: Electronic structure and reactivity of metal surfaces (Derouane, E.G. and A.A. Lucas, eds.) (Plenum Press, New York) pp.45–79.
Allan, G. and P. Lenglart, 1970, J. Phys. (Paris) **31**, Suppl. Cl. 93.
Anderson, P.W., 1961, Phys. Rev. **124**, 41.
Anderson, P.W., 1969, Phys. Rev. **181**, 25.
Appelbaum, J.A. and R.D. Hamann, 1975, Phys. Rev. Lett. **34**, 806.
Appelbaum, J.A. and R.D. Hamann, 1976, CRC Crit. Rev. Solid State Sci. **6**, 357.
Baldock, G.R., 1953, Proc. Phys. Soc. (London) **A66**, 1.
Bassett, D.W., 1975, Surface Sci. **53**, 74.
Bonch-Bruevich, V.L. and S.V. Tyablikov, 1962, The Green function method in statistical mechanics (North-Holland, Amsterdam).
Brenig, W. and K. Schönhammer, 1974, Z. Phys. **267**, 201.
Bullett, D.W., 1977, Surface Sci. **68**, 149.
Bullett, D.W. and M.L. Cohen, 1977, J. Phys. **C10**, 2083.
Burke, N.R., 1976, Surface Sci. **58**, 349.
Caruthers, E., L. Kleinman and G.P. Alldredge, 1973, Phys. Rev. **B8**, 4570.
Casimir, H.B.G. and D. Polder, 1948, Phys. Rev. **73**, 360.
Ciraci, S. and I.P. Batra, 1977, Phys. Rev. **B15**, 3254.
Coulson, C.A., 1970, σ bonds, in: Physical chemistry, vol. 5 (Eyring, H., Henderson, D. and W. Jost, eds.) (Academic Press, New York) pp. 287–367.
Davison, S.G. and J.D. Levine, 1970, Solid State Phys. **25**, 1.
Desjonquères, M.C. and F. Cyrot-Lackmann, 1975, J. Phys. **C6**, 3077.
Devreese, J.T., A.B. Kunz and T.C. Collins, 1974, eds., Elementary excitations in solids molecules and atoms (Plenum Press, London).
Dovesi, R., C. Pisani, F. Ricca and C. Roetti, 1976, J. Chem. Phys. **65**, 3075.
Doyen, G. and T.B. Grimley, 1977, Adsorbate-induced photoemission, in: Photoemission from surfaces (Feuerbacher, B., Fitton, B. and R.F. Willis, eds.) (Wiley, London) to be published.
Edwards, D.M. and D.M. Newns, 1967, Phys. Lett. **A24**, 236.
Einstein, T.L. and J.R. Schrieffer, 1973, Phys. Rev. **B7**, 3629.
Fassaert, D.J.M. and A. van der Avoird, 1976, Surface Sci. **55**, 291.
Forstmann, F., 1977, Electron states at solid surfaces, in: Photoemission from surfaces (Feuerbacher, B., Fitton, B. and R.F. Willis, eds.) (Wiley, London) to be published.
Gell-Mann, M. and K.A. Brueckner, 1957, Phys. Rev. **106**, 364.
Goodwin, E.T., 1939, Proc. Camb. Phil. Soc. **35**, 221, 232.
Grimley, T.B., 1967a, Pontif. Acad. Sci., Scr. Varia **31**, 443.
Grimley, T.B., 1967b, Proc. Phys. Soc. (London) **90**, 751.
Grimley, T.B., 1970, J. Phys. **C3**, 1934.
Grimley, T.B., 1974, Chemisorption theory: the molecular orbital approach, in: Dynamic aspects of surface physics, Int. School of Phys. "Enrico Fermi", Course 58 (Goodman, F.O., ed.) (Editrice Compositori, Bologna) pp. 298–330.
Grimley, T.B., 1975, Progress in surface and membrane science, vol. 9 (Academic Press, New York) pp. 71–161.
Grimley, T.B., 1976a, Chemisorption, electronic structure and reactivity of metal surfaces,

in: Electronic structure and reactivity of metal surfaces (Derouane, E.G. and A.A. Lucas, eds.) (Plenum Press, New York) pp.113–162.
Grimley, T.B., 1976b, CRC Crit. Rev. Solid State Sci. **6**, 239.
Grimley, T.B. and S.M. Walker, 1969, Surface Sci. **14**, 395.
Grimley, T.B. and C. Pisani, 1974, J. Phys. **C7**, 2831.
Grimley, T.B. and E.E. Mola, 1976, J. Phys. **C9**, 3437.
Grimley, T.B. and A.M. Rosales, 1976, unpublished.
Gunnarsson, O., 1975, Thesis, Chalmers University of Technology, Göteborg, Sweden.
Gunnarsson, O. and H. Hjelmberg, 1975, Phys. Scripta (Sweden) **11**, 97.
Gunnarsson, O. and P. Johansson, 1976, Int. J. Quant. Chem. **10**, 307.
Gunnarsson, O. and B.I. Lundqvist, 1976, Phys. Rev. **B13**, 4274.
Gunnarsson, O. and B.I. Lundqvist, 1977, Surface Sci. **68**, 158.
Gunnarsson, O., B.I. Lundqvist and S. Lundqvist, 1972, Solid State Commun. **11**, 159.
Gunnarsson, O., H. Hjelmberg and B.I. Lundqvist, 1976, Phys. Rev. Lett. **37**, 292.
Gurman, S.J., 1976, Surface Sci. **55**, 93.
Hall, G.G., 1950, Prox. Roy. Soc. **A202**, 336.
Haydock, R., V. Heine and M.J. Kelly, 1972, J. Phys. **C5**, 2845.
Haydock, R., V. Heine and M.J. Kelly, 1975, J. Phys. **C8**, 2591.
Hedin, L., 1974, Exchange and correlation effects in band-structure, in: Elementary excitations in solids, molecules and atoms (Devreese, J.T., Kunz, A.B. and T.C. Collins, eds.) (Plenum Press, London) pp. 189–229.
Heitler, H. and F. London, 1927, Z. Phys. **44**, 455.
Hohenberg, P. and W. Kohn, 1964, Phys. Rev. **136**, B864.
Kelly, M.J., 1974, Surface Sci. **43**, 587 (The precise resonance we need was not calculated here, but is similar in width to those published).
Kittel, C., 1963, Quantum theory of solids (Wiley, New York) p. 360.
Kohn, W. and L.J. Sham, 1965, Phys. Rev. **140**, A1133.
Koopmans, T.A., 1933, Physica **1**, 104.
Koster, G.F. and J.C. Slater, 1954, Phys. Rev. **95**, 1167.
Koutecký, J., 1965, Advan. Chem. Phys. **9**, 85.
Lafon, E.E. and C.C. Lin, 1966, Phys. Rev. **152**, 579.
Lang, N.D., 1976, Density functional approach to electronic structure, in: Electronic structure and reactivity of metal surfaces (Derouane, E.G. and A.A. Lucas, eds.) (Plenum Press, New York) pp. 81–111.
Lang, N.D. and W. Kohn, 1970, Phys. **B1**, 4555.
Lang, N.D. and A.R. Williams, 1976a, Phys. Rev. Lett. **34**, 531.
Lang, N.D. and A.R. Williams, 1976b, Phys. Rev. Lett. **37**, 212.
Lennard-Jones, J.E. and J.A. Pople, 1950, Proc. Roy. Soc. **A202**, 166.
Lifshitz, I.M., 1948, Zh. Ekspim. Theoret. Fiz. **18**, 293.
Lifshitz, I.M., 1964, Advan. Phys. **23**, 483.
Melius, C.F., Moskowitz, J.W., Mortola, A.P. and M. B. Baillie, 1976, Surface Sci. **59**, 279.
Mott, N.F., 1958, Nuovo Cim. Suppl. **7**, 312.
Muscat, J.P. and D.M. Newns, 1977, to be published.
Nerbrant, P-O., 1975, Int. J. Quant. Chem. **9**, 901.
Pendry, J.B., 1974, Low energy electron diffraction (Academic Press, London).
Roothaan, C.C.J., 1960, Rev. Mod. Phys. **32**, 179.
Sachtler, W.M.H., 1969, in: Molecular processes on solid surfaces (Drauglis, E., Gretz, R.D. and R.I. Jaffee, eds.) (McGraw-Hill, New York) p. 313.
Schulte, F.K., 1974, J. Phys. **C7**, L370.

Stoddart, J.C., and N.H. March, 1971, Ann. Phys. (New York) **64**, 174.
Tamm, I., 1932, Z. Phys. **76**, 849.
Veillard, A., 1966, Theoret. Chim. Acta, **4**, 22.
Veillard, A. and E. Clementi, 1967, Theoret. Chim. Acta, **7**, 133.
Wahl, A.C., and G. Das, 1970, Advan. Quant. Chem. **5**, 261.
Wigner, E.P., 1938, Trans. Faraday Soc. **34**, 678.
Zubarev, D.N., 1960, Uspekhi Fiz. Nauk, **71**, 71 (English translation: Soviet Physics Uspekhi **3**, 320).

CHAPTER 2

CLUSTER MODEL THEORY AND ITS APPLICATION TO METAL SURFACE–ADSORBATE SYSTEMS

R.P. MESSMER

General Electric Corporate Research and Development, Schenectady, New York 12301, *USA*

Contents

The nature of the surface chemical bond
Edited by Th.N. Rhodin and G. Ertl

1. Electronic structure and metal–adsorbate systems

In many ways the theory of the electronic structure of surface–adsorbate systems is unique in the way it appears to force us to bring together the theoretical techniques of solid state physics and chemical physics (or quantum chemistry) in order to address the problem. For many years those areas of solid state physics and chemical physics concerned with electronic structure developed rather independently, the former concentrating on the behavior of electrons in solids and the latter primarily on the problems of chemical bonding and the associated behavior of isolated molecules.

During the last five years due to the considerable rekindled interest in surfaces and surface related phenomena, several theoretical techniques of solid state physics have been extended to treat the behavior of electrons at the surfaces of otherwise perfect crystalline materials. It would appear naively that the surface–adsorbate problem would be amenable to theoretical treatment through a combination of a chemical method with such a solid state technique. However, the juxtaposition of techniques, even if possible, may present considerable technical difficulties and result in a scheme which although aesthetically pleasing in a formal sense is computationally intractable. In particular if one looks at the general problem it becomes clear that the commonly used techniques for treating electron correlation effects in solids and molecules are quite different and hence a unified approach which treats the solid and the adsorbate on an equal footing becomes very complex. However, this general problem will not be of concern here. It will be sufficient here to consider some rather more restrictive aspects of electronic structure theory, which nonetheless present non-trivial conceptual and computational problems.

The aspects of electronic structure theory which will be of concern here, are those which may be generically referred to as "independent particle" theories. The particles of interest are electrons and an "independent electron" theory is one in which (for an N electron system) each electron moves in an average potential field produced by the remaining N-1 electrons. To properly carry out calculations using such independent electron theories, the potential field must be computed in an iterative manner so that the proper self-consistent field is generated. Hence, these theories are frequently referred to as self-consistent field (SCF) theories. In particular it should be noted that for a so-called spin-restricted calculation there is one orbital for each pair of electrons, whereas in the spin-unrestricted (or spin-polarized) calculations each electron has its own orbital, i.e. there is one orbital for a spin-up electron and another for the spin-down electron. Generalizations to the case where each electron has its own self-consistently determined orbital, yet the overall wavefunction is an eigenfunction of both spin and angular momenta, are not considered here as such procedures actually incorporate certain electron correlation effects.

In the context of solid state physics, independent electron theory manifests itself as energy band theory. In the last 10–15 years, many band structure calculations using a variety of theoretical techniques have been carried out for a wide range of bulk solids yielding valuable information concerning the electronic properties of these solids. In the area of chemical physics, independent electron theory appears as the ubiquitous molecular orbital theory. Here as well, there have been many calculations. The calculations have been performed on a wide variety of molecules using various molecular orbital techniques – some semi-empirical, some first principles, others, ab initio schemes. Collectively, these calculations have added very significantly to our understanding of the electronic structure and properties of molecules.

The main body of this chapter will be concerned with chemical theory and in particular various approaches to molecular orbital theory and their application to clusters of atoms representing a portion of a surface and the interaction of an adsorbate with such a cluster. It is important, however, to gain some perspective as regards connections between various solid state and molecular methods. In the next few pages an attempt to provide a brief account of these connections between theoretical methods will be made.

To begin it is convenient to divide both solid state and molecular independent particle theories into two classes, which will be referred to as "simple" and "first principles." Here, the term simple models means those in which additional approximations beyond the independent particle framework have been made. For example, self-consistency may be ignored in some methods, semi-empirical parameters used in some others and electron–electron interactions ignored in yet others. The term – first principles methods – is taken to mean those methods which are carried out self-consistently and which include the coulomb interactions between electrons, between nuclei and between electrons and nuclei, and include either a local or non-local exchange potential. In table 2.1, a list of some theoretical methods belonging to each class is presented and the connection between well known molecular and solid state methods is noted. In the table, authors associated with the development of each method are noted first, followed by some examples of recent work associated with the extension or application to metal surface problems. The choice of examples, however, is rather arbitrary. As space does not allow for an exhaustive list, many important contributions have not been included. The methods to the right have been used to investigate metal clusters and/or metal cluster–adsorbate systems, with the exception of those denoted by an asterisk. The methods on the left have been used to investigate metal surfaces and/or surface–adsorbate systems. These latter solid state methods treat the metal substrate in two different ways. Some methods treat the substrate as a semi-infinite solid, others treat it as a "slab." A semi-infinite solid is one in which the solid has infinite extent in the $\pm x$, $\pm y$ and $-z$ directions and has a surface at $z = 0$. A slab is a solid of finite thickness, Δz, but with infinite extent in the $\pm x$ and $\pm y$ directions.

From table 2.1 it can be seen that the solid state model which is frequently referred to in the current literature as cubium, is a generalization to infinite systems of simple Hückel theory. This model was discussed by Goodwin (1939) some time ago and has been used by Kalkstein and Soven (1971) as a model of a simple metal surface and later by Einstein and Schrieffer (1973) to study the interaction of two adsorbed atoms on a surface. The simple theory of Hückel (1931) was used by Blyholder and Coulson (1967) to discuss chemisorption for a linear chain of atoms and for a plane of atoms. Recently, the present author (Messmer 1977a), using simple Hückel theory, has shown that for a simple cubic array of atoms, the eigenvalues and eigenvectors of the problem can be obtained in closed form for any size cluster up to the infinite solid. This model which will be discussed in more detail later, although a very simple one, provides a valuable bridge between finite and infinite systems.

Modified versions of a model first used by Anderson (1961) to investigate magnetic impurities in metals, have been discussed by Grimley (1967) and Newns (1969) as models for chemisorption. Calculations were carried out by Newns (1969) for hydrogen chemisorption. For the non-magnetic problem, this model reduces to one which is essentially

TABLE 2.1
Comparison of theoretical methods

Solid state physics	Molecular physics
Simple models	
Cubium	Simple Hückel theory
Goodwin (1939)	Hückel (1931)
Kalkstein and Soven (1971)	Blyholder and Coulson (1967)
Einstein and Schrieffer (1973)	Messmer (1977)
Modified Anderson model	Hückel ω-technique*
Anderson (1961)	Wheland and Mann (1949)
Grimley (1967)	Streitwieser (1960)
Newns (1969)	
Grimley and Pisani (1974)	
Modified Hubbard model	Pariser–Parr–Pople*
Hubbard (1963)	Pariser and Parr (1953)
Newns (1970)	Pople (1953)
Grimley and Mola (1976)	
Jellium model	
Bardeen (1936)	
Lang and Williams (1975)	
Gunnarsson et al. (1976)	
Parameterized LCAO	Extended Hückel, etc.
Slater and Koster (1954)	Hoffmann (1963)
Fassaert and van der Avoird (1976)	Pople et al. (1965)
	Fassaert and van der Avoird (1976)
First principles models	
Pseudo-potentials	Pseudo-potentials*
Phillips and Kleinman (1959)	Hellmann (1935)
Louie et al. (1976)	Kahn et al. (1976)
$X\alpha$ band theory	$X\alpha$–SW molecular theory
Slater (1951)	Slater (1965)
Kar and Soven (1975)	Johnson (1966)
Kohn (1975)	Johnson and Messmer (1974)
Kasowski (1976)	Batra and Robaux (1974)
Ab initio Hartree–Fock*	Ab initio Hartree–Fock
Harris et al. (1973)	Roothaan (1951, 1960)
	Marshall et al. (1976)
	Bauschlicher et al. (1975)

equivalent to a scheme used for conjugated organic molecules by Wheland and Mann (1949) and later developed by Streitwieser (1960). The difference in the methods arises in the details of the treatment of on-site coulomb repulsions. The molecular model is known as the Hückel ω-technique, and was developed to achieve a partial self-consistency by considering on-site coulomb repulsions in conjugated molecules. For the chemisorption application of Newns, the self-consistency was achieved only for the adsorbate atom.

A model proposed by Hubbard (1963) to investigate electron correlation in narrow bands may be viewed within the cluster context as analogous to the ω-technique or as an approximation to the model proposed by Pariser and Parr (1953) and Pople (1953) for treating planar unsaturated molecules. The latter can clearly be seen when the Hamiltonian of this model is written in second quantized form as in Linderberg and Öhrn (1968). Recently, the Hubbard model has in fact been applied to finite systems as a chemisorption model in the work of Grimley and Mola (1976), which presents both an approximate and a full implementation of this model for clusters. This appears to be the only application of these models to the chemisorption problem within the framework of independent particle theory. Newns (1970) has treated the case of a diatomic molecule, i.e. one surface atom and one adsorbate atom, including some correlation effects, as a model chemisorption system.

The jellium model, first discussed by Bardeen (1936), has no direct molecular analog. The model treats the solid as an electron gas neutralized by a uniform background of positive charge. Hence, the atomic nature of the solid is not taken into account. The most sophisticated calculations using this model are those of Lang and Williams (1975) and Gunnarsson et al. (1976), the former considering the chemisorption of O, Li and H, and the latter the chemisorption of H. The electron density of the jellium model was chosen in both applications to represent an aluminum substrate.

Many parameterized linear combination of atomic orbitals (LCAO) band structures have been published, for insulators and semiconductors as well as metals. Most of the approximations of the band theory methods are similar in philosophy, although considerably different in detail, to such parameterized molecular orbital methods as the extended Hückel theory of Hoffman (1963) and the complete neglect of differential overlap (CNDO) and related methods of Pople et al. (1965) and Pople and Segal (1965). Recently, however, Fassaert and van der Avoird (1976a,b) have used an extended Hückel parameterization to carry out a solid state calculation for slabs of Ni with three different exposed surfaces as well as cluster calculations. The chemisorption of H atoms on these surfaces and clusters was studied. A discussion of semi-empirical molecular orbital studies related to chemisorption will be discussed in a later section.

The first principles methods will now be considered. The first of these listed in table 2.1 is the class of techniques denoted by the name pseudo-potential. Most of the applications of pseudo-potential techniques have been for solids. The principal advantage is due to the fact that the core electrons are not explicitly treated, and hence the computational effort can be greatly reduced for large systems. These techniques have not found much use so far for considering chemisorption on metals; although a recent calculation, which also used the $X\alpha$ exchange approximation, made by Louie et al. (1976) considered the surface of a Nb slab. The pseudo-potential techniques used in solid state and molecular applications are very different, as a comparison of the recent papers by Louie et al. (1976) and Kahn et al. (1976) will quickly show. Thus, these methods are not likely in the immediate future to provide a useful bridge between finite and infinite systems.

The need for such a bridge is not a trivial consideration. Suppose for example, that the methods for treating an isolated molecule and those for treating an isolated metal surface were compatible, i.e. the physical approximations are the same and the mathematics of implementation, such as the nature of the expansion functions and the size of the basis, are compatible. Then it would be possible to compare the isolated systems and the interacting system within the same physical and mathematical framework. This allows meaningful comparisons to be made. The alternative of having to draw conclusions from calculations based on different procedures is at best unsatisfying and at worst misleading. Fortunately, the $X\alpha$ theory of Slater (1972), implemented by the scattered-wave (SW) formalism of Johnson (1966, 1973) for molecules and clusters of atoms is compatible with the formalism of Korringa (1947), and Kohn and Rostoker (1954), the KKR method, for the energy bands of solids. Likewise, other mathematical implementations of the $X\alpha$ theory exist for both the molecular and solid state cases. For example, the LCAO formalism has been used for molecules by Sambe and Felton (1975) and for bulk metals by Callaway and Wang (1973). Recently, the KKR theory of energy bands for bulk solids has been extended by Kar and Soven (1975) and Kohn (1975) to treat metal surfaces by considering slabs. Kasowski (1974, 1976) has also used a related method to investigate metal surfaces and chemisorption. Applications of the $X\alpha$ theory will be considered in more detail in a later section.

The final theoretical method in table 2.1 is the ab initio Hartree–Fock method. The LCAO Hartree–Fock approach as formulated by Roothaan (1951, 1960) has been the mainstay of ab initio quantum chemistry. A survey of applications to molecules may be found in the book by Schaefer (1972). The method is not normally used in the theory of energy bands for metals because of the well known failure of Hartree–Fock theory to describe the electron gas, as discussed, for example, by Kittel (1963). However, the Hartree–Fock band calculations of Harris et al. (1973) have provided useful information with respect to possible improvements in local exchange potentials. Recently, Hartree–Fock calculations have been carried out for metal clusters; for Li clusters by Marshall et al. (1976) and for Be clusters interacting with a H atom by Bauschlicher et al. (1975).

In this section an attempt has been made to provide some perspective as regards the general problem of determining the electronic structure of metal–adsorbate systems as well as trying to relate various solid state and molecular approaches where possible. In the next section, the philosophy of cluster models will be discussed, and in ch. 2 sect. 3, the molecular orbital approaches which have been used to study clusters, will be considered. The remaining two sections will be devoted to discussing applications of cluster model theory and assessing what is being learned from such calculations about the nature of the surface chemical bond.

2. The philosophy of cluster models

Before proceeding on, in the next section, to describe the various quantum chemical methods which have been applied to the description of metal clusters and chemisorption thereon, it is important to consider some general issues regarding clusters as models of solid surfaces. It is important to realize both the advantages and disadvantages of using clusters as model substrates. Furthermore, it is desirable to have in mind a number of criteria for assessing the usefulness of various quantum chemical methods for discussing adsorbate–cluster bonding. A short list of advantages and disadvantages associated with the use of clusters as surface models is given in table 2.2. The comments made in this table will become clear on reading the discussion below.

One question which immediately comes to mind with regard to clusters representing a metal substrate is: how many atoms are needed in the cluster to "describe the metal"? The answer to this question will

TABLE 2.2
Clusters as surface models

Advantages	Disadvantages
Localized effects readily treated	Long range effects may require clusters too large to be readily handled
Concepts and computational methods developed in theoretical chemistry can be used	Must investigate convergence of results as a function of cluster size
Accurate results are possible as long as the cluster size needed for the property of interest is not too large	Unwanted "surface effects" may arise, i.e. a cluster has surfaces on all sides not just the side the adsorbate interacts with
Computational requirements are much less than for the corresponding semi-infinite treatment using a method of comparable physical sophistication	

obviously depend upon which properties of the metal one wishes to describe. For example, a cluster consisting of 50 atoms or less, which is already quite a computational feat for most molecular orbital methods, would be sorely inadequate for the discussion of a number of solid state aspects of the substrate. Such aspects as accurate values of bulk cohesive energies, work functions and any effects associated with the Fermi surface are clearly outside the scope of these clusters. Furthermore, in general, any property which depends upon the knowledge of a distribution of one-electron states with an energy resolution smaller than the average orbital energy difference of a given cluster size cannot be dealt with accurately by such a cluster. However, the interest here is not in reproducing bulk properties with clusters, but in understanding adsorbate bonding at metal surfaces using clusters. This is a very different task and one which should be accessible through cluster studies, if the chemist's view of bonding as a local phenomenon is correct.

The fundamental information obtained from electronic structure calculations on a substrate–adsorbate system is information regarding orbital energies and the nature of the corresponding orbitals (i.e. the one-electron wave functions). The basic premise is that by relating these orbitals (both their energies and wave functions) to those from the isolated systems, understanding regarding metal–adsorbate bonding can be obtained. Imagine that one has a single crystal of a metal with one surface which can interact with an adsorbate. Suppose that this metal with a surface is adequately described by an independent particle model and in particular by the hypothetical "ABC" method. Further, suppose that it has a band width of W which is partially occupied and whose

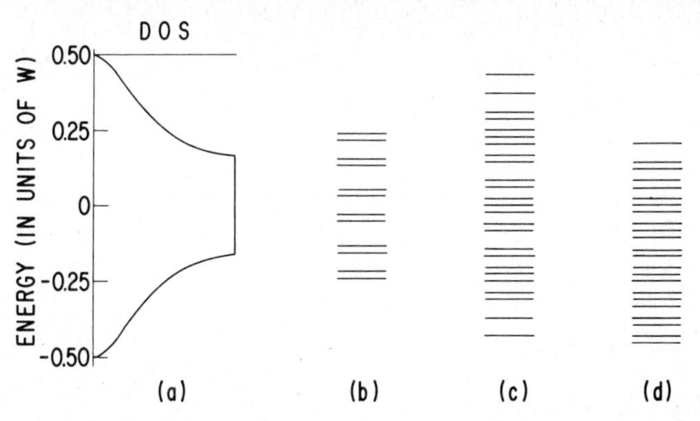

Fig. 2.1. Energy level distributions from two hypothetical theoretical methods. (a) Density of states (DOS) using ABC method for semi-infinite substrate; (b) energy levels using ABC method for small cluster representing the substrate; (c) energy levels using ABC method for larger cluster representing the substrate; (d) energy levels using XYZ method for same cluster as in (c), representing the substrate.

density of states (DOS) is as given in fig. 2.1a. Assume that two clusters (one small and the other larger), which are to provide approximations to the metal, have energy levels calculated by the ABC method as shown in fig. 2.1b and fig. 2.1c respectively. If an adsorbate interacts with the metal and produces shifts in orbital energies in the metal which are small on the scale of the level spacings of the clusters (figs. 2.1a,b) then the clusters will not provide a very accurate description of this interaction. Furthermore, if the adsorbate interacts strongly with the states at the bottom of the band, it may be that the cluster "band width" is not large enough to represent this interaction correctly. In either case, however, it may well be that a larger cluster with more energy levels (as in fig. 2.1c) than a smaller one (as in fig. 2.1b) is adequate to represent the interaction with sufficient accuracy. For orbital interactions associated with chemical bonding, orbital shifts are typically tenths of eV to a few eV. Thus, the bonding should be reasonably well described by a cluster whose average orbital energy difference is, roughly speaking, ~ 0.1 eV to a few tenths of eV, and whose band width, W_c, is large enough to represent all adsorbate induced level shifts.

In the above discussion the ABC method was used both for the infinite case and for the cluster case. Suppose, however, that another method – say the "XYZ" method – had been applied only to clusters and gave the energy levels shown in fig. 2.1d instead of those in fig. 2.1c for the larger cluster. It is not a straightforward task to compare the distribution of energy levels for a finite cluster such as those given in figs. 2.1b–d with the DOS of the semi-infinite case and conclude whether the comparison is good or poor. Such a comparison is made all the more difficult by the fact that many of the discrete energy levels for a symmetrical cluster are highly degenerate. In order to proceed with this discussion, therefore, it is necessary to make a brief digression to consider the construction of DOS curves for clusters.

A convenient way to represent the electronic structure of a cluster which facilitates comparison with the metal substrate results is by constructing DOS curves for the metal clusters. The total density of states for the bulk metal is given by

$$\rho(E) = \sum_k \delta(E - \varepsilon_k) \tag{2.1}$$

where the sum is over all one-electron states. This expression will also be a very good approximation for a solid with a surface, as the relative number of surface states is very small. For the cluster where there is a finite number of discrete states, the delta functions in eq. (2.1) can be replaced by Gaussians and the DOS written as

$$\rho^{c}(E) = (2\pi\sigma^2)^{-1/2} \sum_l \exp[-(E - \varepsilon_l)^2/2\sigma^2], \tag{2.2}$$

where the sum is over all discrete states and σ is the Gaussian width parameter, which is related to the full width at half maximum (FWHM) by FWHM = 2.355σ. As pointed out above, the total density of states of the solid will be dominated by the bulk contributions as the ratio of surface to bulk states is extremely small. In order to gain information about the surface electronic structure or about the electronic structure in a local region of the solid it is customary to define a local density of states (LDOS) or projected density of states (PDOS) in which local information is projected out of the total DOS. If, for example, one assumes a one band solid in an LCAO framework then the total DOS can be written as

$$\rho(E) = \sum_m \sum_k C^*_{km} C_{km} \delta(E - \varepsilon_k) \tag{2.3}$$

where the C_{km} are the atomic orbital coefficients and the sum on k is over all states and the sum on m is over all sites. Note that for a bulk solid with Bloch functions as eigenfunctions, eq. (2.3) reduces to eq. (2.1). The LDOS or PDOS on atom m is given by

$$\rho_m(E) = \sum_k C^*_{km} C_{km} \delta(E - \varepsilon_k) \tag{2.4}$$

and the corresponding expression for the cluster is

$$\rho^{c}_{m}(E) = (2\pi\sigma^2)^{-1/2}\sum_{l}|C_{lm}|^2 \exp[-(E-\varepsilon_l)^2/2\sigma^2]. \tag{2.5}$$

With these quantities in mind, the discussion which was interrupted above may now be completed.

The cluster DOS curves, representing the clusters whose discrete energy levels were shown in figs. 2.1b–d, are presented in fig. 2.2 which may be compared with the metal DOS in fig. 2.1a. The value of σ used for all the clusters is $\sigma = 0.033\,W$. It is clear that the larger cluster results using the ABC method (fig. 2.2b) are in better agreement with the metal DOS curve (fig. 2.1a) than are the results of the smaller cluster DOS curve (fig. 2.2a). The DOS curve obtained from the XYZ method for the larger cluster is rather different from any of the other DOS curves. If the only results available were those of fig. 2.1a and fig. 2.2c, one might arrive at two possible conclusions on a comparison of these curves. First, one might conclude that clusters have a rather different electronic energy level distribution than the real metal and are, therefore, incapable of providing an approximate model to discuss surface bonding. Second, one might conclude that the XYZ method is not very good for describing the electronic structure of metal clusters. Certainly, one cannot logically arrive at the first conclusion, without eliminating the second possibility. The best way to resolve the problem is to use the XYZ method to calculate the DOS for the semi-infinite case. This points to the im-

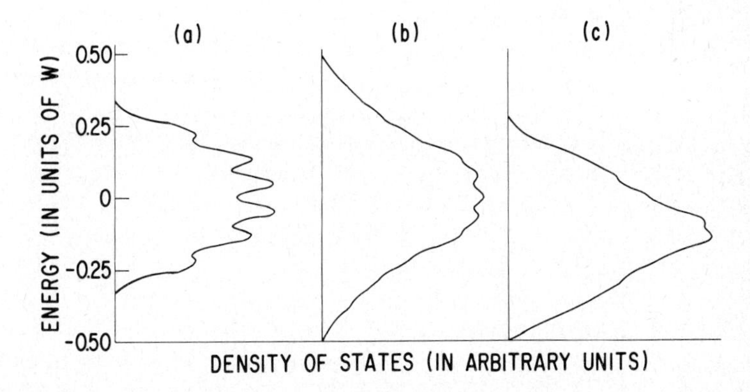

Fig. 2.2. Density of states curves for clusters of atoms. (a) DOS curve for small cluster from ABC method (energy levels shown in fig. 2.1b); (b) DOS curves for larger cluster from ABC method (energy levels shown in fig. 2.1c); (c) DOS curve for larger cluster from XYZ method (energy levels shown in fig. 2.1d).

portance of having methods which can treat both the semi-infinite and finite cluster limits.

As mentioned above, the cluster calculations produce one-electron orbital energies and wave functions from which information concerning bonding can be obtained. To the degree that independent particle theory is accurate, the ionization potentials of a molecule or cluster can be identified with the energies needed to remove electrons from the individual orbital wave functions (MO's). In order to obtain reasonably accurate ionization potentials, it is necessary to take account of orbital relaxation effects which take place due to the remaining electrons readjusting to a new potential field. A detailed discussion of how these effects can be computed and the relationship of orbital energies to ionization potentials within the Hartree–Fock and Xα theories is given in subsect. 3.3. Experimentally, the most direct technique for determining ionization energies is photoemission spectroscopy. As an example, the orbital ionization energies for CO as deduced from its photoemission spectrum are listed in table 2.3 and compared to an Xα–SW molecular orbital calculation of Rösch et al. (1973). Photoemission spectra can also yield information regarding orbital wave functions as will be discussed in more detail later. However, this information is very much more difficult to unravel, as it involves the behavior of peak intensities as a function of photon energy and electron ejection angles. The main point here is that the preoccupation with photoemission studies in the remainder of the chapter is due to the fact that such experiments offer the most direct point of comparison with calculated DOS curves, orbital energies, orbital energy shifts, etc.

TABLE 2.3
Ionization potentials of CO (in eV), a comparison of theory and experiment

Orbital	Theory[a]	Experiment[b]
5σ	14.1	14.5
1π	17.5	17.2
4σ	20.1	20.1
3σ	34.5	38.3
2σ	298.9	295.9
1σ	544.0	542.1

[a] Rösch et al. (1973) using Xα–SW method with overlapping spheres.
[b] Siegbahn et al. (1969) X-ray photoemission spectrum.

It is useful to consider in some detail the concepts which have been introduced so far by studying a concrete example. The model which will be used as an introduction to clusters is the Hückel/cubium model. This is due to its simplicity and as discussed in the last section its advantage of being a technique which provides a bridge between the finite cluster and the infinite solid.

Consider a simple cubic array of atoms in a cluster with one atomic orbital χ_μ per site. In the simple Hückel LCAO approximation the Hamiltonian is completely specified by its matrix elements over these atomic orbitals. These are $\langle\chi_\mu|H|\chi_\nu\rangle = \varepsilon_0$ for the diagonal elements and for the off-diagonal elements, $\langle\chi_\mu|H|\chi_\nu\rangle = t$ if μ and ν are nearest neighbors, otherwise the off-diagonal elements are zero. As is usual in this simple method overlap integrals are neglected, i.e. $\langle\chi_\mu|\chi_\upsilon\rangle = \delta_{\mu\nu}$. If the cluster consists of N_A, N_B and N_C atoms in the x, y and z directions respectively, the resultant secular equation of dimension $N_A N_B N_C \times N_A N_B N_C$ is

$$(\boldsymbol{H} - \boldsymbol{\varepsilon})\boldsymbol{C} = 0. \tag{2.6}$$

The matrix $\boldsymbol{H}$ contains the elements discussed above, $\boldsymbol{\varepsilon}$ is a diagonal matrix and contains the eigenvalues, and $\boldsymbol{C}$ is the eigenvector matrix, which contains the coefficients of the molecular orbitals (MO's). The MO's are given by

$$\psi_\rho = \sum_\mu c_{\mu\rho}\chi_\mu, \qquad \text{with } \rho = 1, \ldots, N_A N_B N_C. \tag{2.7}$$

It has recently been shown by the author (Messmer 1977) that this problem can be solved in closed form for arbitrary values of N_A, N_B and N_C. Actually, a more general case was proven, in which the off-diagonal elements in the x, y and z directions can be arbitrary as well as the nearest neighbor overlap integrals in the three directions; however, only the more restrictive case outlined above need be considered as an example. The proof will not be repeated here, only the results will be quoted. The eigenvalues of the secular problem, eq. (2.6), are

$$E_\rho = \varepsilon_0 + 2t\{\cos[i\pi/(N_A+1)] + \cos[j\pi/(N_B+1)] + \cos[k\pi/(N_C+1)]\}, \tag{2.8}$$

where $\rho = (k-1)N_A N_B + (j-1)N_A + i$ and $i = 1, \ldots, N_A$; $j = 1, \ldots, N_B$; and $k = 1, \ldots, N_C$. The eigenvector coefficients $c_{\mu\rho}$ of eq. (2.7) are given by

$$c_{\mu\rho} = U_C(k, n)U_B(j, m)U_A(i, l) \tag{2.9}$$

where ρ is as given above and $\mu = (n-1)N_A N_B + (m-1)N_A + l$, with $l = 1, \ldots, N_A$, $m = 1, \ldots, N_B$ and $n = 1, \ldots, N_C$. The functions $U_\alpha(\beta, \gamma)$ are given by

$$U_\alpha(\beta, \gamma) = [2/(N_\alpha + 1)]^{1/2} \sin[\beta\gamma\pi/(N_\alpha + 1)] \tag{2.10}$$

TABLE 2.4
Convergence of some cluster results

N	No. of atoms	E_{min}	Band width	Fraction of surface atoms
2	8	−1.500000	3.000000	1.000000
3	27	−2.121320	4.242641	0.962963
4	64	−2.427051	4.854102	0.875000
5	125	−2.598076	5.196153	0.784000
6	216	−2.702907	5.405813	0.703704
7	343	−2.771639	5.543277	0.635569
8	512	−2.819078	5.638156	0.578125
9	729	−2.853170	5.706339	0.529492
10	1000	−2.878479	5.756958	0.488000
11	1331	−2.897778	5.795555	0.452292
12	1728	−2.912825	5.825651	0.421296
13	2197	−2.924784	5.849568	0.394174
14	2744	−2.934443	5.868886	0.370262
15	3375	−2.942356	5.884712	0.349037
16	4096	−2.948919	5.897839	0.330078
17	4913	−2.954423	5.908847	0.313047
18	5832	−2.959084	5.918168	0.297668
19	6859	−2.963065	5.926130	0.283715
20	8000	−2.966493	5.932985	0.271000
21	9261	−2.969464	5.938929	0.259367
22	10648	−2.972058	5.944116	0.248685
23	12167	−2.974335	5.948669	0.238843
24	13824	−2.976344	5.952688	0.229745
25	15625	−2.978127	5.956253	0.221312
26	17576	−2.979715	5.959430	0.213473
27	19683	−2.981137	5.962273	0.206168
28	21952	−2.982414	5.964828	0.199344
29	24389	−2.983566	5.967131	0.192956
30	27000	−2.984608	5.969216	0.186963
40	64000	−2.991197	5.982395	0.142625
50	125000	−2.994310	5.988620	0.115264
60	216000	−2.996022	5.992045	0.096704
70	343000	−2.997064	5.994127	0.083289
80	512000	−2.997744	5.995488	0.073141
90	729000	−2.998212	5.996425	0.065196
100	1000000	−2.998549	5.997098	0.058808

with $\alpha = A, B, C$, $\beta = i, j, k$ and $\gamma = l, m, n$; these indices take on the range of values specified above. This analytic cluster model (ACM) allows one to calculate various electronic properties of clusters and determine how quickly they converge as a function of cluster size to the infinite or semi-infinite cases (i.e. bulk metal case or metal with a surface).

In table 2.4 some simple results for clusters of up to a million atoms are presented; the results are for cubic clusters with $N_A = N_B = N_C = N$. The table lists: the value of N, the number of atoms in the cluster, the lowest eigenvalue, the total band width, and the fraction of surface atoms in the cluster. The energies in table 2.4 are given in units of $2t$; note that the total band width in this model is simply $2|E_{min}|$. It can be seen that E_{min} and, therefore, the band width as well, converges smoothly to the semi-infinite value. Note that even for a cluster of a million atoms nearly 6% of the atoms are still surface atoms. The relatively slow convergence of the model can be attributed to at least three factors. First, there is only one orbital per site; second, the clusters chosen are highly symmetric, which leads to fewer unique eigenvalues with larger degeneracies. Third, the model, due to its simple form, introduces a significant amount of "accidental degeneracy". These factors all lead to a rather sparse eigenvalue spectrum. One would expect if a more sophisticated model with many orbitals per site, appropriate to such metals as Mg, Al or to transition metals, could be solved in analytic form, that the convergence would be much more rapid. Thus, although the present model may be of considerable pedagogic value for a number of features, it certainly does not yield results indicative of the rapidity of convergence to be expected in general for real metals.

Grimley and Pisani (1974) have studied chemisorption on the semi-infinite solid using the cubium model for the substrate and treating the adatom and the substrate atom to which it is attached self-consistently. Grimley and Mola (1976) then used the same procedure to treat clusters of up to 153 atoms to study the convergence to the semi-infinite limit. Their results, although over a rather small size range, are in agreement with the results discussed above. The limitation on size was due to the fact that secular equations such as eq. (2.6) had to be diagonalized numerically. Using the analytic cluster results presented above, together with the Green's function techniques discussed by Kalkstein and Soven (1971), Einstein and Schrieffer (1973) and Grimley and Pisani (1974), which can be extended to clusters with obvious modifications, one should in the future be able to discuss chemisorption on cubic clusters of very much greater size, including self-consistency on the adatom.

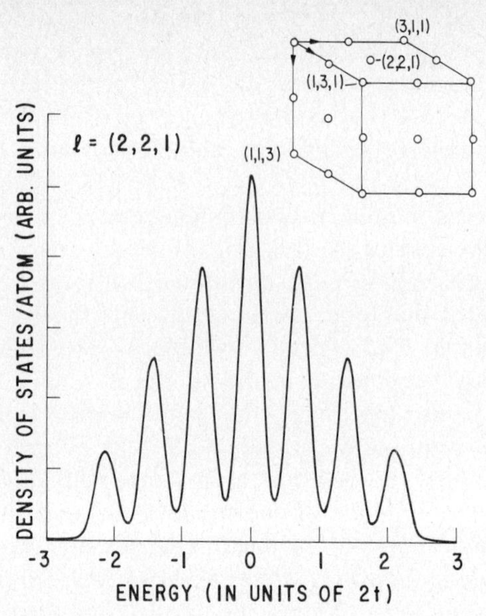

Fig. 2.3. Local density of states from the simple Hückel method for the central atom on a (100) face of a $3 \times 3 \times 3$ atom cluster. The coordinate system is shown in the inset and the coordinates of the central atom are $l = (2, 2, 1)$.

As a final comparison between semi-infinite results and clusters using Hückel theory, the evaluation of the LDOS for clusters of 27 and 729 atoms using the analytic cluster model will be compared to the LDOS results of Kalkstein and Soven (1971) for the semi-infinite case. The 27-atom cluster is a simple array of $3 \times 3 \times 3$ atoms and the 729-atom cluster is a $9 \times 9 \times 9$ array of atoms. The coordinate system used is shown in the inset of fig. 2.3, and in the main figure the LDOS for the surface atom of the 27-atom cluster with coordinates (2, 2, 1) is shown. The Gaussian width is $\sigma = 0.15$ in units of $2t$. The seven unique eigenvalues of this example are clearly seen. The LDOS curves for the 729-atom cluster are shown in fig. 2.4. The labeling scheme is analogous to that used in fig. 2.3. For this cluster there are 129 unique eigenvalues. The curves of fig. 2.4 may be compared to the curves published by Kalkstein and Soven (1971) for the corresponding quantities calculated by the semi-infinite cubium model. These latter curves are shown in fig. 2.5. The notation is the same as that used in the two previous figures, although i and j are arbitrary, as any atom in a given plane parallel to the surface will have the same LDOS in the semi-infinite crystal.

Comparison of figs. 2.4 and 2.5 shows that the cluster results reproduce

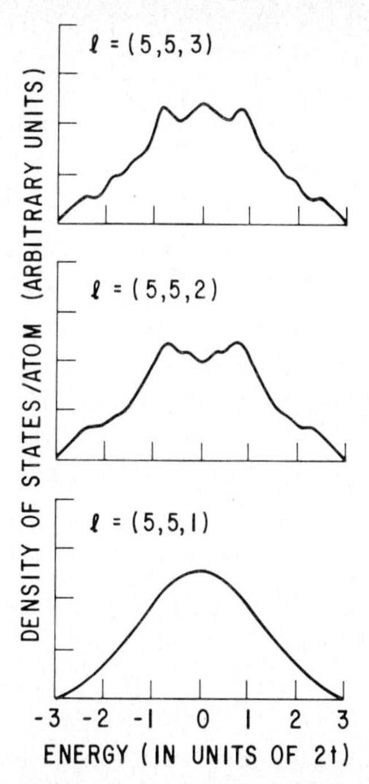

Fig. 2.4. Local density of states from the simple Hückel method for three atoms in a $9 \times 9 \times 9$ atom cluster. The coordinates of the three atoms are: $l = (5, 5, 1)$, central atom on a (100) face of the cluster; $l = (5, 5, 2)$, atom in second plane just beneath the first atom; $l = (5, 5, 3)$, atom in third plane of cluster just beneath the second atom.

the main features of the semi-infinite results, although "fine structure" features can be discerned in some of the curves of fig. 2.4 showing that the discreteness of the eigenvalue spectrum continues to manifest itself. Nonetheless, the agreement is quite striking, and for many purposes the cluster results would be more than adequate. Thus, for this particular property one could conclude that the 729-atom cluster constitutes a "usefully converged" representation of the semi-infinite result, although in an absolute sense this cluster is still quite far from the semi-infinite system having only 129 discrete eigenvalues and 95% of the proper band width.

Thus far, we have stressed the point that in the limit of an infinite cluster the theoretical method chosen to describe the electronic structure of the cluster must yield a reasonable description of the band structure of the bulk metal. Hence, this should be one criterion for the selection of a theoretical method. On the other hand, as the desired

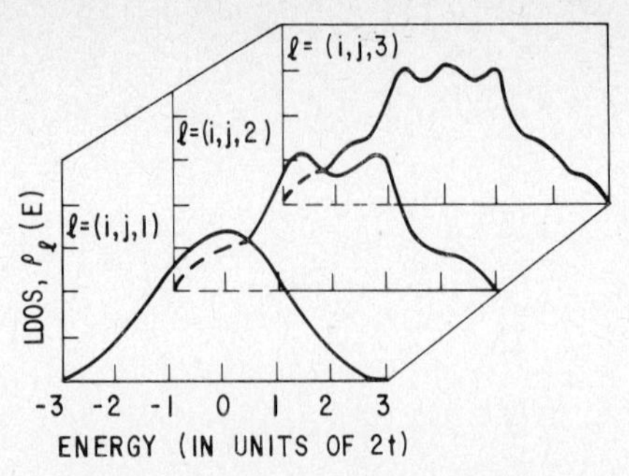

Fig. 2.5. Local density of states from the semi-infinite cubium model for atoms in three different planes. The coordinates of the three atoms are: $l = (i, j, 1)$, atom on a (100) surface; $l = (i, j, 2)$, atom in second layer; $l = (i, j, 3)$, atom in third layer.

information concerns the bonding of an adsorbate to a metal cluster, the theoretical method should also provide reliable information about metal–ligand bonding as, for example, in transition metal complexes. This represents a second criterion in choosing a theoretical method. A third criterion which arises from strictly pragmatic considerations is that the method should be computationally tractable for clusters large enough to allow the investigation of the convergence of the properties of interest as a function of cluster size. Hence, a method which is capable of treating only a very restricted number of metal atoms is likely to be of limited utility in discussing bonding at metal surfaces. From the discussion above concerning the Hückel/cubium model, it is clear that such a simple model can be very instructive for establishing qualitative effects and making connections between semi-infinite and finite systems. However, such models are far too simplified to draw definitive conclusions regarding real systems, especially with regard to bonding. In order to discuss bonding, more sophisticated models are demanded.

3. Survey of theoretical techniques applied to cluster models

The purpose of the present section is to outline the more commonly used theoretical techniques which have been applied to metal clusters. As space does not permit an extensive review of applications using each method, only one example from the recent literature will be cited for each method as an illustration. In sect. 5, a few specific systems will be discussed in considerable detail and comparisons of results among the various theoretical methods and with experiment will be presented.

The techniques to be discussed in this section represent considerable improvements over the simple Hückel model discussed in the last section. With varying degrees of success they have been used to describe bonding in a wide variety of molecules. The class of techniques which have been most widely used are those referred to as semi-empirical methods. These methods will be dealt with first.

3.1. Semi-empirical LCAO–MO theory

The two most widely used semi-empirical methods to treat all valence electrons of a molecule are the extended Hückel and CNDO methods. As discussed in the last section, simple Hückel theory treats one orbital per atomic site, neglects overlap of orbitals and considers only nearest neighbor interactions. In the extended Hückel theory as developed by Hoffmann (1963), several orbitals per site can be treated; all two center interactions are included and all overlap integrals between atomic orbitals are taken into account. To be more explicit, the matrix elements between atomic orbitals for an effective one-electron Hamiltonian, H, are

$$\langle\chi_\mu|H|\chi_\mu\rangle = -I_\mu \tag{2.11}$$

for the diagonal elements and

$$\langle\chi_\mu|H|\chi_\nu\rangle = -\tfrac{1}{2}K(I_\mu + I_\nu)S_{\mu\nu} \tag{2.12}$$

for the off-diagonal elements. In these expressions, the quantities I_μ are energy parameters obtained from experimental atomic ionization data; they are frequently referred to as valence state ionization potentials. The constant K is an empirical factor usually taken as 1.75 and $S_{\mu\nu}$ is the overlap integral between atomic orbitals μ and ν. The atomic orbitals χ_μ, are usually taken as Slater orbitals, although linear combinations of Slater orbitals are sometimes used. The secular equation to be solved is

$$(\boldsymbol{H} - \varepsilon\boldsymbol{S})\boldsymbol{C} = 0 \tag{2.13}$$

where $\boldsymbol{S}$ is the overlap matrix. It should be noted that the extended Hückel (EH) method as outlined above and as usually applied is not a self-consistent procedure.

The CNDO method which is an approximate self-consistent field theory, has for a closed shell system in the CNDO/2 parameterization, diagonal and off-diagonal matrix elements given by

$$H_{\mu\mu} = -\tfrac{1}{2}(I_\mu + A_\mu) + [(P_{AA} - Z_A) - \tfrac{1}{2}(P_{\mu\mu} - 1)]\gamma_{AA} + \sum_{B \neq A} (P_{BB} - Z_B)\gamma_{AB} \tag{2.14}$$

and

$$H_{\mu\nu} = \tfrac{1}{2}(\beta_A + \beta_B)S_{\mu\nu} - \tfrac{1}{2}P_{\mu\nu}\gamma_{AB} \tag{2.15}$$

respectively. In eqs. (2.14) and (2.15) the subscripts μ, ν refer to atomic orbitals and subscripts A, B refer to atoms. Z_A is the core charge on atom A, I_μ has the same meaning as in the EH method, A_μ is the valence state electron affinity of orbital μ; γ_{AA} and γ_{AB} are one and two center electron repulsion integrals evaluated for Slater s-type orbitals; β_A and β_B are empirical parameters determined by comparison with ab initio calculations on diatomic molecules; $P_{\mu\nu}$ and P_{AA} are given by

$$P_{\mu\nu} = 2 \sum_i c_{i\mu} c_{i\nu} \qquad \text{and} \qquad P_{AA} = \sum_\mu^{A} P_{\mu\mu} \tag{2.16}$$

respectively, where the sum on i extends over all doubly occupied MO's. In eq. (2.16), the $c_{i\mu}$ are coefficients of the LCAO expansion of the molecular orbitals. The secular equation which is solved, can be written as

$$(\boldsymbol{H} - \boldsymbol{\varepsilon})\boldsymbol{C} = 0 \tag{2.17}$$

where, due to the complete neglect of differential overlap (CNDO) assumption, the overlap matrix is neglected. The self-consistency comes into eq. (2.17), by the fact that the matrix elements of H given by eqs. (2.14) and (2.15) depend upon the LCAO coefficients which are determined by the solution of eq. (2.17). There are many other semi-empirical methods, some based on modifications of CNDO, which have been applied to molecular problems. A good discussion of these various methods is given in the book by Murrell and Harget (1972). The discussion here has been limited to those methods which have been applied to chemisorption problems.

The studies of the present author and his co-workers (Bennett et al. 1971a,b, Messmer et al. 1972) were the first to use clusters in conjunction with the EH and CNDO methods to investigate the energetics of site preference in chemisorption as well as to employ the semi-infinite slab geometry, in conjunction with the EH method, to discuss chemisorption. These studies considered graphite as the substrate rather than a metal substrate. Since these initial studies, many subsequent investigations using the EH and CNDO methods have been reported which consider chemisorption on metal substrates. Many of these studies have used rather arbitrary choices of parameters and have made little attempt to relate the cluster calculations to the semi-infinite limit within a common framework and using the same set of parameters. The problems associated with such procedures have previously been pointed out by Messmer (1977b) and Messmer et al. (1976).

The recent work of Fassaert and van der Avoird (1976a,b), makes a very significant step to provide a bridge between metal clusters and the semi-infinite limit within the EH framework. These authors have used the same parameterization to discuss chemisorption of hydrogen overlayers on semi-infinite Ni slabs as well as hydrogen adsorption on Ni clusters. They considered H atoms on (100), (110) and (111) surfaces at positions: A, above a Ni atom; B, a bridge site between two nickel atoms; and C, a site centered over a group of nickel atoms. For the (110) surface there are two independent B-sites and for the (111) surface there are two independent C-sites. Nickel clusters of up to 13 atoms were employed in the study. They conclude from their studies that: (1) significant Ni 4s, 4p character as well as Ni 3d character is involved in the chemisorption bond with hydrogen; (2) the chemisorption bond strengths on the three surfaces increase with decreasing number of nearest neighbors to the surface atom, although these differences are not very pronounced; (3) on a given surface the adsorption energies decrease such that, $A > B > C$; and (4) the results of the cluster and semi-infinite calculations agree very well, quantitatively as well as qualitatively.

Blyholder (1974) has applied the CNDO method to Ni clusters of up to 10 atoms to discuss hydrogen atom chemisorption. The parameterization used for the Ni clusters, however, yields cluster DOS curves which are in significant disagreement with the bulk metal DOS as well as cluster DOS curves obtained by other methods, as pointed out previously by Messmer et al. (1976). Hence, one should be somewhat cautious in accepting the conclusions of this work without further substantiation. Nonetheless, the chief conclusions of the work will be summarized. Blyholder concludes from his calculations that hydrogen atoms are more stable in multicenter bonds over hole sites rather than directly over a surface Ni atom and that the hydrogen atoms have the largest binding energy on a (111) surface. Both of these results are contrary to the conclusions of Fassaert and van der Avoird. Blyholder also concludes that the s and p orbitals of Ni are primarily responsible for nickel–hydrogen bonding with the d-orbitals playing a relatively minor role.

Further discussion of the nickel–hydrogen system will be given in sect. 5, where comparisons with experiment and other theoretical methods will be presented.

3.2. AB INITIO LCAO–MO THEORY

The phrase "ab initio calculation," although very commonly used, is perhaps one of the most ill-defined and frequently misunderstood phrases in scientific jargon. The phrase is applied to a very wide variety of methods and calculations, ranging from minimal or sub-minimal basis set SCF–LCAO MO calculations to the most sophisticated calculations which include nearly all electronic correlation effects. Depending upon the method, how it is implemented, and the system being treated, agreement with experiment can range from very poor to almost exact. Hence the phrase, without further qualification, is not very useful as it provides so little information about a calculation.

In the present chapter the primary concern is with ab initio methods within the independent particle context where explicit account of electronic correlation is neglected. These methods are usually referred to as restricted and unrestricted Hartree–Fock theory. The molecular orbitals are approximated by a linear combination of atomic orbitals and unlike semi-empirical methods all the integrals over atomic orbitals are explicitly evaluated rather than being neglected or represented by quantities deduced from atomic data. The matrix elements over atomic orbitals for the restricted Hartree–Fock (RHF) method are

$$H_{\mu\nu} = h_{\mu\nu} + \sum_{\lambda\eta} P_{\lambda\nu}(\gamma_{\mu\nu\lambda\eta} - \tfrac{1}{2}\gamma_{\mu\lambda\nu\eta}), \tag{2.18}$$

where $h_{\mu\nu} = \langle\chi_\mu|h|\chi_\nu\rangle$, and h is the one-electron part of the Hamiltonian; the term

$$\gamma_{\mu\nu\lambda\eta} = \iint \chi_\mu(1)\chi_\nu(1)(1/r_{12})\chi_\lambda(2)\chi_\eta(2)\,\mathrm{d}\tau_1\,\mathrm{d}\tau_2 \tag{2.19}$$

is a two-electron repulsion integral. $P_{\lambda\eta}$ is given by eq. (2.16). The secular equation to be solved is

$$(\boldsymbol{H} - \varepsilon\boldsymbol{S})\boldsymbol{C} = 0. \tag{2.20}$$

A major task in the solution of these equations is the evaluation of the two-electron integrals in eq. (2.19). These integrals are usually more involved than they appear, as each atomic orbital may be expanded in a set of basis functions. These basis functions are usually chosen as

Gaussian functions in calculations for polyatomic systems. There are many procedures which one might apply to choose a basis set for a given system, but the procedures are non-trivial and not without ambiguity. This is clearly demonstrated by the recent discussions of Binkley and Pople (1977) and Brewington et al. (1976) on the choice of Gaussian basis sets for an atom as simple as Be; appropriate basis sets had to be generated for consideration of Be clusters and Be compounds.

Stimulated by the current interest in chemisorption and metal clusters as well as the previous semi-empirical and Xα calculations, a few ab initio Hartree–Fock calculations for metal clusters have recently appeared. The only work to date which considers chemisorption on a cluster of more than one or two atoms is the work of Bauschlicher et al. (1975). They consider the interaction of a hydrogen atom with clusters of up to ten Be atoms. The system was chosen for its relative computational simplicity. It is unfortunate, however, that no experimental data exists for the Be–hydrogen system. The main emphasis of the study was on the binding energy of a hydrogen atom at various sites on the Be clusters. However, no general qualitative conclusions with regard to preferred binding sites were made because the binding energy at various sites was found to depend quite strongly on the size of the Be clusters. Large enough clusters were not considered for the binding energies to converge to a well defined trend.

3.3. Xα–MO THEORY

Unlike the semi-empirical and ab initio molecular orbital methods which invariably employ the LCAO approximation to represent the molecular orbitals, the Xα theory of Slater (1972) has used three different mathematical representations for the molecular orbitals. The most commonly used and flexible method is the scattered wave (SW) method of Johnson (1966, 1973). This Xα–SW method has been applied to a rather wide variety of surface related problems; some of these applications will be discussed later. A second method which has been used to solve the Xα equations is the discrete variational (DV) method of Ellis and Painter (1970). This Xα–DV method has recently been applied to the chemisorption of first row atoms on a nickel cluster and will be discussed below as the example in this subsection. The third method is the LCAO method as used by Sambe and Felton (1975), although this method has not yet been applied to a surface related problem.

There are advantages and disadvantages to each of the methods. The Xα–SW method has the advantages of being computationally fast, and of being able to represent excited states rather easily and naturally. Furthermore, it can be straightforwardly extended to treat photoemission, as shown in the recent work of Dehmer and Dill (1975) and Davenport (1976a). Its chief disadvantage is the inaccuracies introduced by the muffin-tin potentials for molecules or clusters in which considerable charge density resides outside the atomic spheres. Much recent work by a number of investigators has focussed on the correction of these muffin-tin errors. The Xα–DV method is a numerical method whose accuracy depends upon the number of sampling points. Although potentially more accurate than the Xα–SW method because it avoids the muffin-tin approximation, a very large number of sampling points are required in practice to accomplish this. As the computational effort depends upon the number of sampling points, reasonably accurate SCF results require significantly more computational time than for the Xα–SW method. Furthermore, extended sets of orbital functions are necessary to describe excited states and it is not clear if the method can be extended to treat photoemission intensities. The Xα–LCAO method, like the Hartree–Fock LCAO method requires the evaluation of vast numbers of two-electron integrals, although they are fewer and less involved than in the Hartree–Fock method. It is the most computationally demanding of the Xα methods, although less demanding than the Hartree–Fock LCAO method. With large basis sets this method is potentially the most accurate Xα method for ground state properties, however, extensions to excited states and the treatment of photoemission would introduce considerable additional complications. Another promising method is the Xα linear combination of muffin-tin orbitals (LCMTO) approach of Anderson and Kasowski (1971). However, no self-consistent calculations have as yet been carried out, thus, a comparison with the methods above would be premature.

A brief outline of Xα theory and its concepts will be given and compared to the more familiar concepts of Hartree–Fock theory. Due to space limitations, however, the theory will be discussed in terms of molecular orbitals without considering the various mathematical representations of these molecular orbitals. Complete accounts of the SW, LCAO, DV and LCMTO representations of the Xα theory can be found in the references cited above.

For the purposes of the present discussion only the restricted Hartree–Fock (RHF) and restricted Xα methods need be compared. The RHF and restricted Xα total energies are given by

$$E_{\mathrm{HF}} = \sum_i h_i + \tfrac{1}{2} \sum_{ij} (J_{ij} - K_{ij}) \tag{2.21}$$

and

$$E_{X\alpha} = \sum_i n_i h_i + \tfrac{1}{2} \sum_{ij} n_i n_j J_{ij} - \tfrac{1}{2} \sum_i n_i V_i^{X\alpha} \tag{2.22}$$

respectively. The sums in eqs. (2.21) and (2.22) are over all occupied molecular spin orbitals (MSO's), $\{\psi_i\}$; the n_i in eq. (2.22) are the occupation numbers of the MSO's. The various matrix elements over molecular orbitals are given by

$$h_i = \langle \psi_i | h | \psi_i \rangle \tag{2.23}$$

$$J_{ij} = \langle \psi_i \psi_j | 1/r_{12} | \psi_i \psi_j \rangle; \qquad K_{ij} = \langle \psi_i \psi_j | 1/r_{12} | \psi_j \psi_i \rangle \tag{2.24}$$

$$V_i^{X\alpha} = 9\alpha (3/4\pi)^{1/3} \langle \psi_i | \rho^{1/3} | \psi_i \rangle \tag{2.25}$$

where h is the one-electron part of the Hamiltonian, $V_i^{X\alpha}$ is the $X\alpha$ exchange-correlation energy associated with the ith MO and ρ is the total electronic charge density given by

$$\rho = \sum_i n_i \psi_i^* \psi_i. \tag{2.26}$$

The α in eq. (2.25) is the exchange parameter which in the original Slater (1951) work was unity; the later work of Gaspar (1954) and of Kohn and Sham (1965) arrived at a value of $\frac{2}{3}$. Optimum values of α with respect to satisfying the virial theorem or yielding the Hartree–Fock energy have been given by Schwarz (1972, 1974) for each atom in the periodic table, with all values falling between $\frac{2}{3}$ and 1.

One of the chief conceptual differences between the restricted Hartree–Fock theory and $X\alpha$ theory is the physical significance of the orbital energies. The orbital energies, i.e. the one-electron eigenvalues, of the Hartree–Fock theory or of the various semi-empirical methods discussed above, may be identified by the theorem of Koopmans (1933) with the negative of the ionization potentials of the molecule. For the case of $X\alpha$ theory, this identification is not possible and the orbital energies, $\varepsilon_i^{X\alpha}$, are given rigorously by

$$\varepsilon_i^{X\alpha} = \frac{\partial E_{X\alpha}}{\partial n_i}. \tag{2.27}$$

The ionization potentials within $X\alpha$ theory may be obtained however by the "transition state" method of Slater (1972). This is such an important point that it is worthwhile making a digression here to consider it in more detail by studying a definite example.

The simplest possible example is that of a two-electron system, the He atom. The total energies for the ground state of the He atom in the HF and Xα theories are given by

$$E_{\mathrm{HF}}(\mathrm{He}) = 2h_1 + J_{11} \tag{2.28}$$

and

$$E_{\mathrm{X}\alpha}(\mathrm{He}) = 2h_1 + 2J_{11} - V_1^{\mathrm{X}\alpha} = 2h_1 + J_{11} \tag{2.29}$$

respectively. However, if α in the eq. (2.25) is chosen such that $E_{\mathrm{X}\alpha} = E_{\mathrm{HF}}$ for the He atom, as is done in fact in choosing the α values as tabulated by Schwarz (1972), then $J_{11} = V_1^{\mathrm{X}\alpha}$ in eq. (2.29). The total energies for the He ion are given by

$$E_{\mathrm{HF}}(\mathrm{He}^+) = h_1 + \tfrac{1}{2}(J_{11} - K_{11}) = h_1 \tag{2.30}$$

and

$$E_{\mathrm{X}\alpha}(\mathrm{He}^+) = h_1 + \tfrac{1}{2}(J_{11} - V_1^{\mathrm{X}\alpha}) = h_1 \tag{2.31}$$

where it is assumed that the same orbitals are employed for the ion as for the ground state neutral atom, i.e. the orbitals are not re-optimized or are not allowed to "relax". Under these circumstances the first ionization potential of He is given by

$$I_{\mathrm{HF}} = -\varepsilon_1^{\mathrm{HF}} = -[E_{\mathrm{HF}}(\mathrm{He}) - E_{\mathrm{HF}}(\mathrm{He}^+)] = -(h_1 + J_{11}) \tag{2.32}$$

and

$$I_{\mathrm{X}\alpha} = -[E_{\mathrm{X}\alpha}(\mathrm{He}) - E_{\mathrm{X}\alpha}(\mathrm{He}^+)] = -(h_1 + J_{11}). \tag{2.33}$$

As can be easily shown, the Hartree–Fock ionization potential is equal to the negative of the orbital energy. However, as discussed by Slater (1972), $\varepsilon_i^{\mathrm{X}\alpha} \neq -I_{\mathrm{X}\alpha}$ but rather $\varepsilon_i^{\mathrm{X}\alpha}$ is given by eq. (2.27). By explicitly evaluating eq. (2.27) for the case of He, one obtains

$$\varepsilon_i^{\mathrm{X}\alpha} = h_1 + 2J_{11} - \tfrac{1}{2}V_1^{\mathrm{X}\alpha} = h_1 + \tfrac{3}{2}J_{11} \tag{2.34}$$

which clearly shows that $I_{\mathrm{X}\alpha} \neq -\varepsilon_1^{\mathrm{X}\alpha}$. Slater (1972) has shown that if one chooses an occupation of the orbital midway between the initial ground state and the final ionized state, the so-called transition state, then the Xα ionization potential can be derived by a convenient procedure avoiding the difference of total energies as given in eq. (2.33). For the spin–orbital for which the ionization occurs the initial occupancy is 1 and the final occupancy is 0, hence the transition state occupancy is $\frac{1}{2}$, and the transition state energy for this orbital is

$$\partial E_{X\alpha}/\partial n_i\Big|_{n_i=1/2} = \left(h_i + \sum_j n_j J_{ij} - \tfrac{1}{2}V_i^{X\alpha}\right)\Big|_{n_i=-1/2} = \varepsilon_i^{TS}, \tag{2.35}$$

which for the present case gives

$$\varepsilon_1^{TS} = h_1 + \tfrac{3}{2}J_{11} - \tfrac{1}{2}V_1^{X\alpha} = h_1 + J_{11}. \tag{2.36}$$

Thus it is the transition state energy ε_1^{TS} rather than the orbital energy $\varepsilon_1^{X\alpha}$, which is to be identified with the negative of the ionization potential; compare eqs. (2.36) and (2.33). The orbital energies of Xα theory thus contain additional self-energy contributions with respect to Hartree–Fock theory. The transition state procedure cancels these extra self-energy terms allowing the resultant orbital energy, the transition state energy, to be identified with the negative ionization potential. For the case of molecules the cancellation between J_{ii} and $V_i^{X\alpha}$ will not be exact as in eq. (2.29) because atomic α values are used rather than optimal molecular α values, hence the self-energy will not be exactly cancelled.

Most often in Hartree–Fock theory, it is found that the Koopmans' theorem ionization potentials, $I_{HF,i} = -\varepsilon_i^{HF}$, are not very accurate. This is due to the fact that the same orbitals are used for the ground state and for the ionized state, leading to an ionization potential which is too large in magnitude. A better procedure is to allow the orbitals for the ionized state to readjust self-consistently to the new environment created by the loss of an electron in the system. The new ionization potential evaluated by taking the difference between the total energies of these two states for the molecule,

$$\Delta E_i = E(\text{ion}, i) - E\ (\text{neutral}) \tag{2.37}$$

is found to be in better agreement with experiment. In eq. (2.37), $E(\text{ion}, i)$ is the total self-consistent energy of the ion resulting from the loss of an electron from orbital i. The difference between the two quantities, $I_{HF,i}$ and ΔE_i, is often referred to as the relaxation energy of orbital i,

$$\Delta E_{\text{relax},i} = I_{HF,i} - \Delta E_i \tag{2.38}$$

and always has a positive value.

In the discussion of the transition state procedure of the Xα theory above, it was assumed that the orbitals for the ground state, the ionized state and the transition state were all identical, being those appropriate to the ground state. However, in practice this is not the case. The orbitals are determined self-consistently in the Xα transition state procedure, thus taking into account orbital relaxation associated with the ionization process. Note that $|\varepsilon_i^{X\alpha}| < |\tilde{\varepsilon}_i^{TS}| < |\varepsilon_i^{TS}|$, where $\tilde{\varepsilon}_i^{TS}$ is the transition state energy of orbital i determined self-consistently, i.e. it includes orbital relaxation. Hence the relaxation energy

$$\Delta E_{\mathrm{relax},i}^{X\alpha} = |\varepsilon_i^{TS}| - |\tilde{\varepsilon}_i^{TS}| \tag{2.39}$$

is always a positive quantity as in the Hartree–Fock theory. However, confusion often arises on this point, because it is sometimes forgotten that $\varepsilon_i^{X\alpha}$ should not be considered as an ionization potential and if one considers the difference

$$\Delta = |\varepsilon_i^{X\alpha}| - |\tilde{\varepsilon}_i^{TS}| \tag{2.40}$$

it gives a negative value not because of a negative relaxation energy but because of the large self-energy term in $\varepsilon_i^{X\alpha}$ which has not been cancelled. Thus the Slater transition state procedure accomplishes two things simultaneously; it approximately cancels the extra self-energy terms in $\varepsilon_i^{X\alpha}$ and also allows for orbital relaxation.

So far in this discussion it has been pointed out that the orbital energies, $\varepsilon_i^{X\alpha}$, are not to be associated with ionization potentials; however the real physical significance of the $\varepsilon_i^{X\alpha}$ has not been discussed. Slater (1974) has shown that the true significance of the $\varepsilon_i^{X\alpha}$ are as orbital electronegativities. This aspect of Xα theory has recently been developed and applied by Johnson (1977) and his co-workers and will be discussed later, when various applications are considered.

The other chief conceptual difference between the Hartree–Fock and Xα theories is the fact that the Xα theory rigorously satisfies Fermi statistics whereas Hartree–Fock theory does not. This means that in Xα theory the orbitals are filled from the lowest level upward in order, i.e. for the ground state of the system no level j is occupied when an empty level i exists such that $\varepsilon_i < \varepsilon_j$.

As an example of the Xα theory applied to a chemisorption problem, the recent work of Ellis et al. (1976) using the Xα–DV method will be discussed. These workers considered the chemisorption of C, N, and O on a five atom Ni cluster representing a (100) surface. For the case of the oxygen atom, which may be compared to experiment, calculations for several cluster–oxygen distances were presented. Significant shifts of orbital energies were obtained for distances shorter than 1.6 Å, associated with O–Ni bonding. Projected densities of states were presented in order to obtain the energy distribution of the O 2p content of the orbitals. At a distance of 0.9 Å, which is the distance above the

(100) plane for an oxygen atom as deduced from the low energy electron diffraction (LEED) studies of Demuth et al. (1973), two levels associated with oxygen–nickel bonding occur below the Ni "d-band". Using the transition state procedure, extra structure at ~6.7 eV below the Fermi level (E_F) is predicted on the basis of these orbitals. This may be compared with the He I photoemission results of Eastman and Cashion (1971) for oxygen chemisorption on (100) Ni, in which structure centered at ~5.5 eV below E_F is attributed to the oxygen. The valence electronic structure calculated by Ellis et al. for the 0.9 Å distance was found to be "in rather good agreement" with a previous Xα–SW calculation by Batra and Robaux (1975) for this system at the same nickel–oxygen distance. A survey of other oxygen–metal chemisorption systems by the Xα method is given in the recent review of Rösch (1977).

4. Metal complexes as surface models

Chemists, for some time, have held the view that considerable insight into the nature of chemisorption on transition metal surfaces and reactions thereon can be gained by drawing analogies with transition metal inorganic and organometallic molecules and their reactions. This viewpoint is expressed very clearly in two recent reviews by Ugo (1975) and Muetterties (1975) from an experimental perspective, as well as in a review by Messmer and Johnson (1976) from a theoretical perspective. Many molecules containing a single transition metal atom such as ML_n, where M represents the metal atom and L is a ligand, have been characterized and their chemistry well studied. A simple example is $Ni(CO)_4$. Likewise, there are a wide variety of di-nuclear compounds, of which $Mn_2(CO)_{10}$ is a simple example. More intriguing however is the fact that within the last five years or so an increasing number of molecules with three or more transition metal atoms have been prepared and characterized. Molecules which have two or more metal atoms bonded to each other are referred to as "metal cluster compounds". Although mono-nuclear metal complexes are useful as analogs for the bonding of an adsorbate (ligand) to a single surface atom, it is clear that metal cluster compounds can represent a far larger class of surface bonding configurations and reactions. This field of metal cluster compounds is rather in its infancy, however it promises as it matures to have great impact on our understanding of chemisorption and reactivity on metal surfaces.

In this section a few examples of bonding in metal complexes will be considered in order to elucidate some of the types of adsorbate–metal interactions one might expect to encounter in discussing chemisorption on metals. There are a number of interesting and important questions, however, which arise with regard to using transition metal complexes such as $Ni(CO)_4$ as analogs for the chemisorption of CO on Ni. One question is: what is the effect of the other CO ligands on the Ni–CO interaction? For metal surfaces, the ratio of metal atoms to adsorbate molecules is usually much larger than in the case of a simple complex such as $Ni(CO)_4$. Stable molecules such as M_nL do not appear to exist at room temperature, which would enable one to address this question directly. Fortunately, the application of a well-known experimental technique appears to be able to provide an answer to this question as well as to a number of others.

The technique is the co-condensation of reactive species at low temperatures in inert gas matrices, which in the hands of Ozin, Moskovits and their co-workers, for example Hulse and Moskovits (1976) and Ozin and Power (1977), is proving a powerful experimental method indeed. In this work co-condensation of monatomic metal vapor with ligand molecules such as CO, ethylene, etc. at low temperatures in inert gas matrices has been achieved. The usually unstable complexes are trapped by this technique and can be studied by visible, ultra-violet and infra-red spectroscopies. Hulse and Moskovits (1976) studying the (Ni, CO) system by such techniques have found evidence for a number of previously unknown species including Ni_2CO and three forms of Ni_3CO. From infra-red evidence they concluded that for Ni_2CO, CO was bridge bonded to the two Ni atoms. For the Ni_3CO species they concluded that both bridge bonding and three-center bonding of CO occurred and that the nature of the metal–CO bond is essentially that found in stable metal carbonyls.

The isolation of such species containing clusters of metal atoms and an adsorbed molecule should be of considerable interest to theoreticians who use such metal clusters to model chemisorption on metal surfaces. They may represent the experimental realization of their theoretical models. Indeed the present author, Messmer (1975), considered the simple species $Ni(C_2H_4)$, $Pd(C_2H_4)$ and $Pt(C_2H_4)$ as theoretical models to investigate changes in chemisorptive bonding of ethylene by the series of metals Ni, Pd and Pt. The actual species $Ni(C_2H_4)$ and $Pd(C_2H_4)$ have since been reported to be isolated by Huber et al. (1976) and Ozin and Power (1977a) respectively. They are found to be π-bonded to the metal atoms as assumed in the study of Messmer (1975). In addition the cluster $Ni_2C_2H_4$, considered by Rösch and Rhodin (1974) as a model to carry out theoretical calculations, has recently been isolated by Ozin and Power (1977b). Thus in the future, with such matrix isolation methods one may be able to assess by experiment how closely the electronic properties of a cluster of metal atoms with a single ligand correspond to those of the same molecule chemisorbed on a metal surface. In order for this to occur however, additional means of characterizing the matrix isolated species will have to be developed; photoemission experiments would be particularly useful in this regard.

4.1. Ni–CO BONDING IN $Ni(CO)_4$

In order to discuss the bonding in $Ni(CO)_4$, a recent $X\alpha$–SW calculation by Klemperer and Johnson (1975) will be considered. In fig. 2.6 the valence orbital energies of CO, Ni and $Ni(CO)_4$ are shown. Three of the occupied valence orbitals of CO, i.e. the 4σ, 1π and 5σ orbitals, together with the lowest occupied MO of CO, the 2π orbital, are shown in fig. 2.7.

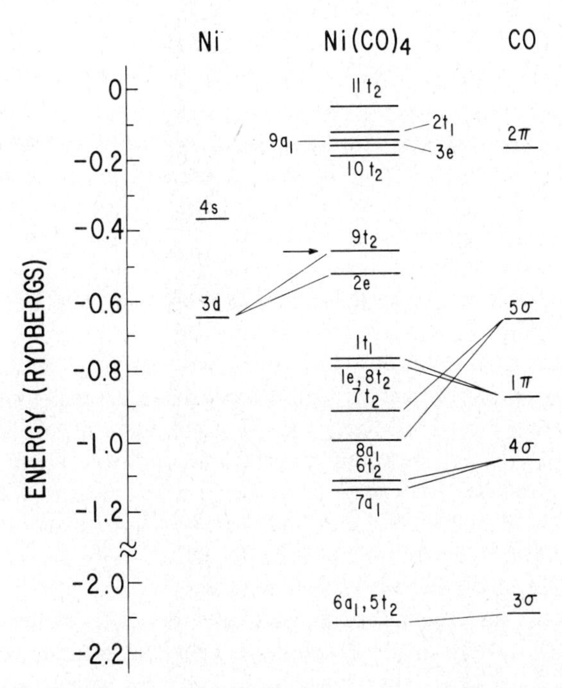

Fig. 2.6. SCF–$X\alpha$ energy levels for the $Ni(CO)_4$ molecule calculated by Klemperer and Johnson (1975). Connecting lines show the dominant CO orbital character contributing to the $Ni(CO)_4$ orbitals. The arrow denotes the highest occupied orbital.

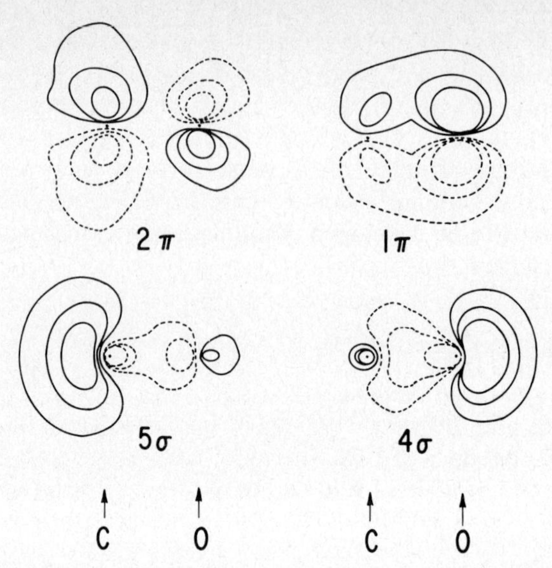

Fig. 2.7. Contour plots of some valence orbitals of the CO molecule. The arrows mark the carbon and oxygen nuclear positions. The 4σ, 1π and 5σ are occupied orbitals; the 2π orbital is unoccupied. (See Johnson and Klemperer 1977).

These orbitals are adapted from the work of Johnson and Klemperer (1977) and use the overlapping sphere parameterization of Rösch et al. (1975). The CO molecules are bonded to the nickel atom through the carbon end, and as can be seen from fig. 2.7 the occupied MO of CO most likely to participate in bonding to Ni is the 5σ. This comes about not only because it is the highest occupied MO but also because it has considerable weight on the carbon atom with a spatial extent pointing away from the CO bond, it is a "lone pair" orbital. The 4σ and 1π orbitals are more concentrated on the oxygen than on the carbon, with the 4σ corresponding to a "lone pair" of the oxygen atom. Hence these orbitals are not likely to be significantly perturbed on forming an Ni–CO bond. The 1π orbital is doubly degenerate and is responsible for the π-bonds of CO, whereas the 3σ orbital, which is not shown in the figure, is responsible for the σ-bond between C and O. The 2π is the lowest unoccupied CO orbital and is anti-bonding. It can overlap with an occupied metal d-orbital of appropriate symmetry to contribute to the metal–carbon bonding. Such a component to the bonding of a ligand with a metal is called "back-bonding". The CO ligand is frequently called a π-acceptor, due to this feature that the 2π orbital can accept electron transfer from the metal.

In fig. 2.6, the lines show the origin of the dominant contributions to the molecular orbitals of $Ni(CO)_4$. The $6a_1$ and $5t_2$ orbitals are largely unperturbed combinations of the 3σ orbitals of CO. The $7a_1$ and $6t_2$ orbitals are mainly derived from combinations of the 4σ orbitals of CO with Ni 4s contributing to the $7a_1$ orbital and Ni 3d and 4p contributing to the $6t_2$ orbital. The $8a_1$ and $7t_2$ orbitals are chiefly responsible for the Ni–CO bonding in $Ni(CO)_4$ and arise primarily from the 5σ orbitals of CO overlapping with Ni 4s and Ni 3d and 4p orbitals respectively. The $1t_1$, 1e and $8t_2$ orbitals arise mainly from combinations of CO 1π orbitals with a small contribution of d character from Ni. The $9t_2$ and 2e orbitals are dominantly Ni 3d in character with mixtures of 2π and 1π orbitals from CO; the $9t_2$ orbital also contains a very small amount of σ character from CO. It should be noted in this regard that for the t_2 representation of the T_d point group, both σ and π character are allowed. In an octahedral molecule such as $Cr(CO)_6$, σ and π orbitals of CO are always separated by symmetry, i.e. they transform according to different irreducible representations.

The $9t_2$ and 2e orbitals are responsible for any of the back-bonding occurring in the molecule, which is concluded to be relatively weak from the calculations. Thus the Ni 3d – CO 5σ interactions are found to be much stronger than the Ni 3d – CO 2π interactions. This is also found to be the case in $Cr(CO)_6$ by Johnson and Klemperer (1977) who present a critical discussion of experimental vibrational data which gives strong support for this conclusion.

When the transition state energies of the Xα–SW method are computed for $Ni(CO)_4$, the resultant ionization potentials follow the order of orbitals shown in fig. 2.6, with the $9t_2$ having the lowest ionization potential, the 2e orbital next, etc. The same thing holds true for the CO molecule. Thus for a qualitative discussion of differences in the photoemission spectra of CO and $Ni(CO)_4$, the levels shown in fig. 2.6 will suffice. The main qualitative feature to be deduced from fig. 2.6 is that the 5σ-derived orbitals are shifted to higher ionization energy with respect to the 1π-derived orbitals. This is due to the fact that the 5σ orbitals of the CO molecules have a strong bonding interaction with the nickel atom. One may now use this result as an analog for the system of CO chemisorption on nickel to anticipate what might happen in that system. Thus, one might expect for CO chemisorbed on Ni that the 5σ-derived levels would shift to higher ionization energies with respect to the 1π-derived orbitals. Further, depending upon the strength of the metal–CO interaction on the surface, the 5σ-derived orbitals might even have higher ionization energies than the 1π-derived orbitals as in the case of $Ni(CO)_4$. A detailed discussion of the chemisorption of CO on Ni will be given in the next section. Suffice it to say here however, that the qualitative aspects discussed above are found experimentally. A recent angle-resolved photoemission study by Williams et al. (1976) has found for CO chemisorbed on Ni (111) that the 5σ-derived peak has higher ionization energy than the 1π-derived peak.

4.2. $Pt-C_2H_4$ BONDING IN $Pt(PH_3)_2C_2H_4$

The recent $X\alpha$–SW calculation of Norman (1977) on the model system $Pt(PH_3)_2C_2H_4$ will be used as a vehicle to discuss the $Pt-C_2H_4$ interaction. In fig. 2.8 the geometry of the $Pt(PH_3)_2C_2H_4$ molecule is shown along with the coordinate system used to label the orbitals. This model system represents a simplification of the actual molecule $Pt(PPh_3)_2C_2H_4$ by replacing the phenyl groups with hydrogen atoms. The planes defined by the CH_2 groups of atoms, are not parallel to the yz-plane, but bent away from the Pt atom by 28°. Such a tilt of the CH_2 groups away from the metal atom is common in a wide variety of metal–olefin systems, with a recognized correlation between the degree of tilt and the C–C distance in the coordinated ethylene molecule. In fig. 2.9, those orbital energy levels in $Pt(PH_3)_2C_2H_4$ and C_2H_4 which are associated with the C–C and C_2H_4–Pt bonds and Pt orbitals are shown. This simplifies the diagram and allows one to focus on those aspects associated solely with

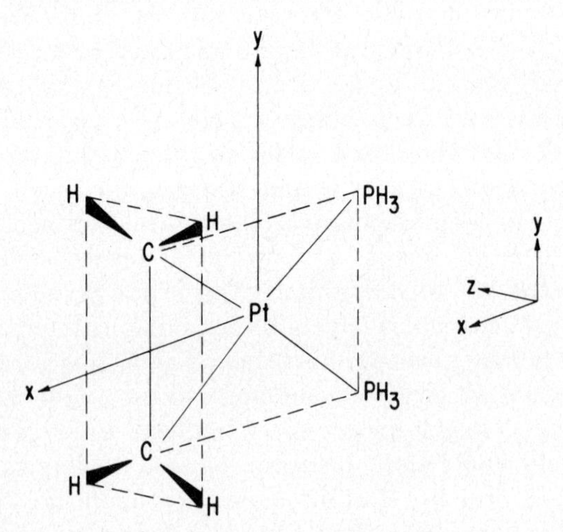

Fig. 2.8. Geometry and coordinate system used to describe the orbitals of the $Pt(PH_3)_2(C_2H_4)$ molecule.

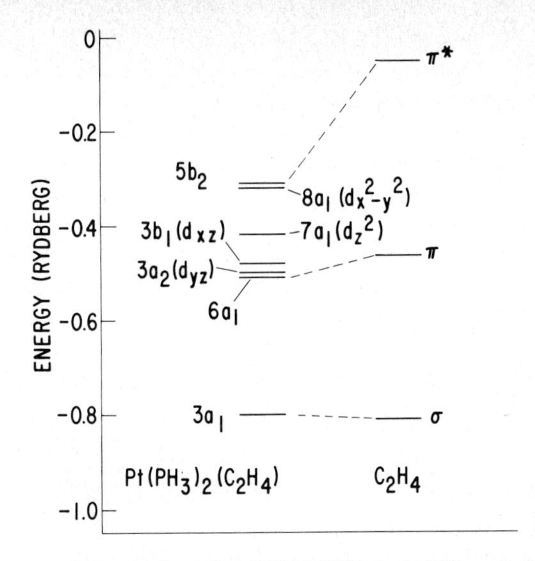

Fig. 2.9. Valence orbital energies of $Pt(PH_3)_2(C_2H_4)$ and C_2H_4 associated with the orbitals involved in $Pt–C_2H_4$ bonding as well as the Pt 5d-derived orbitals. Dashed lines show the dominant CO orbital contributions to the valence orbitals of $Pt(PH_3)_2(C_2H_4)$. (See Norman 1977).

the $Pt–C_2H_4$ bondong. For C_2H_4, the orbitals labelled σ and π are occupied and represent the σ and π bonds between the two carbon atoms of ethylene. The π^* orbital of ethylene is anti-bonding with respect to the two carbon atoms and is unoccupied. As in the case of CO this π^* orbital can act as an electron acceptor by overlapping with occupied Pt orbitals of the appropriate symmetry. Thus back-bonding is of potential importance in describing the $Pt–C_2H_4$ bonding.

In $Pt(PH_3)_2C_2H_4$ the $3a_1$, $6a_1$, $8a_1$ and $5b_2$ orbitals are responsible for the $Pt–C_2H_4$ bonding. The $3a_2$, $3b_1$ and $7a_1$ orbitals are largely Pt d-orbitals (whose atomic symmetries are noted) with contributions from the PH_3 groups. The $5b_2$ level is the highest occupied orbital. The $3a_1$, $6a_2$, $8a_1$ and $5b_2$ orbitals from the work of Norman (1977) are shown in fig. 2.10; they are plotted in the *xy*-plane. The $3a_1$ orbital is predominantly the C–C σ bond of C_2H_4, which interacts somewhat with the Pt atom in a bonding fashion. Due to this interaction a small amount of ethylene π character is also mixed into the orbital. The $6a_1$ orbital consists of the orbital of ethylene interacting with the $5d_{x^2-y^2}$ and $5d_{z^2}$ orbitals of Pt, but there is a significant contribution of C–C σ character responsible for the tilting of the p orbitals on the carbon atoms. The $8a_1$ orbital consists of the ethylene π orbital interacting with the Pt $5d_{x^2-y^2}$

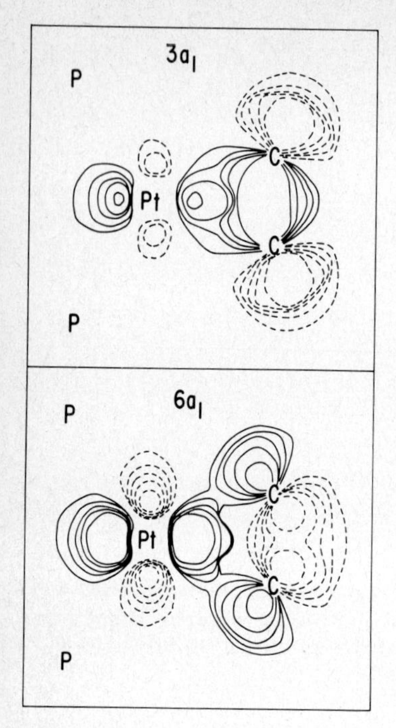

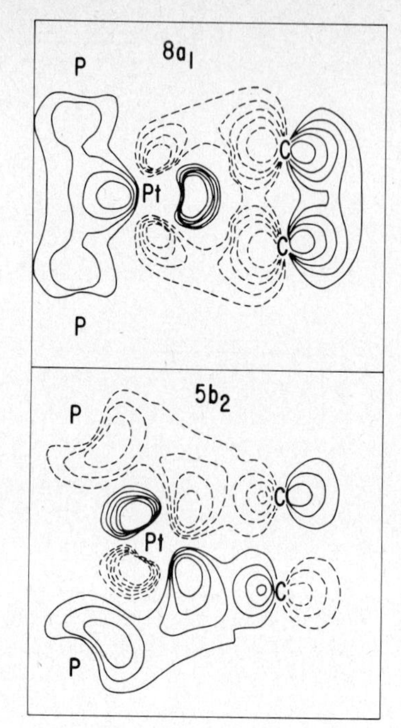

Fig. 2.10. Contour plots of the four valence orbitals of $Pt(PH_3)_2$ (C_2H_4) responsible for the $Pt–C_2H_4$ bonding. The positions of the various nuclei are marked. The plots are in the *xy*-plane (see fig. 2.8). The orbitals are the $3a_1$, $6a_1$, $8a_1$ and $5b_2$; their orbital energies are shown in fig. 2.9. (See Norman 1977.)

and $6p_x$ orbitals. The $5b_2$ orbital arises from the interaction of the ethylene π^* orbital with Pt $5d_{xy}$ and $6p_y$ orbitals and constitutes the back-bonding component of the $Pt–C_2H_4$ bond. It is instructive to compare these results with a few features of the $Pt–C_2H_4$ bonding found for Zeise's anion ($PtCl_3C_2H_4^-$) by Rösch et al. (1974) and for the PtC_2H_4 fragment by Messmer (1975). For example, the back-bonding contribution, as deduced from orbital plots, appears to be less important for $Pt(PH_3)_2C_2H_4$ than for the other two cases. This may well be a reflection of the fact that the $5b_2$ orbital is higher in energy than the $8a_1$ in the present case, whereas in the other two cases the corresponding orbitals are reversed, leading to a more favorable back-bonding stabilization. Another feature of $Pt(PH_3)_2C_2H_4$ which differs from the other two cases, is the interaction of the C–C σ orbital of ethylene with the Pt atom. This interaction can be seen in the $3a_1$ orbital of fig. 2.10. In the case of PtC_2H_4 and Zeise's anion, this orbital is almost exclusively a C–C σ

orbital. Such differences arise from the influence exerted by the other ligands on the orbitals of Pt, e.g. one may view PH_3 as being an electron donor and Cl an electron acceptor with respect to a neutral Pt atom.

Can this information serve to guide one as to what to expect in terms of "orbital positions" for ethylene when it is chemisorbed on Pt or a similar metal? Qualitatively all three of the model complexes, i.e. $Pt(PH_3)_2C_2H_4$, Zeise's anion and PtC_2H_4, predict a stabilization shift of the π-derived orbital ($6a_1$ orbital in fig. 2.9) with respect to the σ-derived orbital ($3a_1$ orbital in fig. 2.9) on going from the isolated C_2H_4 molecule to the complex. One would thus expect the same sort of shift for the case of chemisorbed C_2H_4, and this shift should manifest itself in the photoemission spectrum. This is in fact the case observed experimentally as for example in ethylene chemisorption on Ni (111) as discussed by Demuth and Eastman (1976). These authors also try to correlate the small shifts in ethylene σ orbitals associated with both C–C and C–H bonding with the state of hybridization of the carbon atoms bonded to Ni. They suggest from calculations on the C_2H_4 molecule that the shifts observed in the spectrum indicate a mixing of π and σ C_2H_4 character and that the CH_2 groups are bent out of a plane parallel to the surface. This effect, as mentioned above, is found in a variety of metal complexes containing C_2H_4. However, one cautionary remark is appropriate here. As seen above, the C–C σ orbital can interact with the Pt orbitals ($3a_1$ orbital of fig. 2.10) in some cases. Such an interaction can cause a shift of this σ orbital with respect to those σ orbitals describing C–H bonds. Thus analyzing the shifts in the photoemission spectrum, strictly on the basis of a comparison with various isolated C_2H_4 molecular geometries could be somewhat misleading. However, one would expect this situation to be more of a problem in discussing chemisorption on Pt where the atomic d-orbitals are more diffuse than for the case of chemisorption on Ni. It is however a point which should be kept in mind.

It appears therefore that there is a variety of useful information to be obtained by considering transition metal complexes as analogs of simple chemisorption systems. With closer interaction between workers in these fields, better advantage of such information will undoubtedly be made in the future.

5. Cluster model applications

In this section a few examples of systems in which cluster models have been applied to discuss chemisorption will be studied in some detail.

Where possible a comparison among theoretical methods applied to the same system will be presented and will be accompanied by a review of and comparison with the relevant experimental data.

5.1. THE Ni + CO SYSTEM

An introduction to this system has already been provided in the last section where the $Ni(CO)_4$ molecule was considered as an analog for discussing the bonding of CO to a metal surface. However, the first complete cluster calculation for this system was by Batra and Bagus (1975) who considered a cluster of five metal atoms representing a (100) face of Ni and assumed the CO molecule to bind symmetrically at the 4-fold hole site through the carbon end. The Ni–C distance was assumed to be that known for $Ni(CO)_4$. The method of calculation was the $X\alpha$–SW method, and the transition state procedure was used to calculate the ionization energies of the orbitals. A comparison was then made with the photoemission experiments of Eastman and Cashion (1971) for CO chemisorbed on Ni (100). In order to make meaningful comparisons between an ionization spectrum from a cluster calculation and the experimental photoemission spectrum a constant energy has to be added to all the ionization energies of the cluster so as to bring the

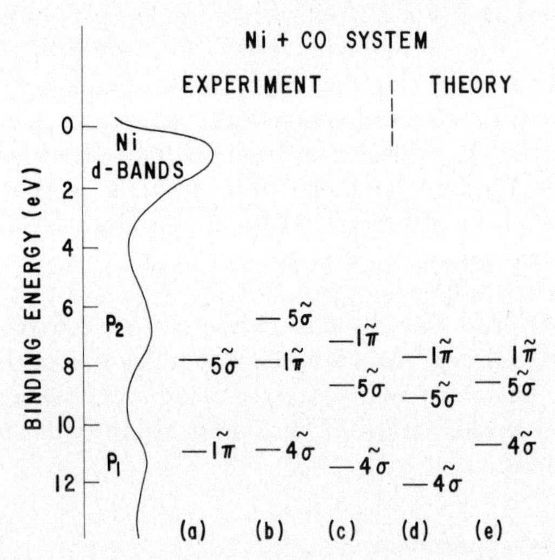

Fig. 2.11. Comparison of theory and experiment for photoemission spectra of CO adsorbed on Ni. (a) Eastman and Cashion (1971); (b) Gustafsson et al. (1975); (c) Williams et al. (1976); (d) Batra and Bagus (1975); (e) Kasowski (1976). See text.

lowest ionization potential of the cluster in coincidence with the Fermi level (E_F), i.e. onset of photoemission, in the experimental spectrum.

In fig. 2.11a the photoemission spectrum at 40.8 eV photon energy is shown with the original assignment of the peaks proposed by Eastman and Cashion; in (d) the Ni_5CO cluster results are shown. Although the calculations were carried out in a spin-unrestricted manner, resulting in slight energy differences (≤ 0.2 eV) between spin-up and spin-down orbitals, an average of the spin-up and spin-down level positions are shown in (b) in order to simplify the discussion. In the region of 0 to ~2.5 eV binding energy (relative to E_F) there are a number of closely spaced levels for the cluster calculation which arise largely from combinations of Ni 3d orbitals. These levels have been omitted from the figure as they do not enter into the discussion of the main bonding effects. The notation used in fig. 2.11, e.g. $\widetilde{5\sigma}$ means 5σ-derived orbital, i.e. a 5σ orbital changed due to the interaction with the metal atoms. It is interesting to note that the original assignment of Eastman and Cashion gives a $\widetilde{5\sigma}$–$\widetilde{1\pi}$ energy separation which is comparable to that found in the gas phase CO spectrum. This is counter to chemical intuition and to what is known regarding the bonding in metal carbonyls as was discussed in the last section using a recent calculation for $Ni(CO)_4$ as an example. On the other hand the predicted order of the orbitals from the Ni_5CO calculation shows a pronounced stabilization of the 5σ-derived orbital and is strikingly similar to the relative level positions obtained for the $Ni(CO)_4$ molecule using the same theoretical methods. This latter point is quite interesting as the geometries of the two situations are very different.

Batra and Bagus on comparing their results with the photoemission spectrum of Eastman and Cashion offered two possible assignments for the peaks, P_1 and P_2, of the spectrum. In the first, they suggested that P_2 might be assigned to $\widetilde{1\pi}$ and that P_1 be assigned to $\widetilde{5\sigma}$ with the calculation giving too small a splitting between them. They then suggested that the $\widetilde{4\sigma}$ orbital might have a very small ionization cross section and thus not be seen by the experiment. The second possibility they presented was that both $\widetilde{1\pi}$ and $\widetilde{5\sigma}$ were responsible for P_2, with the calculations predicting too large a splitting between $\widetilde{1\pi}$ and $\widetilde{5\sigma}$, and that $\widetilde{4\sigma}$ was responsible for P_1.

Almost simultaneously with the publication of the cluster results, new experimental results by Gustafsson et al. (1975) were published. They studied the ionization cross sections of molecular CO and CO adsorbed on Ni as a function of photon energy from 25 eV to 105 eV using synchrotron radiation. The relative intensities of P_1 and P_2 were found to vary with photon energy and by correlating this with the orbital cross section for molecular CO they arrived at a new assignment for P_1 and P_2. P_1 was shown to arise from $\widetilde{4\sigma}$ and the peak P_2 was assigned to both $\widetilde{5\sigma}$ and $\widetilde{1\pi}$ with possibly the $\widetilde{1\pi}$ centered at ~7.8 eV with a weaker $\widetilde{5\sigma}$ centered at ~6.5 eV. This assignment is shown in column (b) of fig. 2.11.

A theoretical calculation by Kasowski (1976) using the Xα–LCMTO approach for a semi-infinite slab geometry studied a CO monolayer adsorbed on a (100) surface of Ni; the calculation however was not carried out self-consistently. His results for the calculated peak positions are shown in column (e). It is interesting to note that the qualitative aspects, i.e. level ordering and relative spacings, are in agreement with the Ni_5CO and $Ni(CO)_4$ results. All of the calculations give the ordering $\widetilde{1\pi} < \widetilde{5\sigma} < \widetilde{4\sigma}$.

The most recent experimental results are for a Ni (111) surface using angle-resolved photoemission. The work was carried out by Williams et al. (1976). In a number of recent studies it has been demonstrated that the photoemission cross sections in surface studies vary quite dramatically as a function of detector angle. A general discussion of photoemission and angle-resolved photoemission is given in ch. 3 by Rhodin. Williams et al. used 40.8 eV radiation and studied the emission as a function of polar angle (θ, measured with respect to the surface normal)

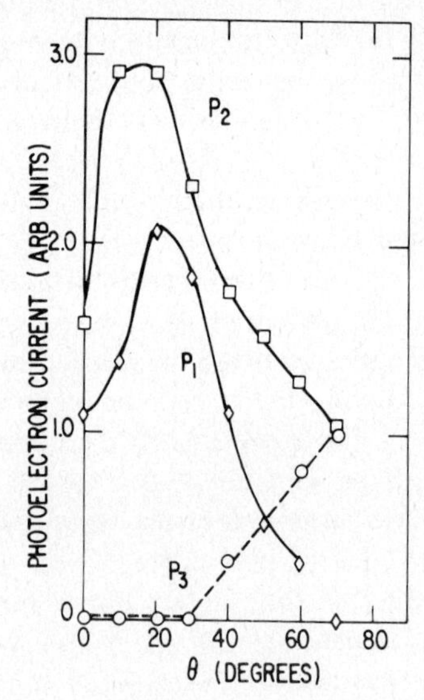

Fig. 2.12. Measured angular distribution of photoelectrons for CO chemisorbed on Ni(111) from Williams et al. (1976).

at a fixed azimuthal angle. At small values of θ, the photoemission spectrum shows only P_1 and P_2 with maxima at slightly different energies than in fig. 2.11a. However, with increasing θ, the peaks P_1 and P_2 lose intensity and a new peak, P_3, centered at ~7.0 eV grows in. At $\theta = 70°$, P_1 has essentially disappeared and P_3 has slightly higher intensity than P_2. In fig. 2.12 the measured angular distributions of the three peaks are shown. On the basis of the observed behavior in angle-resolved emission, the authors provided an assignment for the peaks shown in column (c). Except for a uniform shift in level positions this assignment is in striking agreement with the Xα–SW cluster results of Batra and Bagus. This however, may not be the final word on the assignment of the photoemission spectrum for CO on Ni. Other workers, namely Allyn et al. (1977), have carried out angle-resolved studies on this system and have obtained evidence that P_3 arises from substrate effects and that P_2 contains both $\widetilde{1\pi}$ and $\widetilde{5\sigma}$ with $\widetilde{1\pi}$ at a lower ionization energy than $\widetilde{5\sigma}$.

This example clearly points out the importance of systematic angle-resolved photoemission studies to elucidate the electronic structure of chemisorbed species. Hence experiments at one or two detector angles might completely miss crucial structure in the spectrum necessary to understand the nature of the local electronic structure of adsorbates. Only theoretical calculations of ionization energies have been discussed here so far. But it is apparent from the above discussion that an important theoretical problem is to calculate the photoionization cross sections as a function of photon energy, incident light angle and polarization and detector angles in order to compare and interpret modern photoemission experiments. Fortunately, the recent theoretical work of Davenport (1976a,b) now provides a means, within the Xα–SW cluster framework, to calculate such information.

The observed photoemission intensity is proportional to the photoionization cross section, which in the dipole approximation depends upon a matrix element which couples the initial and final states of the system through the dipole operator. In Davenport's work the initial state is described as in a usual Xα–SW calculation, however in order to calculate the necessary matrix elements, he had to construct a proper treatment of the final state, describing the outgoing photoionized electron. As an example of the applications of this method, a brief discussion of the "oriented CO molecule" will be presented, as it connects with the angle-resolved experiments discussed above.

An important difference between molecules in the gas phase and those chemisorbed on a surface, as far as photoemission is concerned, is the fact that the chemisorbed molecules are oriented in space. This is of course what makes angle-resolved spectra important in surface studies, as an average over orientations is not obtained as in the gas phase. Davenport, as a first approximation to obtaining angular distributions of photoelectrons for CO chemisorbed on Ni, carried out calculations using an oriented CO molecule as his model. According to the previous discussions above, the $\widetilde{1\pi}$ and $\widetilde{4\sigma}$ orbitals of CO, assuming the carbon atom bonds to the surface, are not much perturbed by the bonding. It is the 5σ which is significantly changed on bonding. Thus, one might expect the oriented molecule model to provide a reasonable approximation for photoemission associated with $\widetilde{1\pi}$ and $\widetilde{4\sigma}$ but be less reliable for $\widetilde{5\sigma}$. In fig. 2.13 are Davenport's calculated angular distributions which may be compared to the experimental findings in fig. 2.12. There is clearly good agreement between P_1 and 4σ and between P_3 and 1π, however, as expected, the agreement is not as good for 5σ and P_2. While these results tend to support the assignment of Williams et al., it must be kept in mind that no account of the Ni substrate has been included in the calculations of Davenport for the oriented CO molecule. In fact, Davenport (1977) has recently carried out similar calculations for an NiCO molecule, in order to include some aspects of the substrate. He finds that the photoelectron intensity of the 5σ-derived orbital increases considerably, relative to $\widetilde{4\sigma}$, as compared to the 5σ and 4σ intensities of the isolated CO molecule. Furthermore, the calculated angle dependence of photoemission for the combination of $\widetilde{5\sigma}$ and $\widetilde{1\pi}$ (of the oriented NiCO molecule) is in good agreement with P_2 of fig. 2.12, which is consistent with the experimental evidence of Allyn et al. Such calculations in conjunction with $X\alpha$–SW studies on clusters promise to offer a closer interaction between theory and experiment in future photoemission studies of chemisorption.

5.2. THE Al + O SYSTEM

Some of the results of extensive cluster studies by Messmer and Salahub (1977a,b) and Salahub and Messmer (1977) directed at understanding oxygen chemisorption on aluminum, will be discussed. The choice of aluminum as a substrate for the theoretical investigation of chemisorption was motivated by four factors. First, as aluminum is traditionally thought of as a prime example of a free electron metal, it provides a stringent test of the cluster approach. Second, a comparison can be made with the previous theoretical work of Harris and Painter (1976) and of Lang and Williams (1975), the latter of which has employed a solid state viewpoint – namely the atom–jellium model.

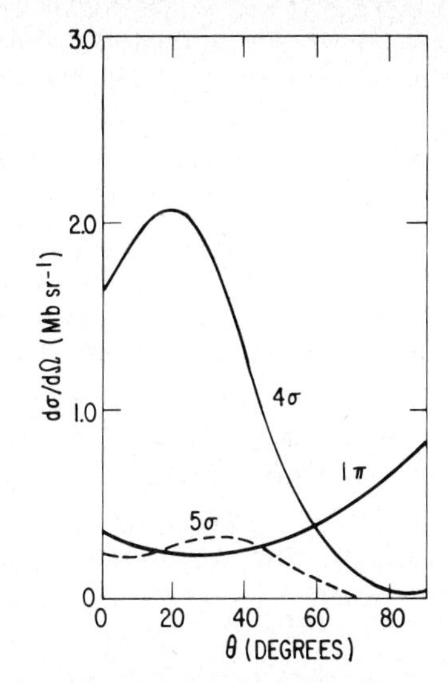

Fig. 2.13. Calculated angular distribution of photoelectrons for "oriented" CO from Davenport (1976b).

Third, unlike the transition metals, only s and p electrons need to be considered for aluminum, hence larger clusters can be conveniently treated and the convergence of the results as a function of cluster size can be studied. Fourth, recent and on-going photoemission work for aluminum and for oxygen chemisorption on aluminum by Flodström et al. (1976a,b) allows a comparison with experiment and the possibility of a fruitful interplay between theory and experiment.

The theoretical method used to obtain the electronic structures of the clusters is the SCF–Xα scattered wave method. Although this method has been used to investigate metal cluster–adsorbate systems previously, the calculations for the Al + O system consider the largest clusters used to represent the substrate to date.

The clusters of 5, 9 and 25 Al atoms used to represent the (100) surface of the metal exhibit C_{4v} symmetry. The 25 atom cluster includes up to third nearest neighbors of an adsorbate atom situated in the central 4-fold site of the first surface plane. It consists of 12 atoms in this plane, nine in the second plane and four in the third plane. For the five and nine atom clusters, calculations were carried out for an oxygen atom above the 4-fold hole site at three distances from the first atom plane, namely at $Z_O = 0.0$, 2.0 and 4.0 bohrs. For the 25 atom cluster, calculations for $Z_O = 0.0$ and 2.0 bohrs were performed for the 4-fold site. The $Z_O = 0.0$

calculations represent a model for the incorporation of oxygen into the first atomic layer of aluminum.

The convergence of the electronic structure of the aluminum clusters as a function of cluster size will not be discussed here, but it has been taken up in detail by Salahub and Messmer (1977). Suffice it to say that in so far as the distribution of energy levels and character of orbitals is concerned, the cluster of 25 atoms represents a very reasonable description of the local electronic structure of the substrate.

In order to compare results obtained from cluster calculations with those obtained by solid state approaches or with experiment, it is convenient to construct density of states (DOS) curves of various types for the clusters, as described in sect. 2, by replacing each discrete energy level of the cluster by a Gaussian. For all of the aluminum calculations the Gaussian width parameter has been taken as $\sigma = 0.05$ rydberg. Curves which display the changes in density of states (ΔDOS) upon chemisorption are quite instructive and are constructed by performing a point by point subtraction of the DOS curves obtained for the individual clusters, i.e. $Al_n + O$ and Al_n, after aligning the Fermi levels (energies of highest occupied orbitals).

In fig. 2.14, ΔDOS curves are presented for the case of an oxygen atom situated at three distances ($Z_O = 4.0$, 2.0 and 0.0 bohr) above the four-fold symmetric hole site of a cluster of five aluminum atoms. The structure in the ΔDOS curves arises mainly from the 2p orbitals of oxygen as a result of their interaction with the aluminum. Adsorbate atomic levels broadened by the adsorbate–substrate interaction are commonly referred to as "orbital resonances". At $Z_O = 4.0$ bohr a rather well-defined orbital resonance is found at ~2 eV below the Fermi level (E_F). This result is in substantial agreement with the ΔDOS curve derived from the atom–jellium model of Lang and Williams (1975) as well as the result deduced for $Z_O = 3.0$ from a previous cluster calculation for oxygen on aluminum by Harris and Painter (1976). A similar orbital resonance is also observed for $Z_O = 4.0$ on the 9-atom cluster. Hence, for large adsorbate–substrate separations all of the models mentioned above give essentially the same result which is quite insensitive to the detailed nature of the substrate. For short distances, however, and particularly for $Z_O = 0.0$ not considered in the previously mentioned work, increased structure in the ΔDOS occurs and a simple orbital resonance no longer results. Also for shorter distances the ΔDOS curves are quite sensitive to the substrate model chosen. Thus it is important to use a reasonably large cluster; in the present case, 25 atoms.

It should be pointed out that the results in fig. 2.14 are based on Xα orbital energies. However, when based on the Slater transition state energies which should be used in a comparison with photoemission results, the peaks in the ΔDOS curves derived from the transition states exhibit a maximum shift of 0.3 eV (with respect to E_F) from those shown in fig. 2.14. Therefore, in the following only the results based on orbital energies need be discussed.

The ΔDOS curves for the Al_{25} cluster with an oxygen atom at $Z_O = 2.0$ and 0.0 bohr above the center of the (100) face are presented in fig. 2.15. The ΔDOS curve for $Z_O = 2.0$ bohr is very similar to the corresponding

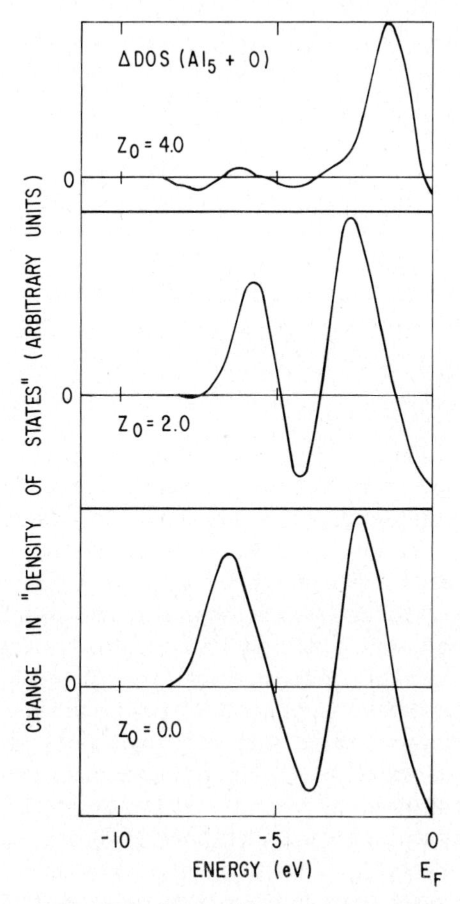

Fig. 2.14. ΔDOS curves for Al_5 cluster with an oxygen atom at $Z_O = 4.0$, 2.0 and 0.0 bohr (1 bohr ≡ 0.52917 Å) above the midpoint of the plane determining the (100) face.

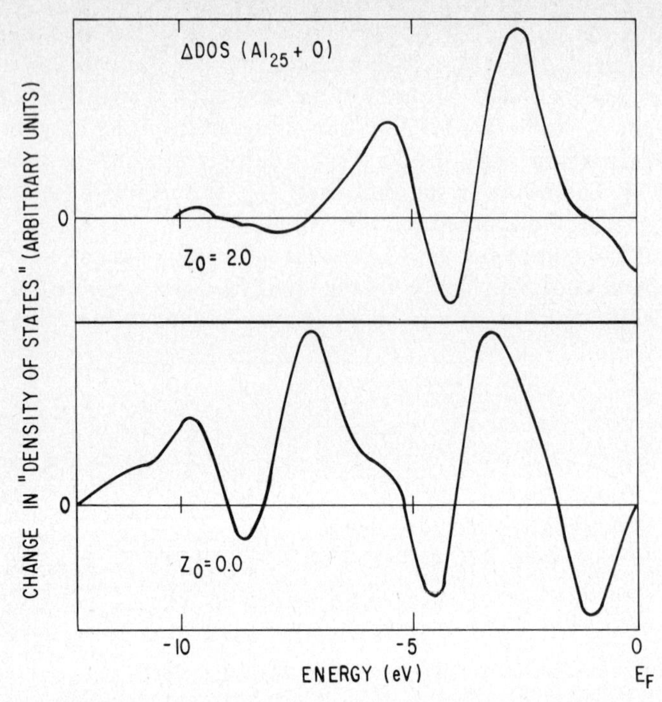

Fig. 2.15. ΔDOS curves for Al_{25} cluster with an oxygen atom at $Z_O = 2.0$ and 0.0 bohr (1 bohr ≡ 0.52917 Å) above the (100) face of the cluster.

curve for the 5-atom cluster shown in fig. 2.14. The curves for $Z_O = 0.0$ bohr are notably different however, especially as regards the appearance of a third peak at ~9.5 eV below E_F for the $Al_{25} + O$ case. The 5-atom cluster is incapable of reproducing this structure at −9.5 eV since it yields a band width which is too narrow and thus contributes no states in this region with which the oxygen atom can interact. Thus the five atom cluster might be adequate to roughly represent a situation where the oxygen atom is at $Z_O > 2.0$ bohr, but it is inadequate to represent the situation where the atom penetrates the lattice (i.e. $Z_O = 0.0$ bohr). This conclusion is important because recent experimental evidence of Flodström et al. (1976a) and Dawson (1976) suggests rather strongly that oxygen atoms are incorporated into the aluminum lattice, even at low exposures.

Recently some photoemission results have been published for oxygen chemisorption on aluminum by Flodström et al. (1976a,b). Although these experiments have been for evaporated polycrystalline films rather

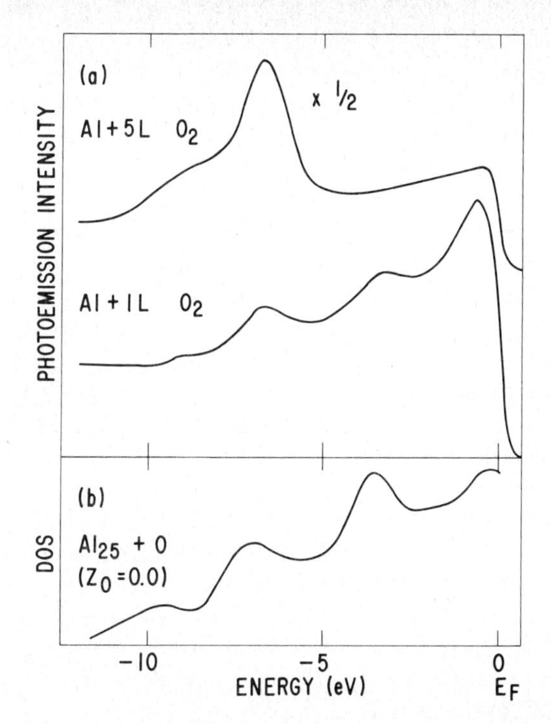

Fig. 2.16. (a) Photoemission spectra for oxygen chemisorbed on aluminum at two different oxygen exposures. The intensity of the curve for the 5L exposure has been reduced by $\frac{1}{2}$ to aid in displaying the curves. Data is from Flodström et al. (1976a). (b) Calculated DOS curve for a cluster of Al_{25} with an oxygen atom incorporated in the surface plane.

than for single crystal faces, they do allow some point of comparison between theory and experiment. In fig. 2.16a the results of Flodström et al. using He II (40.8 eV) radiation are shown for two different exposures, 1 L and 5 L; 1 L = 1 langmuir = 10^{-6} torr · sec. Notice the increase in the intensities of the two peaks at ~7 eV and ~9.5 eV below E_F with increased exposure to oxygen. Photoemission intensity difference curves which could be compared to our calculated ΔDOS curves were not generated by the experimenters. Thus in order to provide a comparison with experiment, the calculated DOS curve for $Al_{25}O$ ($Z_O = 0.0$) is shown in fig. 2.16b. An Xα–DOS curve, even though based on transition state energies which take account of certain relaxation effects attendant to the photoemission process, is not strictly comparable to an experimental photoemission curve. This is because matrix element effects can modify dramatically the intensities based on the simple DOS curve. As discussed above, these effects have recently been shown by Davenport (1976a) to be accessible within the $X\alpha$–SW framework, and will undoubtedly be included in future calculations. However, the work presently available proceeded in the spirit of previous theoretical work which also neglected these effects, and thus the comparison in fig. 2.16 is presented. The similarity between the calculated curve and the experimental curves, especially the lower coverage result, is quite striking. The fact that the only calculated curves which exhibit the two peaks at −9.5 and −7.0 eV are those for which the oxygen is incorporated in the lattice, is strong support for the oxygen penetrating the lattice. Besides the results shown in fig. 2.15b for oxygen at $Z_O = 0.0$, similar results showing these two peaks have been obtained for five oxygen atoms at $Z_O = 0.0$ and for four oxygen atoms at octahedral sites inside the Al_{25} cluster, corresponding to incorporation below the surface.

It should be mentioned that the atom–jellium model of Lang and Williams has been approximately corrected in order to investigate lattice penetration by Yu et al. (1976). The chief result is that as the oxygen atom approaches the surface and is moved into the metal, the O 2p resonance quickly drops from 2 eV below E_F (previous result) to an interior limit of ~10 eV below E_F. Evidently no structure in the orbital resonance is found, which can accommodate the structure found experimentally in the region 5–10 eV below E_F.

The Al 2p level shifts which take place on interaction with an oxygen atom have also been calculated, using the $Al_{25}O$ clusters. The results are shown in table 2.5. Again the results for lattice incorporation ($Z_O = 0.0$) are consistent with the experimental findings of Flodström et al. (1976), for less than a monolayer coverage. In this cluster model each aluminum atom listed in table 2.5 has a single oxygen atom as a nearest neighbor. As the number of oxygen atoms increases the core shifts will also

TABLE 2.5
Calculated Al 2p level shifts due to chemisorption of oxygen (in eV)

Atom[a]	$Z_O = 2.0$	$Z_O = 0.0$	Expt.[b]
Al (1)	0.30	1.21	1.30
Al (4)	0.65	1.05	

[a] Al (1) – atom just beneath the oxygen atom; Al (4) – one of the four atoms on the surface plane closest to the oxygen atom.
[b] See Flodström et al. (1976c).

increase as has been found for the models involving four or five oxygen atoms mentioned above. The experimental shift of 1.3 eV is for intermediate exposure of oxygen (much less than a monlayer) and so it is reasonable to assume that most of the surface aluminum atoms have only one adjacent oxygen and hence the comparison made is valid. At higher exposures a shift of 2.6 eV is observed which corresponds to an aluminum oxide type surface.

The discussion thus far has centered on the calculated peaks at ~7 eV and 9.5 eV below E_F. However, there is a third peak at −3.5 eV in the ΔDOS of fig. 2.15 which also appears in projected DOS curves in which O 2p character only is projected out of the total DOS of fig. 2.15. Hence this third peak arises from O 2p character according to the calculations. In fig. 2.16a there appears to be some change in the photoemission intensity in this energy region as a function of oxygen exposure. The change is however much less dramatic than that observed below −5 eV. Moreover further recent photoemission studies by Flodström et al. (1976c), Yu et al. (1976), and Martinsson et al. (1977), have appeared, which have studied much higher exposures. Even at 100 L it appears that a complete monolayer has not formed. At these higher exposures the oxygen structure below −5 eV is clearly seen above a very small Al background and there is no indication of oxygen related structure above −5 eV.

It should be emphasized however, that the experimental results have been obtained using a rather limited number of photon energies and also a very limited number of incident light and electron collection angles. There is ample theoretical and experimental evidence, as discussed above for the Ni + CO system, to show the extreme sensitivity of surface photoemission results to all of these parameters. There are thus two possible situations which may be present for the part of the spectrum above −5 eV. First, there may be no (or very little) oxygen related structure in this region, in which case our theoretical model is inadequate in this respect. Second, as systematic angle and photon energy dependent photoemission studies have not yet been carried out, the structure may not yet have been detected in a convincing manner. The choice between these two possibilities clearly requires further study both experimentally and theoretically.

One thing is clear, the aluminum + oxygen system is not so simple as it had at first appeared. A close interaction between theory and experiment will be necessary in order to fully understand this system, and the calculation of photoemission intensities as discussed for the CO molecule above will undoubtedly play a key role in this interaction.

5.3. Ni, Pd, Pt + H SYSTEMS

Calculations for four-atom tetrahedral and six-atom octahedral clusters of Ni, Pd and Pt containing hydrogen at the interstitial sites have been carried out by Messmer et al. (1977), using the Xα–SW method. The metal–metal distances were constrained to be equal to the corresponding crystalline metals. The cluster configurations chosen have the advantage that they are large enough to represent the local effects on electronic structure of embedding hydrogen atoms in a surface or bulk interstitial environment, yet small enough to permit resolution of individual metal–hydrogen bonding orbitals. The work also includes the first application of the relativistic Xα–SW formalism developed by Yang and Rabii (1975) to metal clusters.

The non-relativistic orbital energies for the tetrahedral clusters with and without interstitial atomic hydrogen are shown in fig. 2.17. Also shown for comparison is the SCF–Xα 1s orbital energy for the isolated hydrogen atom. It should be recalled that the Xα orbital energies are to be associated with orbital electronegativities as discussed above. Thus for the hydrogen atom the orbital energy is approximately $\frac{1}{2}(I + A)$, where I and A are the ionization potential and electron affinity, respectively. These quantities indidually may be calculated by Slater's transition state procedure.

Johnson (1977) has given a lucid discussion of the conceptual and

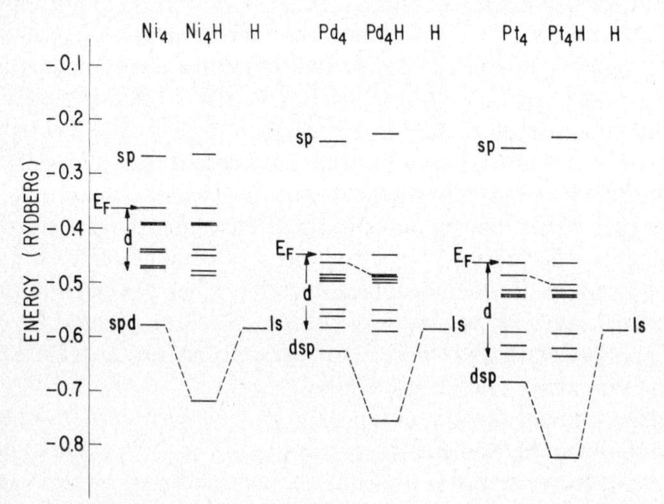

Fig. 2.17. Non-relativistic SCF–Xα orbital energies of tetrahedral group VIII transition metal clusters with and without interstitial atomic hydrogen.

practical advantages of the fact that the Xα eigenvalues can be rigorously treated as orbital electronegativities, thus extending the electronegativity concept of Mulliken (1949). In the present case, the relative positions of the SCF–Xα orbital energies for the Ni_4, Pd_4 and Pt_4 clusters with respect to the orbital energy for atomic hydrogen, as shown in fig. 2.17, are a measure of the differences in orbital electronegativity and chemical potential between these metal aggregates and atomic hydrogen.

The electronic energy levels in fig. 2.17 are labeled according to the principal atomic s, p, d character of the associated molecular orbitals, with the highest occupied orbital in each case indicated by the "fermi level" (E_F). As these clusters are intended to simulate the local interstitial bonding configurations of isolated hydrogen atoms embedded in an otherwise perfect bulk or surface lattice, the energy levels of the clusters containing hydrogen have been shifted with respect to those of the corresponding hydrogen-free clusters so that the respective Fermi levels line up.

The electronic structure of each cluster shown in fig. 2.17 is characterized by a manifold of closely spaced d-levels bracketed by s,p- or s,p,d-hybrid levels. This is similar to the results of Messmer et al. (1976) obtained for larger Ni, Pd and Pt clusters and is analogous to the overlap of the "d-band" by the "s,p-like conduction band" in the respective bulk metals. There are several additional features previously observed in the case of the larger metal clusters, which again appear for the present clusters. For example, although the calculated d-band width of each of the metal clusters is less than that of the corresponding bulk metal, there is a trend of increasing band width and a downward trend of the energy levels from Ni_4 to Pd_4 to Pt_4 which parallels the trends of the crystalline metals. Furthermore, the electronic structures of the Pd_4 and Pt_4 clusters are more similar to each other than they are to the Ni_4 cluster, consistent with the behavior of the bulk band structures.

The deepest energy levels shown in fig. 2.17 for Pd_4 and Pt_4 are associated with orbitals having a_1 symmetry and are predominantly d-like with a small amount of s,p-hybridization. In contrast, the deepest energy level for the Ni_4 cluster shown in fig. 2.17, also associated with an a_1 molecular orbital, is predominantly s-like, but with significant d-orbital hybridization and some p-like character. The partial-wave decomposition of these cluster orbitals having a_1 symmetry is summarized in table 2.6. These differences between the electronic structures of nickel aggregates and the electronic structures of palladium and platinum aggregates are crucial to understanding the differences in the photo-

TABLE 2.6
Partial-wave decomposition of the metal-atom contributions to the a_1 molecular-orbital charge distributions of tetrahedral clusters with and without interstitial atomic hydrogen. The orbital energies correspond to levels shown in fig. 2.17.

Cluster	Orbital energy ($-\varepsilon$ in Rydberg units)	Fractional partial-wave character s	p	d
Ni_4	0.580	0.63	0.11	0.26
Ni_4H	0.724	0.47	0.18	0.35
Pd_4	0.628	0.05	0.03	0.92
Pd_4H	0.758	0.15	0.10	0.75
Pt_4	0.681	0.04	0.03	0.93
Pt_4H	0.821	0.12	0.09	0.79

emission spectra for hydrogen chemisorbed on these metals as discussed below, and can also be related to the differences among these metals with respect to hydrogen solubility and catalytic reactivity as discussed in the paper of Messmer et al. (1977).

The bonding of atomic hydrogen to the cluster interstices arises from the overlap of the a_1 orbitals near the bottom of the Ni_4, Pd_4 and Pt_4 d-bands with the H 1s orbital, with the result that a hydrogen–metal bonding energy level of a_1 orbital symmetry is split off from the bottom of the d-band of each cluster, accompanied by much smaller level shifts within the d-manifolds. This is indicated in fig. 2.17 by the orbital energies for the Ni_4H, Pd_4H and Pt_4H clusters and the connecting dashed lines. The metallic 4s-like component of this a_1 orbital is largely responsible for the bonding of hydrogen to the nickel aggregate, as indicated by the partial-wave decomposition for this a_1 orbital in table 2.6. However, the contribution of the 3d-like component to the bonding is not negligible, amounting to 35% of the Ni_4H a_1 bonding orbital charge. The re-hybridization of 4s- and 3d-like orbital components upon interaction of hydrogen with the Ni_4 cluster is evident in the somewhat different mixtures of s and d partial-wave components for Ni_4 and Ni_4H shown in table 2.6. In contrast to the results for nickel, the metal d-orbital components almost exclusively dominate the bonding of hydrogen to palladium and platinum aggregates, the contributions of Pd 5s- and Pt 6s-like components amounting to only 15% and 12% for Pd_4H and Pt_4H respectively. These results underscore the danger of making general conclusions about the dominance of s-orbitals over d-orbitals in determining the chemisorption and catalytic reactivity of hydrogen on transition metals exclusively on the basis of theoretical studies of first-row transition metals, as done in the recent work of Kunz et al. (1975) and Melius et al. (1976). The Xα findings are essentially unaltered by hydrogen interacting with the larger nickel, palladium and platinum aggregates having octahedral configurations. The Xα results for the s, p, and d character of bonding between a hydrogen atom and nickel can be compared with the semi-empirical results discussed in sect. 3. The CNDO results of Blyholder (1974) predict much too large an s,p-like contribution with respect to d in comparison with the Xα results. The extended Hückel results of Faessert and van der Avoird (1976b) are however in reasonable agreement with the Xα results in terms of s, p and d contributions to the bonding.

In order to determine the effects of relativity on the electronic structures of the transition-metal cluster results shown in fig. 2.17, calculations were carried out using the relativistic version of the Xα scattered-wave method recently developed by Yang and Rabii (1975). The results of these calculations, which included the relativistic energy-level shifts and spin–orbital splittings, are shown in fig. 2.18. The manifolds of energy levels in fig. 2.18 including the hydrogen–metal bonding orbital split off from the bottom of each d-band, are qualitatively similar to the non-relativistic results of fig. 2.17. The principal effect of relativity is to widen the Pt_4 d-band and to lower its center of gravity with respect to the H 1s orbital energy. The latter result implies a relativistic increase

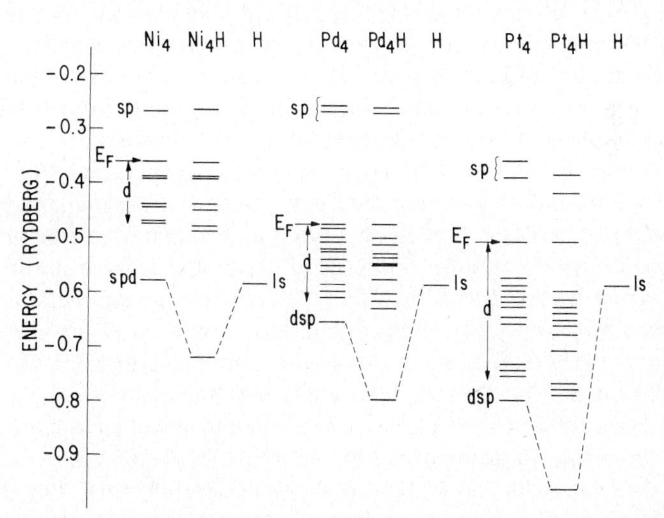

Fig. 2.18. SCF–Xα orbital energies of tetrahedral group VIII transition metal clusters with and without interstitial atomic hydrogen, including relativistic corrections for the Pd_4, Pd_4H, Pt_4 and Pt_4H clusters.

in the effective group orbital electronegativity of the platinum cluster, making it somewhat of an electron acceptor with respect to hydrogen, rather than an electron donor as implied by the respective positions of the non-relativistic Pt_4 energies and H is level in fig. 2.17. This finding is consistent with the slight decrease in the work function of platinum upon hydrogen chemisorption observed in the photoemission measurements of Demuth (1977), whereas the slight increases in work function observed for hydrogen chemisorption on nickel and palladium are explicable in terms of the relative positions of the corresponding energy levels in figs. 2.17 and 2.18. The relativistic effects on the electronic structures of Pd_4 and Pd_4H are qualitatively similar to those for Pt_4 and Pt_4H respectively, although they are too small to significantly affect the chemistry and the interpretation of the photoemission data. The relativistic contributions to the electronic structure of Ni_4 and Ni_4H are negligible.

The most striking confirmation of these theoretical results are the photoemission spectra recently measured by Demuth (1977) for hydrogen chemisorbed on the (111) faces of nickel, palladium and platinum. For each metal, the data clearly show a chemisorption-induced photoemission peak at an energy slightly higher than the metal d-band photoemission peaks, suggestive of a hydrogen–metal bonding state or "resonance" split off from the manifold of d-orbitals as predicted in figs. 2.17 and 2.18. On the basis of the intensity and width of the chemisorption-induced photoemission peak as a function of incident photon energy, Demuth (1977) concludes that hydrogen chemisorption on Ni (111) occurs primarily via the s-orbitals (with some d-orbital participation), whereas the metal d-orbitals dominate hydrogen chemisorption on Pd (111) and Pt (111). This interpretation is completely consistent with the partial-wave decomposition of the hydrogen–metal bonding orbitals of a_1 symmetry described above and summarized in table 2.6. Furthermore, the chemisorption-induced spectral modification of d-band photoemission from Pd (111) and Pt (111) surfaces is explicable in terms of the calculated shifts of d-manifold energy levels in going from Pd_4 to Pd_4H (fig. 2.18). On the other hand, the uniform enhancement of the d-band photoemission from the Ni (111) surface upon hydrogen chemisorption is consistent with the almost negligible shifts calculated for the d-orbital manifold in going from Ni_4 to Ni_4H. Note that the hydrogen-induced d-level shifts for Pt_4H and Pd_4H are significantly larger in the relativistic limit (fig. 2.18) than in the non-relativistic limit (fig. 2.17), showing the role of relativistic contributions to the electronic structure in elucidating certain details of the chemistry of heavy transition-metal aggregates. The close correspondence between the theoretical results for hydrogen bonded in the interstices of tetrahedral Ni_4, Pd_4 and Pt_4 clusters and the photoemission spectra for hydrogen chemisorbed on Ni (111), Pd (111) and Pt (111) strongly suggests that such chemisorption may lead to incorporation of hydrogen atoms in the tetrahedral interstices bounded by the (111) surfaces.

6. Summary

In this chapter an attempt has been made to place cluster model theory in proper perspective with respect to the solid state viewpoint. The importance of having bridges between the two points of view has been stressed. The analytic cluster model which uses simple Hückel theory was shown to provide a bridge between the cluster model and the semi-infinite solid model and to provide some useful qualitative information about changes in electronic structure as a function of cluster size. Future work which would compare results from the cluster model viewpoint and the semi-infinite solid or slab viewpoint for given chemisorption systems using realistic, closely related, theoretical approaches would be particularly valuable. As an example, comparisons between KKR slab calculations and $X\alpha$-SW cluster calculations on common chemisorption systems would be highly desirable.

The utility of metal complexes and organometallic molecules, as well as matrix isolated metal cluster–ligand fragments as analogs for understanding adsorbate bonding on metal surfaces, has been discussed. Future surface work will undoubtedly take increasing advantage of the knowledge being generated by inorganic chemists with respect to metal–ligand bonding.

Of the various theoretical approaches which have been employed to treat cluster models, the $X\alpha$–SW approach has been the most widely applied first principles method. Hence, the discussions of applications which have been presented have had a definite emphasis on systems which have employed this method. Nonetheless, a survey of other independent particle theories was presented and examples of chemisorption applications discussed. Some inter-comparisons of theoretical results were also presented.

In discussing applications, the philosophy adopted was that it would be of greater value to present a few examples in some detail, than to present a cursory discussion of many examples.

In the next few years, there will undoubtedly appear many new examples of cluster model calculations with applications to chemisorption systems. It is the author's hope that the present chapter may serve as a brief introduction to the philosophy and utility of such calculations.

Acknowledgements

The author gratefully acknowledges many useful discussions with Professors K.H. Johnson and D.R. Salahub. He wishes to thank Professors K.H. Johnson, W.G. Klemperer, J.G. Norman and E.W. Plummer and Dr. J.W. Davenport for permission to quote unpublished work. Finally, he especially wants to thank Professor D.R. Salahub for reading the manuscript and making many valuable suggestions.

References

Allyn, C.L., T. Gustafsson and E.W. Plummer, 1977, to be published.
Anderson, O.K. and R.V. Kasowski, 1971, Phys. Rev. **B4**, 1064.
Anderson, P.W., 1961, Phys. Rev. **124**, 41.
Bardeen, J., 1936, Phys. Rev. **49**, 653.
Batra, I.P. and P.S. Bagus, 1975, Solid State Commun. **16**, 1097.
Batra, I.P. and O. Robaux, 1975, Surface Sci. **49**, 653.
Bauschlicher, C.W. Jr., D.H. Liskow, C.F. Bender and H.F. Schaefer III, 1975, J. Chem. Phys. **62**, 4815.
Bennett, A.J., B. McCarroll and R.P. Messmer, 1971a, Surface Sci. **24**, 191.
Bennett, A.J., B. McCarroll and R.P. Messmer, 1971b, Phys. Rev. **B3**, 1397.
Binkley, J.S. and J.A. Pople, 1977, J. Chem. Phys. **66**, 879.
Blyholder, G., 1974, Surface Sci. **42**, 249.
Blyholder, G. and C.A. Coulson, 1967, Trans. Faraday Soc. **63**, 1782.
Brewington, R.B., C.F. Bender and H.F. Schaefer III, 1976, J. Chem. Phys. **64**, 905.
Callaway, J. and C.S. Wang, 1973, Phys. Rev. **B7**, 1096.
Davenport, J.W., 1976a, Phys. Rev. Lett. **36**, 945.
Davenport, J.W., 1976b, Ph.D thesis, Phys. Dept., U. of Pennsylvania.
Davenport, J.W., 1977, to be published.
Dawson, P.H., 1976, Surface Sci. **57**, 229.
Dehmer, J.L. and D. Dill, 1975, Phys. Rev. Lett. **15**, 213.
Demuth, J.E., 1977, to be published.
Demuth, J.E. and D.E. Eastman, 1976, Phys. Rev. **B13**, 1523.
Demuth, J.E., D.W. Jepsen and P.M. Marcus, 1973, Phys. Rev. Lett. **31**, 540.
Eastman, D.E. and J.K. Cashion, 1971, Phys. Rev. Lett. **27**, 1520.
Einstein, T.L. and J.R. Schrieffer, 1973, Phys. Rev. **B7**, 3629.
Ellis, D.E. and G.S. Painter, 1970, Phys. Rev. **B2**, 2887.
Ellis, D.E., H. Adachi and F.W. Averill, 1976, Surface Sci. **58**, 497.
Fassaert, D.J.M. and A. van der Avoird, 1976a, Surface Sci. **55**, 291.
Fassaert, D.J.M. and A. van der Avoird, 1976b, Surface Sci. **55**, 313.

Flodström, S.A., L.G. Petersson and S.B.M. Hagström, 1976a, J. Vac. Sci. Technol. **13**, 280.
Flodström, S.A., L.G. Petersson and S.B.M. Hagström, 1976b, Solid State Commun. **19**, 257.
Flodström, S.A., R.Z. Bachrach, R.S. Bauer and S.B.M. Hagström, 1976c, Phys. Rev. Lett. **37**, 1282.
Gaspar, R., 1954, Acta Phys. Acad. Sci. Hung. **3**, 263.
Goodwin, E.T., 1939, Proc. Cambridge Phil. Soc. **35**, 232.
Grimley, T.B., 1967, Proc. Phys. Soc. **90**, 751.
Grimley, T.B. and E.E. Mola, 1976, J. Phys. C: Solid State Phys. **9**, 3437.
Grimley, T.B. and C. Pisani, 1974, J. Phys. C: Solid State Phys. **7**, 2831.
Gunnarsson, O., H. Hjelmberg and B.I. Lundqvist, 1976, Phys. Rev. Lett. **37**, 292.
Gustafsson, T., E.W. Plummer, D.E. Eastman and J.L. Freeouf, 1975, Solid State Commun. **17**, 391.
Harris, J. and G.S. Painter, 1976, Phys. Rev. Lett. **36**, 151.
Harris, F.E., L. Kumar and H.J. Monkhorst, 1973, Phys. Rev. **B7**, 2850.
Hellmann, H., 1935, Acta Physicochimica URSS **1**, 913.
Hoffmann, R., 1963, J. Chem. Phys. **39**, 1397.
Hubbard, J., 1963, Proc. Roy. Soc. (London) **A276**, 238.
Huber, H., G.A. Ozin and W.J. Power, 1976, J. Am. Chem. Soc. **98**, 6508.
Hückel, E., 1931, Z. Physik **70**, 204.
Hulse, J.E. and M. Moskovits, 1976, Surface Sci. **57**, 125.
Johnson, J.B. and W.G. Klemperer, 1977, to be published.
Johnson, K.H., 1966, J. Chem. Phys. **45**, 3085.
Johnson, K.H., 1973, Scattered-wave theory of the chemical bond, in: Advances in quantum chemistry, vol. 7 (Löwdin, P.O., ed.) (Academic Press, New York) pp. 143–186.
Johnson, K.H., 1977, Intern. J. Quantum Chem. **11S**, 39.
Johnson, K.H. and R.P. Messmer, 1974, J. Vac. Sci. Technol. **11**, 236.
Kahn, L.R., P. Baybutt and D.G. Truhlar, 1976, J. Chem. Phys. **65**, 3826.
Kalkstein, D. and P. Soven, 1971, Surface Sci. **26**, 85.
Kar, N. and P. Soven, 1975, Phys. Rev. **B11**, 3761.
Kasowski, R.V., 1974, Phys. Rev. Lett. **33**, 83.
Kasowski, R.V., 1976, Phys. Rev. Lett. **37**, 219.
Kittel, C., 1963, Quantum theory of solids (Wiley, New York).
Klemperer, W.G. and K.H. Johnson, 1975, unpublished calculations.
Kohn, W., 1975, Phys. Rev. **B11**, 3756.
Kohn, W. and N. Rostoker, 1954, Phys. Rev. **94**, 1111.
Kohn, W. and L.J. Sham, 1965, Phys. Rev. **140**, A1133.
Koopmans, T., 1933, Physica **1**, 104.
Korringa, J., 1947, Physica **13**, 392.
Kunz, A.B., M.P. Guse and R.J. Blint, 1975, J. Phys. **B8**, L358.
Lang, N.D. and A.R. Williams, 1975, Phys. Rev. Lett. **34**, 531.
Linderberg, J. and Y. Öhrn, 1968, J. Chem. Phys. **49**, 716.
Louie, S.G., K.M. Ho, J.R. Chelikowsky and M.L. Cohen, 1976, Phys. Rev. Lett. **37**, 1289.
Marshall, R.F., R.J. Blint and A.B. Kunz, 1976, Phys. Rev. **B13**, 3333.
Martinsson, C., L.G. Petersson, S.A. Flodström and S.B.M. Hagström, 1976, Photoemission from surfaces, European space agency report ESA SP-118 Noordvijk, Holland, p. 117.

Melius, C.F., J.W. Moskowitz, A.P. Mortola, M.B. Baille and M.A. Ratner, 1976, Surface Sci. **59**, 279.
Messmer, R.P., B. McCarroll and C.M. Singal, 1972, J. Vac. Sci. Technol. **9**, 891.
Messmer, R.P., 1975, Theoretical Studies of Metal Aggregates and Organometallic Complexes Relevant to Catalysis, in: Drauglis, E. and R.I. Jaffee, eds., The Physical Basis for Heterogeneous Catalysis, (Plenum Press, New York), pp. 261–275.
Messmer, R.P., S.K. Knudson, K.H. Johnson, J.B. Diamond and C.Y. Yang, 1976, Phys. Rev. **B13**, 1396.
Messmer, R.P. and K.H. Johnson, 1976, Quantum Chemistry and Catalysis, in: Franklin, A.D., ed., Electrocatalysis on Non-Metallic Surfaces, Spec. Publ. 455 (National Bureau of Standards, Washington, D.C.) pp. 67–86.
Messmer, R.P. and K.H. Johnson, 1976, Quantum Chemistry and Catalysis, in: Franklin, A.D., ed., Electrocatalysis on Non-Metallic Surfaces, Spec. Publ. 455 (National Bureau of Standards, Washington, D.C.) pp. 67–86.
Messmer, R.P., D.R. Salahub, K.H. Johnson and C.Y. Yang, 1977, Chem. Phys. Letters **51**, 84.
Messmer, R.P. and D.R. Salahub, 1977a, Theoretical Studies of Metal Clusters as Models for Surface Phenomena, in: Ludeña, L.V., N. Sabelli and A.C. Wahl, eds., Computers in Chemical Research and Education, (Plenum Press, New York), p. 156.
Messmer, R.P., 1977a, Phys. Rev. **B15**, 1811.
Messmer, R.P., 1977b, The Molecular Cluster Approach to Some Solid State Problems, in: Segal, G., ed., Modern Theoretical Chemistry, Vol. 8 (Plenum Press, New York), pp. 215–246.
Messmer, R.P. and D.R. Salahub, 1977b, Chem. Phys. Lett. **49**, 59; Phys. Rev. B **16**, 3415.
Muetterties, E.L., 1975, Bull. Soc. Chim. Belg. **84**, 959.
Mulliken, R.S., 1949, J. Chem. Phys. **46**, 497.
Murrell, J.N. and A.J. Harget, 1972, Semi-Empirical Self-Consistent Field Molecular Orbital Theory, (Wiley-Interscience, New York).
Newns, D.M., 1969, Phys. Rev. **178**, 1123.
Newns, D.M., 1970, Phys. Rev. Lett. **25**, 1575.
Norman, J.G., Jr., 1977, to be published.
Ozin, G.A. and W.J. Power, 1977a, Inorg. Chem. **16**, 212.
Ozin, G.A. and W.J. Power, 1977b, to be published.
Pariser, R. and R.G. Parr, 1953, J. Chem. Phys. **21**, 466.
Phillips, J.C. and L. Kleinman, 1959, Phys. Rev. **116**, 287.
Pople, J.A., 1953, Trans. Faraday Soc. **49**, 1375.
Pople, J.A. and G.A. Segal, 1965, J. Chem. Phys. **43**, S136.
Pople, J.A., D.P. Santry and G.A. Segal, 1965, J. Chem. Phys. **43**, S129.
Roothaan, C.C.J., 1951, Rev. Mod. Phys. **23**, 69.
Roothaan, C.C.J., 1960, Rev. Mod. Phys. **32**, 179.
Rösch, N., 1977, The SCF–Xα scattered wave method with applications to molecules and surfaces, in: Electrons in finite and infinite systems (Pharisean, P., ed.) (Plenum Press, New York).
Rösch, N. and T.N. Rhodin, 1974, Phys. Rev. Lett. **32**, 1189.
Rösch, N., W.G. Klemperer and K.H. Johnson, 1973, Chem. Phys. Lett. **23**, 149.
Rösch, N., R.P. Messmer and K.H. Johnson, 1974, J. Am. Chem. Soc. **96**, 3855.
Salahub, D.R. and R.P. Messmer, 1977, Phys. Rev. **B16**, 2526.
Sambe, H. and R.H. Felton, 1975, J. Chem. Phys. **62**, 1122.

Schaefer, H.F., 1972, The electronic structure of atoms and molecules – a survey of rigorous quantum mechanical results (Addison-Wesley, Reading, Mass.).
Schwarz, K., 1972, Phys. Rev. **V5**, 2466.
Schwarz, K., 1974, Theoret. Chim. Acta **34**, 225.
Slater, J.C., 1951, Phys. Rev. **81**, 385.
Slater, J.C., 1965, J. Chem. Phys. **43**, 5228.
Slater, J.C., 1972, Statistical exchange-correlation in the self-consistent field, in: Advances in quantum chemistry, vol. 6 (Löwdin, P.O., ed.) (Academic Press, New York) pp. 1–93.
Slater, J.C., 1974, Quantum theory of molecules and solids, vol. 4, The self-consistent field for molecules and solids (McGraw-Hill, New York) p. 94.
Slater, J.C. and G.F. Koster, 1954, Phys. Rev. **94**, 1498.
Streitwieser, A., 1960, J. Am. Chem. Soc. **82**, 4123.
Ugo, R., 1975, Catal. Rev. – Sci. Eng. **11**, 225.
Wheland, G.W. and D.E. Mann, 1949, J. Chem. Phys. **17**, 264.
Williams, P.M., P. Butcher, J. Wood and K. Jacobi, 1976, Phys. Rev. **B14**, 3215.
Yang, C.Y. and S. Rabii, 1975, Phys. Rev. **A12**, 362.
Yu, K.Y., J.N. Miller, P. Chye, W.E. Spicer, N.D. Langard and A.R. Williams, 1976, Phys. Rev. **B14**, 1446.

CHAPTER 3

ELECTRON SPECTROSCOPY AND SURFACE CHEMICAL BONDING

T.N. RHODIN*

School of Applied and Engineering Physics, Cornell University, Ithaca, N.Y. 14853, USA

and

J.W. GADZUK

Surface Science Division, National Bureau of Standards, Washington, D.C. 20234, USA

*Supported by NSF-DMR77-05078-A02 and the Cornell Materials Science Center NSF-DMR76-81083-A01.

Contents

The nature of the surface chemical bond
Edited by Th.N. Rhodin and G. Ertl

1. Introduction

1.1. General background

That a chapter of this nature exists is an indication of the intense activity throughout the past decade which has brought us to a closer understanding of the microscopic basis of the nature of the surface chemical bond. A number of different forces have propelled us towards this new insight. On the one hand, the development of new ultra-high vacuum techniques enabled preparation and preservation of well defined surfaces, which could be characterized in a reproducible way. This situation, in turn, has led to the discovery of a multitude of experimental methods for probing atomic scale phenomena at solid–vacuum interfaces. Of particular interest in this chapter are the electron spectroscopies which have, in our opinion, proven most useful in studies of chemisorptive bonding. Consideration has been limited here to just a few of the many spectroscopic techniques currently being employed by researchers throughout the world. The overwhelming number of these techniques as well as an introduction to the acronym-collage of experimental surface science techniques are summarized in appendix A where the row and column indices of the matrix label, respectively, the incident and detected particles of the particular "scattering" experiment (Powell 1977). Since a number of books, edited proceedings, special journal issues, and review articles on various combinations of electron spectroscopies (atoms, molecules, solids and surfaces) (Shirley 1972, Koch et al. 1974, Carlson 1975, Siegbahn 1976, Castex et al. 1977, Ibach 1977a, Feuerbacher et al. 1978) as well as surface physics and chemistry (Ertl and Küppers 1974, Gomer 1975a, Hannay 1976, Ibach 1977b, Dobrozemsky et al. 1977) in general are available, no attempt will be made here to provide a critical review of the field. Instead the focus will be on the basic physical features of four commonly used techniques in order to gain the insight needed to proceed from raw experimental data to system information, devoid of measurement-process-induced features.

A common element of most surface sensitive spectroscopies is the utilization of charged particles or photons for either the incident probe, detected signal, or both. Most common is an electron beam, which in the energy range 10–1000 eV with respect to the Fermi level of the solid, undergoes inelastic collisions in which plasmons, surface plasmons, and/or electron–hole pairs (including core states) are created within ~10 Å of the surface (Powell 1974, Penn 1976). Thus any process known to involve unscattered electrons necessarily drops off exponentially going into the solid, with a decay or attenuation length that is a function of electron energy, substrate material, and surface conditions. For the $\lesssim$1000 eV domain, it usually remains $\lesssim$10 Å, thus guaranteeing that a large component of the total output originates in the first layer or two. With this in mind, it is apparent why the techniques discussed in this chapter are so particularly well suited for interpretative studies of chemisorptive systems.

1.2. Solid state electron spectroscopies

Specifically, we will consider ultraviolet photoelectron spectroscopy (UPS) (Feuerbacher and Fitton 1977b, Brundle 1975, Derouane and Lucas 1976, West 1977, Gadzuk 1976, Plummer 1975, Feuerbacher et al. 1978), X-ray photoelectron spectroscopy (XPS) (alternatively called electron spectroscopy for chemical analysis (ESCA) (Brundle 1975b, Siegbahn 1977, Feuerbacher et al. 1978, Martin and Shirley 1977), Auger electron spectroscopy (AES) (Hawkins 1977, Gallon 1978), and electron energy loss spectroscopy (EELS) (Froitzheim 1977, Ibach et al. 1977). The general plan of these techniques is shown in fig. 3.1. The left half of each column depicts the relevant energy level diagrams and currents involved in each technique. The UPS, XPS, and AES diagrams show the substrate energy levels as a broadened conduction band (cross-hatched) and some discrete atomic-like core states. As drawn, the doublet structure might correspond to spin–orbit split $P_{1/2}$ and $P_{3/2}$ states whereas the single level might be an s-state. The vacuum level just outside the surface ("just" is a distance which is long compared to atomic dimensions but no larger than the size of a single crystal facet, in order to avoid transverse fringing fields from other crystal facets which are required to equalize the electrostatic potential "far" from the crystal) (Herring and Nichols 1949, Lang 1973), is $e\phi_e$ above the Fermi level, with ϕ_e the work function. The possibility of a chemisorbed atom or molecule is signified both by the localized broadened resonance within the conduction band (Newns 1969, Gomer 1975b, Gadzuk 1975b,

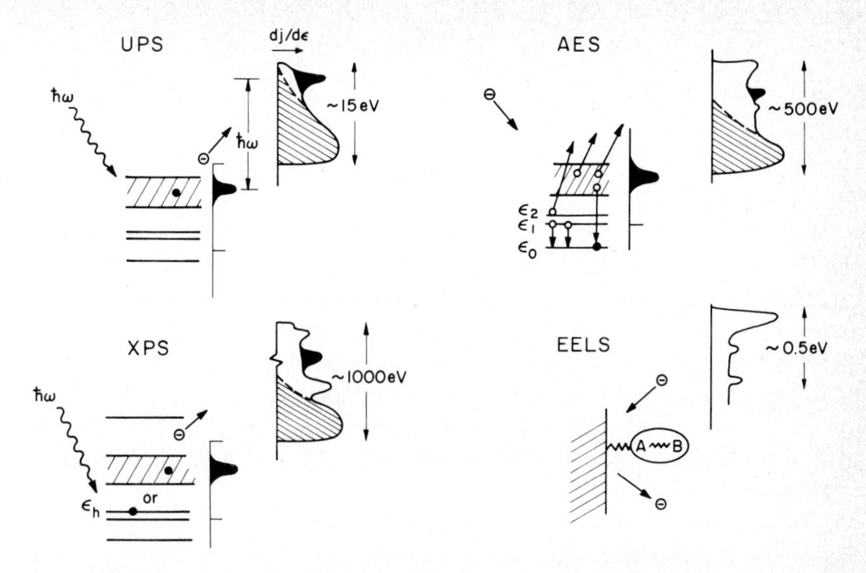

Fig. 3.1. Overview of the measurement processes considered in this chapter. On the left of each pair are shown the energy level diagrams for the 3-emission spectroscopies (UPS, XPS, AES) and the mechanical analog for the absorption spectroscopy (EELS). On the right are schematic forms of the raw energy distribution data obtained with each technique with the characteristic span of energies so marked.

Muscat and Newns 1978) and also by core states at different energies than for the substrate. The valence electron resonances usually, but not always, bear a one-to-one correspondence with the electron orbitals of the adparticle, some of which are shifted and/or rehybridized due to the formation of the surface chemical bond. This is discussed in greater detail in both Grimley's and Messmer's chapters in this volume.

The most common mode of electron spectroscopy is the recording of the energy shifts (both absolute and differential) of the electrons emitted from the various adparticle states relative to their gas phase counterparts. If one can properly separate out both the relaxation shifts and shake-up satellite structure (to be discussed later in this chapter) inherent in the measurement process and also the ubiquitous "matrix element effects", one has a direct measure of the energy level structure of the chemisorption complex which includes both the electrostatic chemical shifts championed by Siegbahn et al. (1976) and the shifts due to chemical bond formation. In actual practice, this separation is not straightforward and the uncovering of means to this end is one of the challenges actively being addressed today by many researchers. Another increasingly popular mode of surface electron spectroscopies is to record the angular distribution of electrons ejected at a specified energy; in other words, the momentum distribution. In many cases, the momentum distribution of photoejected electrons is simply related to the Fourier transform of the initial state wave function and thus such measurements can, in principle, provide a picture of the shapes of the localized chemisorption states.

The EELS diagram is quite different from the previous three, since this techniques as most often used, probes the vibrational rather than electronic excitations of the chemisorption complex (Froitzheim 1977, Ibach et al. 1977; also Andersson and Jostell 1974, Schröder and Höltz 1977, Lindgren and Walldén 1978). The energy losses suffered by a highly monochromatic incident electron beam, backscattered by the surface, are related to gas phase intramolecular vibrational excitation and to possible new modes formed between the substrate and the adparticle bonded to the surface. It is in this way that one hopes to ascertain the state of dissociation and/or adsorption site location with respect to the substrate atoms.

The second column shows prototypical energy distributions (ED). Take UPS first: ultraviolet photons ($\hbar\omega = 16.85, 21.2, 40.8$ eV with discrete line sources are popular and available) do not provide sufficient energy to remove electrons from anything but the relatively weakly bound valence electron states of the substrate and chemisorption system (with a few exceptions of high lying core electrons such as 2s in O, or 4p in Rb). The ED of the emitted electrons, divided into three segments in this diagram, have the universal features that the maximum allowed energy for the photoelectron is $\hbar\omega$ above the Fermi level and the width of the ED is $\hbar\omega - \phi_e$. For UPS this width is in the range of $10 \sim 30$ eV. In this diagram, the unmarked area corresponds to photoelectrons from the substrate which have not undergone inelastic scattering on their way out. The blacked-in portion is emission from the chemisorption complex. It is these types of features which, upon proper interpretation, yield direct information on the chemisorption bond. Typically an admolecule will have some orbitals involved in bonding to the surface and some which remain non-bonding. Differential shifts in peak positions and shapes are observed which correspond to photoionization from various ones of these orbitals. Finally, the cross-hatched area are those secondary electrons which have been involved in the inelastic scattering of the primary photoelectrons (Chung and Everhart 1977, Tung and Ritchie 1977, Sickafus 1977). This portion of the spectrum is an annoyance and is often removed by a procedure which is given the dignified name of "background subtraction". As usually performed, a well trained eye is most important, although some quasi-rigorous precedures are being established (Houston 1975).

XPS is in principle no different from UPS, so much of the previous paragraph still applies. However, X-rays (typically Mg- or Al-K_α radiation with $\hbar\omega = 1253.6$ or 1486.6 eV respectively) have sufficient energy to photoionize not only the more or less extended valence electrons, but also some core electrons which are localized on specific atoms. Photoelectrons from the narrow core states appear in the spectrum as discrete peaks, asymmetrically broadened on the low kinetic energy side by core hole lifetimes (Citrin et al. 1974) and electron–hole pairs (Doniach and Šunjić 1970, Gadzuk and Šunjić 1975, Minnhagen 1976a, b, 1977) or phonon (Citrin et al. 1974, Šunjić and Lucas 1976, Minnhagen 1976, Gadzuk 1976a, Citrin and Hamann 1977, Mahan 1977, Almbladh and Minnhagen 1978) creation. To a first approximation, the binding energy (and hence kinetic energy) of a core electron depends solely upon the atomic species, which makes XPS a viable method of nondestructive elemental analysis. The presence of neighboring atoms, in the case of molecules, solids, or adsorption systems, causes small (~1–10 eV) shifts in the binding energies (~100–1000 eV), which are due to changes in the electrostatic potential at the core site. These environmentally induced shifts in the initial state binding energies have appropriately been termed chemical shifts (Siegbahn et al. 1976) and it is with this in mind that XPS and ESCA holds promise for study of the surface chemical bond.

Unfortunately, life is not so simple though, for upon creation of a localized core hole, the mobile electrons of the substrate are attracted to the extra positive charge. The attractive Coulomb interaction between this induced screening charge and the core hole charge results in a lowering of the total energy of the $N-1$ electron ionized system with respect to its total energy when the core state is occupied. In order that energy be conserved in the photoemission process, the photoelectron emerges at a kinetic energy which has been upshifted by an equal amount, call it $\Delta\varepsilon_r$ often referred to as the extra atomic relaxation, screening, or polarization energy. Typically this shift is ~1–10 eV and always positive in sign. The experimentally observed shifts are a sum of the initial state chemical and final state relaxation shifts, which requires that there be an independent determination of the relaxation energy before XPS can be used routinely for the type of chemical analysis first hoped for. The importance of realizing this hope has spurred on considerable work on relaxation effects, some of which will be discussed later. As a result of one-electron photoexcitation matrix elements, the relative intensity of emitted electrons from various surface-modified Bloch states within the conduction band, or molecular orbital states from the chemisorption complex, may be quite dependent on the photon frequency and hence final state energy of the photoelectron. Depending upon ones point of view, this is regarded in principle as an easily dealt with single electron effect presenting no fundamental problem, thus allowing one to drop further considerations of it, or alternatively the frequency dependence is thought of as a useful property, permitting assignment of initial state orbital labels to various features in the ED on the basis of the $\hbar\omega$-dependence of their cross sections (Gustafsson 1975, Feuerbacher et al. 1978, Feuerbacher and Fitton 1977b). Examples of this will also be discussed later.

Moving on to AES, we note that the fundamental nature of this process is quite different from that of the photoelectron spectroscopies for the following reason. In UPS and XPS, the system, initially in its ground state, is lifted to an excited state (one with a hole) by absorption of a photon. A possible product of the excitation process is a photoelectron, which is then sensed by a detector. In contrast, the Auger emission is usually considered as follows. Rigorously the initial state is also the system ground state. A high energy excitation beam (usually electrons but sometimes X-rays) (Fuggle et al. 1976a, b) creates a low-lying core hole. The exciting and excited electrons then disappear, leaving behind a prepared or initially excited one-hole state which decays. To the extent that the excitation process does not interfere with the decay process, that is no memory or incomplete relaxation processes (Watts 1972, Yue and Doniach 1973, Almbladh 1974), the initial state in AES can then be regarded as this prepared state with a core hole. For elements with $Z \leqslant 30$, radiative decay (at least for K shell vacancies) is unlikely, in which case two electrons from higher lying occupied states interact, with the end results that the initial core hole is occupied, an Auger excited electron with energy above the Fermi level is produced, and two holes are left behind. It is the Auger electron which then may be emitted and detected. At least three different types of spectral features are possible, depending upon the types of electrons involved in the Auger decay process. The highest energy ejected electrons will be those originating from processes involving two valence electrons. By energy conservation, the maximum energy for the Auger excited electrons is ε_b above the Fermi level (neglecting relaxation shifts for the time being). A rather broad feature (twice the conduction bandwidth) which is approximately the self-convolution of a surface density of states (Powell 1973, Gadzuk 1974a, b, Jackson et al. 1975, Madden and Houston 1977) if matrix element effects are neglected

(Feibelman et al. 1977, 1978), follows from the core–valence–valence (CVV) transition. Related to transitions involving valence electrons is the fate of the conduction band hole or holes created in the Auger decay. Do they remain localized on the atom with the core hole, as might happen for narrow band materials such as noble metals, or do they delocalize in a bandlike way? Localized holes would suggest that CVV transitions in solids would more closely resemble atomic-like processes with characteristically narrow spectral features and reduced ejected electron kinetic energies due to the higher energy of the localized two-hole final state of the target (a consequence of the intra-atomic Coulomb repulsion between the two holes). Such questions define some areas of current activity (Rojo and Baro 1976, Cini 1976, 1977, 1978, Sawatzki 1977). Another possibility involves one core and one valence electron (CCV), in which case the lower lying broad structure in the ED was first thought to replicate the conduction band density of states. Inclusion of matrix-element-derived selection rules (Feibelman et al. 1977, 1978) and considerations from the localized–delocalized valence hole make such expectations far too simple. The final possibility in AES is transitions involving only core states. These CCC transitions show up as sharp peaks in the spectrum at characteristic and well documented energies for a given Auger transition in a specific element. These fingerprints are the ones commonly used for elemental analysis, which is perhaps the most important current application of AES. As usually practiced, neither resolution nor data analysis procedures (including quantitative theoretical understanding) are at present sufficiently precise in AES to extract the sort of chemical shift information obtained in XPS.

Finally, the high resolution energy loss spectrum is shown. The large high energy peak is the elastically scattered primary beam, typically ~10 meV in width. The lower energy peaks are due to to energy loss processes in which the localized vibrational modes in the chemisorption complex are excited and the displacement below the no-loss peak is the information that is most commonly being used to understand the molecular nature of bonding at the surface. Studies involving angular dependencies or relative peak intensities will certainly provide additional useful information, but at present they are in their infancy (Newns 1977, Šokčević et al. 1977, Delanaye et al. 1978, Persson 1977, Davenport et al. 1978).

1.3. Theoretical background

From a theoretical point of view, the principal problems fall into two broad areas, one of which focuses on the single electron matrix elements and the other on many-body relaxation and shake-up effects. The basic similarities, differences, complications, and simplifications of each spectroscopy can be succinctly illustrated in terms of Feynman-like diagrams. Current conceptual theoretical work (as opposed to advanced numerical work on conceptually well-established models) seems to favor an approach in which transition probabilities, cross sections, and the like are expressed in terms of time-dependent correlation functions linking the system ground or initial state with various possible real and virtual excited states (Doniach and Sondheimer 1974). This is reflected in the so-called three-point diagrams shown in fig. 3.2, in which some of the physical features of the measurement process to be dealt with are indicated. The far left column shows the diagrammatic representation of the simplest independent particle description for the so-labeled spectroscopy, with time moving from down (the past) to up (the future). Some of the additional processes such as many-body relaxation, inelastic scattering, and initial excitation which influence the observations are then shown on the right. Let us consider each one in detail.

Conceptually, UPS and XPS are equivalent; however different aspects of the interacting system are emphasized in photoemission from localized (core states or chemisorption resonances) as compared to delocalized states. It is only with X-rays that the core states are accessible. The upper left diagram shows the bare-bones photoemission process (Caroli et al. 1973, Feibelman and Eastman 1974). A photon with energy $\hbar\omega$ is absorbed at the space–time point $\boldsymbol{x}_1$, creating an electron–hole pair. The electron is excited to a state with energy ε_{f} (with respect to a convenient and consistent energy zero, usually the Fermi-level, in which case ϕ_{e} must be subtracted when talking of a free space electron kinetic energy) and momentum $\boldsymbol{k}$ far out of the solid or surface. A hole state characterized by an energy ε_{h} and some remaining quantum numbers n ($\boldsymbol{k}$-vector for Bloch states, angular momentum or multiplets for core states, etc.) is also necessarily created. Both the electron and hole propagate forward in time, the electron ultimately arriving at a detector at $\boldsymbol{R}$, $t = 0$, where the current is measured. Thus far we have described the amplitude for the photoemission process. Completion of the diagram to $\boldsymbol{x}_2 \rightarrow t = +\infty$ is equivalent to taking the modulus squared, which is proportional to the intensity. Single electron matrix elements tell us which hole states are compatible with a given photon and final state

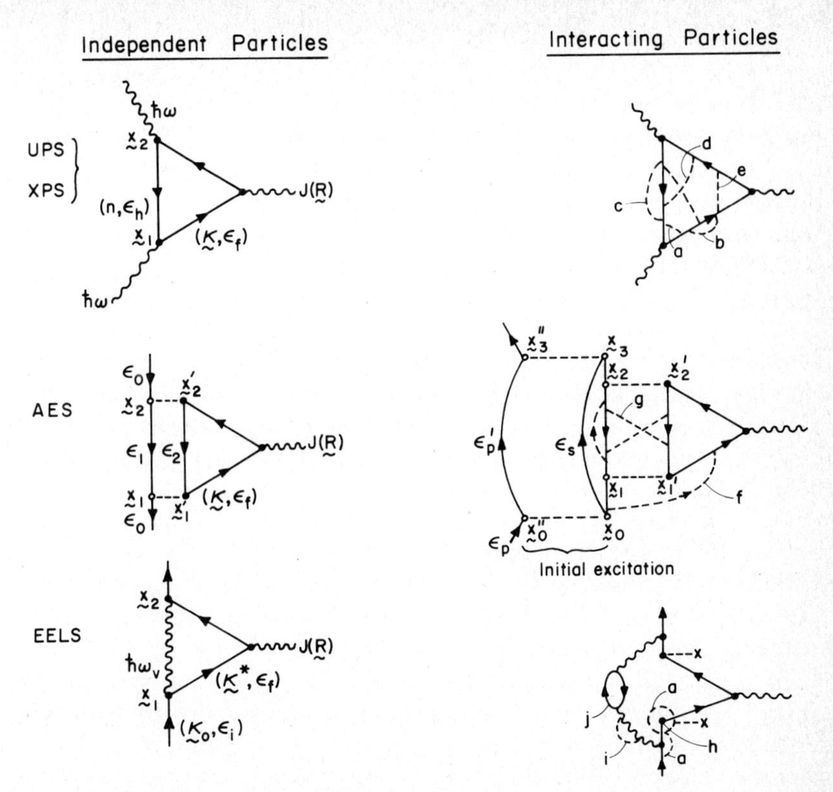

Fig. 3.2. Diagramatic representation of the spectroscopic techniques. On the left are the processes involved, within an independent particle picture, in which times moves upward. Some typical corrections to the single electron picture are shown on the right. These include: (a) vertex renormalization, (b) electron propagator renormalization or self energy corrections (c) hole renormalization and intrinsic satellites, (d) Boson induced electron–hole interferences, (e) extrinsic Boson satellites or inelastic transport loss, (f) Plasmon gain, (g) hole–hole interactions, (h) high momentum transfer elastic collision with surface, (i) vibrational renormalization, (j) vibrational decay into electron–hole pairs.

electron (selection rule) and the relative intensities of various electron and hole combinations. They are to be associated with the heavy dots. For the non-interacting system, conservation of energy requires that $\hbar\omega = \varepsilon_f + \varepsilon_h$, with the excited hole state quite positive in energy.

A number of interesting new possibilities open up when dynamic interactions amongst the particles of the target system are allowed for. The system can now readjust to the presence or absence of the excited electron and hole, thus changing its total energy (relaxation shift) and also can be left in one of its excited states (shake-up). For present purposes, the excitations of the solid are electron–hole pairs, plasmons, surface plasmons, and phonons, all of which can be modeled as boson-fields (Müller-Hartman et al. 1971). It is the interactions with the appropriate bosons, of which some prototypical ones are shown on the right, that must be understood. Upon photon absorption, a bare hole with energy determined by some mean-field procedure such as Hartree–Fock, propagates forward in time polarizing the boson-field along the way. In the real world picture, hole-induced density fluctuations in the electron gas model of the conduction band occur. This induced charge can be expressed in terms of plasmons or surface plasmons which are represented in the diagram by dotted lines. (Phonons are also quite important, particularly in ionic solids.) Some typical hole–plasmon interactions are shown in which the hole emits and reabsorbs arbitrary numbers of real and/or virtual plasmons. These processes renormalize the hole energy by adding a self-energy to ε_h and are most important for spatially localized hole states. Plasmons represented by lines which exist at $t = 0$, the time of the measurement, are considered to be real shake-up excitations (intrinsic plasmons), which require that the photoelectron energy be reduced by the appropriate integral number of plasmon energies.

The structures appearing in a spectrum due to intrinsic plasmon creation are referred to as intrinsic shake-up satellites. A similar energy-dependent renormalization of the electron propagator must be considered. Those plasmon lines which originate and end on the electron line going into and out of the J-vertex give rise to the so-called extrinsic or transport loss satellites characteristic of any electron beam travelling through a metal. They have nothing to do with the actual photon absorption process and as such can in principle be subtracted out of the measured spectrum (Langreth 1971, Chang and Langreth 1972, Pardee et al. 1975, Penn 1977). In addition, the electron and hole can still exchange energy via plasmon-mediated final state interactions as represented by the plasmon lines connecting electrons and holes. If both connections are in the same time epoch, the process is referred to as vertex renormalization. On the other hand, when the final state interaction plasmon lines cross the $t = 0$ level, the result is an interference between intrinsic and extrinsic plasmon satellite production, which makes a rigorous separation of the two in data-analysis meaningless from a quantum mechanical point of view (Šunjić and Šokčević 1974, 1976, Šunjić et al. 1977). In actual practice, final state interactions are often neglected (mostly out of convenience), in which case all relaxation processes are associated with independently propagating electrons and holes. Aside from exciting extrinsic plasmons, the photoelectron is merely a messenger which brings to the detector information about the relaxation which has occurred around the hole. Thus it is usually assumed that there is a one-to-one correspondence between the ED and the spectral function (imaginary part of the Green's function) (Doniach and Sondheimer 1974) for a hole in which certain time-orderings are honored, namely that intrinsic plasmon production cannot start until after the photon has been absorbed. It is this time-ordering which makes the problem non-trivial in subtle ways which show up as the difference between adiabatic and sudden creation of the hole (Meldner and Perez 1971, Müller-Hartman 1971, Gadzuk and Šunjić 1975, Gadzuk 1977a, b).

The zero order picture of AES is shown in row 2, left. Initially the prepared hole state with energy ε_0 propagates forward in time until it reaches $\boldsymbol{x}_1$ where it interacts with the system via the electromagnetic field, creating an electron–hole pair at $\boldsymbol{x}_1'$. The two dots and dashed line represent the two-electron matrix element. Henceforth two holes and an Auger electron with energy ε_1, ε_2, and ε_f respectively, propagate until the Auger electron enters the detector at $\boldsymbol{R}$. Completion of the round trip yields the Auger equivalent of the photoemission three-point diagram. Conservation of energy demands that $\varepsilon_1 + \varepsilon_2 + \varepsilon_f = \varepsilon_0$. Strictly speaking, calling this an independent particle picture is a misnomer as the perturbation giving rise to the Auger transition is in fact an interaction among particles. However, aside from this one interaction, all others are lumped into a static, mean-field potential and thus present no problems. With regards to additional complications, all those processes which were brought up in photoemission persist in AES in addition to several new ones. To begin with we should consider the total process including initial excitation, shown in its most elementary form on the right in fig. 3.2. As drawn, a primary electron with energy ε_p creates the deep core hole ε_0 and secondary electron ε_s, itself dropping to energy $\varepsilon_p' = \varepsilon_p - \varepsilon_s - \varepsilon_0$. However, either electron could go into an intermediate state of energy ε_f where it would appear in the detector as the Auger electron. In this case, the standard Auger electron line would have to end at $\boldsymbol{x}_3$ or $\boldsymbol{x}_3''$ which then couples the Auger decay with the initial excitation processes. Yet another coupling occurs when plasmons, emitted between $\boldsymbol{x}_0$ and $\boldsymbol{x}_1$ (intrinsic to the initial excitation process) are reabsorbed between $\boldsymbol{x}_1$ and $\boldsymbol{x}_2$, after the Auger transition. Some attention has also been directed towards the possibilities of plasmon gain-satellites which might occur when an intrinsic initial excitation plasmon is absorbed by the Auger electron (Jenkins and Zehner 1973). In general, plasmon mediated coupling between the initial excitation and Auger decay processes is enhanced as the magnitude of the core hole width increases towards the plasmon energy. Lastly, dynamically screened interactions in the two-hole state must be included. The importance of the types of processes just mentioned is currently under study (Watts 1972, Yue and Doniach 1973, Almbladh 1974, Rojo and Baro 1976, Cini 1976, 1977, 1978, Sawatzky 1977). Suffice to say that most data interpretation is done on models which do not require details of the initial excitation, which sometimes include renormalized hole propagators (at least with respect to relaxation or self-energy shifts), and models which are beginning to consider dynamic hole–hole interactions. Even without an in-depth understanding of the dynamics of the total Auger process, the technique has certainly proved its pragmatic worth in elemental surface analysis of both clean and chemisorbed surfaces.

The final technique, EELS, presents the fewest complications in understanding its gross characteristics. This is in large part due to the nature of the absorption experiment itself in which no secondary particles are generated. Consequently the charge state of the surface molecules remains constant throughout the scattering event and process-induced-coupling with the electron density fluctuations (plamsons, pairs) of the surface is small. In its simplest form we might take the EELS event to be equivalent to a scattering from an oriented gas phase molecule (Davenport and Ho 1978). As shown in fig. 3.2, bottom left, the incident electron with energy ε_i and wave vector $\boldsymbol{k}$ interacts with the surface molecule at space–time point $\boldsymbol{x}_1$, creating a vibrational quanta $\hbar\omega$, and itself scattering into a new state ε_f, $\boldsymbol{k}'$ where it moves on to the detector. Completing the circuit to $\boldsymbol{x}_2$ yields the intensity diagram. The gas phase scattering matrix element, to any degree of sophistication desired, is represented by the big dots. Note that as drawn, a fairly large momentum transfer scattering event is required.

Although the role of particle interactions is not as severe in EELS as in the emission spectroscopies, some interesting possibilities do exist which must be considered if relative intensities of various loss peaks are to be interpreted. Consider first the fact that the chemisorption complex is more than an oriented molecule; in fact it is an oriented molecule sitting on a massive solid which can provide a lot of high momentum transfer, elastic scattering. Thus the actual sequence of events could be first inelastic forward scattering by the surface molecule followed by elastic specular scattering off the surface (or reversed in order), as represented by the dashed line and x in fig. 3.2 bottom right. The electron is necessarily near the surface when either of these events occurs and it is at this time that virtual surface plasmon effects are most probable. Of particular interest is the vertex renormalization of the gas phase matrix element as they (unrenormalized vertices) are presumably quite accurately known. Small shifts in the energy of the scattering electron could possibly pull the scattering electron into or out of a resonance condition. Two new effects are included on the vibronic line. First since the oscillating dipole of the molecule sets up image charges which must oscillate, a vibron-surface plasmon coupling is set up which appears in the diagram as a vibrational renormalization. Secondly, the lifetime of the vibrationally excited state should be much less than in the gas phase due to the new channels for vibrational relaxation and decay (Fong 1976). One such channel into an electron–hole pair is shown. The popularity of EELS has grown immensely in the past year or two, in part due to the ease in reducing some parts of the collected data into useful form. The one-to-one relation between electron energy losses and vibrational energies makes this rather trivial. On the other hand, angular dependences and branching ratios provide still more data. This aspect of EELS is still to be developed.

Having discussed the qualitative features of the most informative surface spectroscopies, with relatively little attention directed to the fact that our main concern is the surface chemical bond, we here try to point out that connection. In fact, most of what has been said so far is quite general and applies equally well to electronic states of the substrate and of the chemisorption complex. Since the energy of the detected particle is directly related to either the hole state or vibrational energies, those processes known to create holes or vibrations in the chemisorption complex provide information related directly to the surface chemical bond. As a first approximation, two different types of states, for both electrons and vibrations are usually assumed to exist. In the first case one considers those orbitals or vibrational modes which do not participate in the chemical bond. Electronic core levels, lone pair molecular orbitals pointing away from the surface, or intramolecular stretching modes for diatomics are some examples of this type. In some senses they are spectators at the chemisorption event and thus provide a measure of initial state electrostatic or final state relaxation shifts, devoid of the bonding shifts. Obtaining estimates of the non-bonding shifts from spectroscopic probes of the spectator states then enables one to determine a zeroth order bonding shift associated with the participant states. The reliability of this procedure remains to be adequately established or discredited by detailed theoretical and/or computational studies. Finally, angle-resolved measurements are best done on the spectator states, as their wave functions are known, thus permitting the calculation of the appropriate single electron matrix elements which contain the most significant angular dependences (Dill 1976, Dill et al. 1976, Wallace et al. 1978, Davenport 1976, 1978).

1.4. Experimental background

The present experimental status of electron spectroscopy looks promising from the viewpoints of scope and versatility. It must be considered cautiously, however, in full recognition of its limitation in providing basic information critical to the interpretation of chemical bonding of a general nature. The major impact of recent effort has been on bonding interactions associated primarily with extended single crystal surfaces in ultra-high vacuum environments, e.g. conditions chosen to facilitate both the instrumentation of the measurement as well as the preparations of a well defined surface. A great need exists to expand the measurements to include non-UHV conditions involving both gaseous and liquid environments. Measuring approaches which feature absorption or fluorescence behavior, which emphasize, for example, atomistic properties rather than those of two-dimensional single crystals, is desirable. In a similar vein, with some exceptions, the whole area of the solid–solid surface, e.g. that of *interfaces*, lies largely unexplored in terms of relating microscopic electron characterization to chemical bonding. A logical and inevitable extension of this extrapolation is, of course, to the microscopic characterization of dispersed systems, eventually to very small atom clusters and ultimately to single atoms. The present state of the art in the microcharacterization of interfaces and particles already points in these directions although the achievements to date are mainly suggestive.

Extended X-ray absorption fine structures (EXAFS) (Stern et al. 1977) can be effectively applied to the elucidation of bonding and orientation in a region highly localized to a reference atom in a bulk matrix. Extension of this approach using Auger emission (Kincaid et al. 1977) or other effects to enhance signal-to-noise to make it sufficiently surface-sensitive are needed to provide a practical probe of the surface. They are being explored, but the applicability to chemical bonding is largely undeveloped at this time (Rhodin and Brucker 1977). It lends itself well to dispersed systems since it is primarily an absorption effect independent of the constraints associated with long range order. The tool of photoacoustic spectroscopy (Rosencwaig 1975), sensitive as it is to electromagnetic effects in the infrared wavelength range, has been mainly demonstrated to be effective for bulk systems. It suffers from a similar need to achieve adequate signal-to-noise response to make it effective as a surface-sensitive technique. On the other hand, infrared absorption spectroscopy which has been effectively applied for a long time to the vibrational mode study of both dispersed systems (Hair 1967) and single crystal surfaces (Tompkins 1975) suffers from the need to develop supplementary methods for both localized surface structure characterization and for sensitivity. In the area of infrared vibrational study of extended crystal surfaces of metals and semiconductors, new measurements of surface-localized electromagnetic wave phenomena is a promising approach. In this regard the noise-to-signal enhancement problem also exists but may be solved using high intensity laser stimulation to generate surface electromagnetic waves at the appropriate wavelengths (Chabal and Sievers 1978). The use of laser stimulation in general to excite selectively the vibrational behavior of atoms and molecules in or at surfaces with reference to enhancing specific physical or chemical phenomena provides an intriguing tool in general for surface chemical diagnostics.

There are many important heterogeneous forms of matter where the surface probe must be sensitive to the atomic, electronic, and compositional character of a solid interface (Hobson 1974). The usefulness of reflection ion mass spectroscopy for this purpose in special cases to explore subsurface interfaces is well established (McHugh 1975). In a similar fashion, the use of very fine scanning Auger electron beam microprobes (1000 Å diameter) combined with ion milling to achieve subsurface compositional profiling is also effective under special conditions (MacDonald and Hovland 1977). In a similar vein, the development of electron loss measurements using finely collimated electron beam microscopy in transmission offers in principle a promising developmental approach to the compositional and electronic characterization of internal surfaces (Silcox 1977). The use of very high sensitivity dispersive energy analyzers for the microdetection of characteristic X-radiation is an obvious corollary to both electron and X-ray stimulated emission from point, line and surface imperfections characteristic of the interior of solids. The potential usefulness of the latter as a microanalytical probe of the interior bulk matter expands greatly with imminent availability of intense sources of high energy X-rays from unique synchrotron radiation facilities now under development (Hodgson et al. 1976).

These examples are but a few possibilities of new directions for surface electron spectroscopy which are likely to be developed in the forseeable future. They are noted in order to illustrate the broad potential for expansion of the electron spectroscopy approach to a new generation of interface and surface chemical bonding phenomena. The second important point is the necessity to make measurements of combined microscopic, electronic, atomistic, and structural parameters in a correlated manner on the same well defined atomic or molecular interface in the interior of a solid. It is necessary finally to be able to interpret the measurements in terms of electron interactions related to chemical bonding in terms of localized effects.

This chapter will be directed to the use of the specific electron spectroscopies (UPS, XPS, Auger and EELS) in conjunction with supportive data obtained from other relevant measurements such as LEED, TDS, and EID*. It will be directed mainly to well defined single crystal surfaces of metals in low pressure gaseous or UHV environments. Many of the principles discussed apply equally well with some modifications to both semiconductor and ionic solid surfaces. The choice of these particular spectroscopies and conditions of surface and environment is dictated mainly by the present state of the art in surface science. It happens at this time that the state of instrumentation, the potential for correlated measurements and the stage of theoretical understanding all combine to present a substantial body of new and significant information on the nature of chemical bonding for gas-surface metal surfaces. It appears both timely and expedient to attempt to gather together this body of important new experimental information into an organized format. It is intended to develop from it a set of cogent interpretations and principles to facilitate both an improved understanding of the present array of information on bonding at free solid surfaces and to present a basis for future development. It is likely and desirable that a subsequent overview will be constructed at a later date with reference to chemical bonding at interior interfaces when the field is sufficiently advanced in that direction. Electron features characteristic of chemical bonding primarily on external metal surfaces will be emphasized in the subsequent discussion.

2. Theoretical formulation and interpretation

2.1. Concepts and ansatzes

At this point, the time has come to add mathematical substance to the preceeding qualitative discussion. A subsection will be devoted to each

*See appendix A for definition of other acronyms. LEED = low-energy-electron-diffraction, TDS = thermal desorption spectroscopy and EID = electron impact desorption.

spectroscopy. Upon writing the general expressions for the electron current at the detector, the limiting cases in which single electron matrix elements, local density of independent-particle states, and many-body effects will be taken up, if and when appropriate. We will try to point out the approximations necessary to get to each of the limiting cases.

First, recall a few general ideas from quantum mechanics which will serve as a unified starting point. The golden rule expression for the number of electrons per unit time emitted or scattered by a photon or vibrational quantum of energy $\omega(\hbar \equiv 1)$ is

$$\dot{N} = 2\pi \sum_F \delta(E_i \pm \omega - E_F)|\langle \mathrm{FIN}|\tau|\mathrm{IN}\rangle|^2, \tag{3.1}$$

where the $+(-)$ sign in the energy conserving delta-function applies to processes in which the quantum is absorbed (emitted). The detailed nature of the operator τ, which connects the many-body initial ground (or prepared) state $|\mathrm{IN}\rangle$ of total energy E_i with any of the possible final states $|\mathrm{FIN}\rangle$, depends upon the specific spectroscopy under consideration. Contact is made with the Van Hove description in terms of ground state time correlation functions by introducing the Fourier representation of the delta-function (Doniach and Sondheimer 1974, Langreth 1970, Hedin 1973). The sum on the complete set $|\mathrm{FIN}\rangle$ can be eliminated since $\Sigma_F|\mathrm{FIN}\rangle \exp(-\mathrm{i}E_F t)\langle \mathrm{FIN}| = \exp(-\mathrm{i}H_f t)$ where H_f is the system Hamiltonian operator in the final state. Consequently, eq. (3.1) can be cast into the form

$$\dot{N} = \int_{-\infty}^{\infty} \mathrm{d}t\, \mathrm{e}^{\pm \mathrm{i}\omega t}\, \langle \mathrm{IN}|\mathrm{e}^{\mathrm{i}H_i t}\, \tau^\dagger\, \mathrm{e}^{-\mathrm{i}H_f t}\, \tau|\mathrm{IN}\rangle, \tag{3.2}$$

with H_i the initial state Hamiltonian. As will be discussed in quite some detail later, when relaxation effects are unimportant $H_i \simeq H_f$ (this in fact could be regarded as one of many equivalent criterion of their importance) in which case $\mathrm{e}^{\mathrm{i}Ht}\tau^\dagger \mathrm{e}^{-\mathrm{i}Ht} = \tau(t)$ is a Heisenberg operator and eq. (3.2) takes the nice form

$$\dot{N}_{\mathrm{nr}} = \int_{-\infty}^{\infty} \mathrm{d}t\, \mathrm{e}^{\pm \mathrm{i}\omega t}\langle \mathrm{IN}|\tau^\dagger(t)\tau(0)|\mathrm{IN}\rangle, \tag{3.3}$$

which correlates the influence of the excitation process (represented by τ at some point in time ($t = 0$ for convenience) with the conjugate de-excitation at time t returning the system to its ground state.

Further generalization follows if it is assumed that the participating or active electron in the process can be regarded as "special", in the sense that the initial state can be written as a product $|\mathrm{IN}\rangle =$

$|N-1; \text{in}; 0\rangle|\text{el}; \text{in}\rangle$. Here $|N-1; \text{in}; 0\rangle$ is the ground state of the possibly interacting $N-1$ electrons plus ion cores, described by some Hamiltonian $H_0(N-1)$ when $|\text{el}; \text{in}\rangle$, the initial state of the participant electron is occupied. Upon ejection of this electron the many-body Hamiltonian becomes $H_0(N-1)+\Delta V$. The ΔV term can be regarded as an effective potential of the hole created by removal of the active electron, as seen by the many-body system. The electron state is an eigenfunction of H_{el}, some one-body operator. The separability assumption is very crucial in permitting further theoretical development and thus it is almost always invoked. Certain exchange and interference processes are neglected within this ansatz, so caution must be exercised in data interpretation if such effects are thought to be operative. Another consequence of the separability ansatz is that the operator τ is a function of the participant electron operators only, which implies that $[H(N-1), H_{\text{el}}+\tau]=0$ provided $H(N-1)$ has built into it a proper time ordering in which ΔV is not turned on until after τ has acted to eject the electron (create the hole).

The result of the preceding discussion is that eq. (3.2) can be written as

$$\dot{N} = \int_{-\infty}^{\infty} \mathrm{d}t \, \mathrm{e}^{\pm \mathrm{i}\omega t} \langle N-1; \text{in}; 0|\mathrm{e}^{\mathrm{i}H_0 t} \, \mathrm{e}^{-\mathrm{i}(H_0+\Delta V)t}|N-1; \text{in}; 0\rangle \times \langle \text{el}; \text{in}|\mathrm{e}^{\mathrm{i}H_{\text{el}}t} \tau^\dagger \, \mathrm{e}^{-\mathrm{i}H_{\text{el}}t} \, \tau|\text{el}; \text{in}\rangle, \tag{3.4}$$

which is the Fourier transform of the product of two functions of time. Thus it can be expressed as a convolution of the individual Fourier transforms.

Using standard methods of many-body theory (Doniach and Sondheimer 1974, Langreth 1970, Hedin 1973), a spectral function for the many-body part of the problem can be identified as

$$D(\omega) = \int_{-\infty}^{\infty} \mathrm{d}t \, \mathrm{e}^{\pm \mathrm{i}\omega t} \langle N-1; \text{in}; 0|\mathrm{U(t)}|\mathrm{N}-1; \text{in}; 0\rangle, \tag{3.5}$$

where

$$U(t) = \mathrm{e}^{\mathrm{i}H_0 t} \, \mathrm{e}^{-\mathrm{i}(H_0+V)t} = T \exp\left(-\mathrm{i}\int_0^t \Delta V(t') \, \mathrm{d}t'\right),$$

T is a time ordering operator, and $\Delta V(t) = \exp(\mathrm{i}H_0 t)\Delta V \exp(-\mathrm{i}H_0 t)$. Further reduction of eq. (3.5) depends upon specific details of the models which will shortly be taken up.

With regard to the one-electron part, it is convenient to insert a summation over the complete set of electron final states (a valid procedure *only* because of the separability ansatz), in which case a golden-rule-like transition probability emerges from eq. (3.4)

$$\dot{N}_{\text{el}}(\omega) = \int_{-\infty}^{+\infty} \mathrm{d}t\, \mathrm{e}^{\pm i\omega t} \sum_{\mathrm{f}} \langle \mathrm{el;in}|\mathrm{e}^{iH_{\text{el}}t}\tau^{\dagger}\,\mathrm{e}^{-iH_{\text{el}}t}|\mathrm{el;f}\rangle\langle \mathrm{el;f}|\tau|\mathrm{el;in}\rangle$$
$$= 2\pi \sum_{\mathrm{f}} \delta(\varepsilon_{\text{in}} \pm \omega - \varepsilon_{\mathrm{f}})|\langle \mathrm{in}|\tau|\mathrm{f}\rangle|^2, \qquad (3.6)$$

where it is to be emphasized that the lower case indices refer only to the state of the participant electron. Again, the precise form and meaning of each term in eq. (3.6) depends on the specific technique considered. Equations (3.4)–(3.6) combine as

$$\dot{N} \equiv J = \int \mathrm{d}\omega'\, D(\omega')\, \dot{N}_{\text{el}}(\omega - \omega'), \qquad (3.7)$$

which is one of the fundamental results of this section. The sum on observed electron final states in eq. (3.6) can be replaced by the multiple integral $\Sigma_{\mathrm{f}} \Rightarrow \int \rho(\varepsilon_{\mathrm{f}})\, \mathrm{d}\varepsilon_{\mathrm{f}}\, \mathrm{d}\Omega$ where $\rho(\varepsilon_{\mathrm{f}})$ is the free space density of states at kinetic energy ε_{f}. Consequently the angle-resolved electron energy distribution is

$$\frac{\mathrm{d}^2 J}{\mathrm{d}\varepsilon_{\mathrm{f}}\mathrm{d}\Omega} \equiv J'' = \rho(\varepsilon_{\mathrm{f}}) \int \mathrm{d}\omega' D(\omega')\, \dot{N}_{\text{el, in}\to\mathrm{f}}(\omega - \omega'), \qquad (3.8)$$

where the in→f subscript signifies a transition from a given initial electron state to the final state at energy ε_{f} directed into the solid angle $\mathrm{d}\Omega$. In the extreme one-electron limit (no many-body or relaxation effects) $D(\omega') = \delta(\omega')$, in which case J'', from eqs. (3.6)–(3.8), is

$$J''_{\text{oe}} = \rho(\varepsilon_{\mathrm{f}}) 2\pi \sum_{\text{in}} f(\varepsilon_{\text{in}})\delta(\varepsilon_{\text{in}} \pm \omega - \varepsilon_{\mathrm{f}})|\langle \mathrm{in}|\tau|\mathrm{f}\rangle|^2. \qquad (3.9)$$

The sum on initial states is introduced because there are many different occupied single electron states connected by τ to the given final state. Subscript oe significes the one-electron limit and $f(\varepsilon_{\text{in}})$ is a Fermi function. The other extreme limit is that one in which the variation of the matrix element with energy is negligible, in which case it is just a multiplicative scale factor $\equiv m_k$ and the spectrum becomes

$$J''_{\text{mb}} = 2\pi\rho(\varepsilon_{\mathrm{f}}) \sum_{\text{in}} |m_k|^2\, D(\varepsilon_{\text{in}} \pm \omega - \varepsilon_{\mathrm{f}}), \qquad (3.10)$$

with subscript mb signifiying many-body limit. As a general rule (although not necessarily always so), the angular variations in the spectrum will be derived from the one-electron matrix elements and resulting selection rules whereas additional satellite peaks in the energy distribution will originate from many-body relaxation and shake-up phenomenon.

2.2. One-electron aspects

Photoemission

Perhaps the most widely used spectroscopic technique for the study of chemisorption has been photoemission. In theoretical considerations we distinguish three types of electronic states which serve as the initial state; band states typically spread over an energy range of 5–10 eV, chemisorption resonances of width $\lesssim 2$ eV which are accessible with either UV photons or X-rays, and localized core states which require X-rays for their removal. In all cases photoemission from an array of atoms in the condensed phase can be viewed as coherent emission from a distribution of atomic sources and it is the interferences between waves emitted from the various source atoms which are responsible for many of the characteristic solid state features in the observed currents. In the case of core states this point of view can still be held, but due to the narrow bandwidth of the core states one must sum over a filled band, in which case the solid state interference effects average to zero and atomic-like emission from incoherent sources is restored. This has been discussed in great detail elsewhere (Gadzuk 1976a). Working within this philosophy, it is then convenient to express the initial one-electron state as a linear combination of molecular-type functions

$$\psi_{\text{in}}(\boldsymbol{r};\varepsilon_{\text{in}}) = \sum_j c_j(\varepsilon_{\text{in}})\psi_j(\boldsymbol{r}), \tag{3.11}$$

where the coefficients $c_j(\varepsilon_{\text{in}})$ are determined from solutions to the relevant Schrödinger equation. The molecular orbitals, taken in the LCAO form are

$$\psi_j(\boldsymbol{r}) = \sum_n Q_n\, \mathrm{e}^{\mathrm{i}\delta_n}\phi_{j,n}(\boldsymbol{r}-\boldsymbol{R}_n), \tag{3.12}$$

where the n sum is over site locations $\boldsymbol{R}_n$ of the chosen array, and $\phi_{j,n}(\boldsymbol{r}-\boldsymbol{R}_n)$ are localized atomic-like functions centered on site $\boldsymbol{R}_n$. The real coefficient Q_n and phase factor δ_n are chosen out of convenience. Some illustrative examples include (a) atoms: $Q_n = 1$ for $n = 0$, $= 0$ otherwise, δ_n irrelevant; (b) H_2 molecule: $Q_n = 1\sqrt{2}$ for $n = 0, 1, = 0$ otherwise $\delta_0 = \delta_1$ for bonding state, $\delta_0 = \delta_1 + \pi$ anti-bonding state; (c) periodic solids: $Q_n = 1\sqrt{N}$ for all n(N the number of atoms in the solid), $\delta_n = \boldsymbol{k} \cdot \boldsymbol{R}_n$ with k the wave vector in the energy band. Localized chemisorption states will have large Q_n values only for site locations $\boldsymbol{R}_n$ near the surface. Specific values for Q_n and δ_n depend upon the system in question and presumably can be extracted from cluster calculations of the type discussed by Messmer in chapter 2.

The interaction of the radiation field with the array of atoms is given by

$$\tau = \tfrac{1}{2}(\boldsymbol{P}_{\text{op}} \cdot \boldsymbol{A}(\boldsymbol{r}) + \boldsymbol{A}(\boldsymbol{r}) \cdot \boldsymbol{P}_{\text{op}}) \tag{3.13}$$

with $\boldsymbol{P}_{\text{op}}$ the electron momentum operator and $\boldsymbol{A}(\boldsymbol{r})$ the total vector potential of the incident plus induced radiation field. Considerable effort is now being invested in the study of both local and non-local field corrections to the vector potential of the incident light, but at present these studies are still in the clean jellium phase of development (Feibelman 1975a, b, 1976, Bagchi 1977, Kliewer 1976, Forstmann and Stenschke 1977). Roughly these corrections give rise to a rapid decrease, on an atomic length scale, of A_z, the component of $\boldsymbol{A}$ perpendicular to the surface, when the frequency is less than ω_p, the plasmon frequency of the substrate. Until a better understanding is at hand, this complication can be side-stepped by working with s-polarized light at frequencies greater than ω_p. When this is done, it is then usually assumed that the z dependence of $\boldsymbol{A}$ is slowly varying and can be neglected in matrix element calculations. It is also usually assumed that the photon momentum can be neglected, although this is not so true for X-rays.

The final state $\psi_f(\boldsymbol{r}; \varepsilon_f) \equiv \phi_>(\boldsymbol{r}; \hat{R}; \varepsilon_f)$ is an outgoing wave satisfying incoming wave boundary conditions, the time-reversed familiar scattering wave functions (Bethe et al. 1953) and is given by

$$\phi_>(\boldsymbol{r}; \hat{R}; \varepsilon_f) = e^{i\boldsymbol{k} \cdot \boldsymbol{r}} + \int d^3r' \, e^{i\boldsymbol{k} \cdot \boldsymbol{r}'} V(\boldsymbol{r}') G^r(\boldsymbol{r}', \boldsymbol{r}; \varepsilon_f) \tag{3.14}$$

with $\boldsymbol{k} = -(2m\varepsilon_f/\hbar^2)^{1/2} \hat{R}$, $\hat{R}$ a unit vector pointing towards the detector, $V(\boldsymbol{r}')$ the complete semi-infinite potential of the solid including the chemisorbed objects, and G^r the retarded electron Green's function.

Subtleties in the physical interpretation of $\phi_>$ as it relates to the photoemission process have been discussed in quite some detail elsewhere (Caroli et al. 1973, Feibelman and Eastman 1974, Bethe et al. 1953, Mahan 1970, Schaich and Ashcroft 1971). For the present case in which the dominant scattering is due to the lattice potential of the substrate and/or assumed ordered adlayer, the final state given by eq. (3.14) is a time-reversed LEED wave function of the type discussed by

Van Hove in chapter 4. As it stands $\phi_>$ represents a state in which at time $t \to \infty$ a plane wave of energy ε_f is moving in the direction $\hat{R}$. This is what is observed in the detector. The second term in eq. (3.14) allows for proper inclusion of final state diffraction effects.

The fact that inelastic scattering (Powell 1974, Penn 1976) of the photoexcited electron can occur is usually accounted for in a nonrigorous way by inserting a phenomenological damping term, either into $V(\boldsymbol{r}')$, $G^r(\boldsymbol{r}', \boldsymbol{r}; \varepsilon_f)$, or the photoexcitation matrix element. In all cases, photoexcitation near the surface is weighted most strongly. The one electron expression for the photoemission spectrum, which follows from eqs. (3.9)–(3.14) is

$$J''_{oe} \sim \rho(\varepsilon_f)|A|^2 \sum_{in} f(\varepsilon_{in})\, \delta(\varepsilon_{in} + \omega - \varepsilon_f) \times \Big|\sum_{j,n} c_j(\varepsilon_{in})\, Q_n\, e^{i\delta_n}\, \hat{\varepsilon} \times \int d^3r\, e^{(z-z_0)/2\lambda}\, \phi^*_{\gtrless}(\boldsymbol{r}; R; \varepsilon_f)\, \nabla\phi_{j,n}(\boldsymbol{r} - \boldsymbol{R}_n)\Big|^2, \tag{3.15}$$

with $\hat{\varepsilon}$ the photon polarization vector and λ the electron attenuation length which is operative in the range $-\infty < z \leq z_0$. That 2λ rather than λ appears in the matrix element is because the attenuation length is usually defined in terms of currents rather than wave functions.

Density of states limit

If both the energy dependence and selection rules implicit in the matrix element are neglected, then since $\Sigma_{in}\delta(\varepsilon - \varepsilon_{in}) = \rho_{in}(\varepsilon)$, the one-electron energy distribution from eq. (3.15) is

$$\frac{dJ_{oe}}{d\varepsilon_f} \sim \rho(\varepsilon_f)\rho_{in}(\varepsilon_f - \omega),$$

which gives a one-to-one correspondence between the energy distribution and initial density of states. An improvement, although still quite approximate, is to neglect matrix element effects other than the preferential weighting they give to the surface form of the initial state wave function. This is effected by replacing the coherent sum on l by an incoherent sum of layers, each of whose emitting strength varies with position due to the attenuation factor and the probability density of the initial state wave function at layer l. Models have been proposed (Einstein 1974, 1975) which assume something equivalent to approximating the modules squared quantity

$$\Big|\sum_n \cdots\Big|^2 \approx \sum_{n_z} e^{(z_n - z_0)/\lambda} \Big|\sum_{j,n_\parallel} c_j(\varepsilon_{in}) Q_n\, e^{i\delta_n}\, \phi_{j,n}(\boldsymbol{r} - \boldsymbol{R}_n)\Big|^2 = \sum_{n_z} e^{(z_n - z_0)/\lambda}\, |\bar{\psi}_{in}(z_n; \varepsilon_{in})|^2, \tag{3.16}$$

where $\bar{\psi}_{\text{in}}$ denotes the fact that ψ_{in} has been summed over all $n_{\|}$ sites in the z_n layer. Using eq. (3.16) and the definition of the area averaged local density of state at depth z

$$\rho_{\text{in}}(\varepsilon\,; z) = \sum_{\text{in}} |\bar{\psi}_{\text{in}}(z;\, \varepsilon_{\text{in}})|^2 \delta(\varepsilon - \varepsilon_{\text{in}}).$$

Equation (3.15) reduces to

$$\frac{\mathrm{d}\,J_{\text{oe}}}{\mathrm{d}\,\varepsilon_{\text{f}}} \sim \rho(\varepsilon_{\text{f}}) \sum_{n_z} \mathrm{e}^{(z_n - z_0)/\lambda} \rho_{\text{in}}(\varepsilon_{\text{f}} - \omega\,; z_n). \qquad (3.17)$$

For better or worse, this exercise has demonstrated the sort of approximations one has to live with in order to declare that photoemission measures the local surface density of states and its change upon chemisorption.

Of particular interest in the study of the surface chemical bond is an example of the change in the local density of states, due to chemisorption of an s-like adsorbate on s-band cubium presented by Einstein (1974, 1975) and shown in fig. 3.3a. For purposes of illustration, parameters were chosen which yielded a local density of states on the adatom dominated by resonances at the bottom and top of the substrate band, corresponding to bonding and anti-bonding orbitals. Changes of local density of states for substrate layers were at most ~20% of that at the adatom site and in addition, oscillated rapidly (and out of phase, layer upon layer) with energy across the band. The accompanying change in the energy distribution (difference spectra) obtained from eq. (3.17) was calculated for various mean free paths and the results are shown in fig. 3.3b. At least for the case of localized surface complexes all the curves are qualitatively the same, which suggests that analysis of photoemission difference spectra of adsorbate covered surfaces in terms of local density of states on the adsorbate is "qualitatively not bad at all' (Einstein 1974, 1975).

Matrix elements

It is not by accident that we chose the LCAO–MO type of initial state wave function. With them the connection between atomic and solid state effects can be readily demonstrated. This is a big aid in extracting the chemical information from spectroscopic data.

Consider the matrix element $\equiv m_k$ in eq. (3.15), neglecting electron attenuation. If $\phi_>$ and ψ_{in} are eigenstates of the same Hamiltonian, it is easy to show the matrix elements of $\boldsymbol{P}_{\text{op}} = -(m\omega/\text{i})\boldsymbol{x} = \text{i}\nabla V/\omega$. Translat-

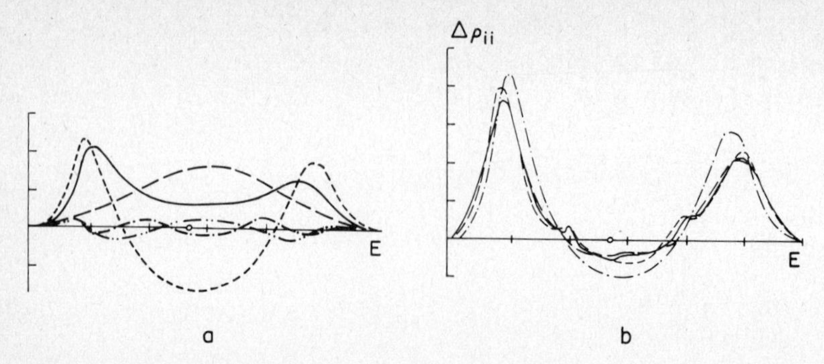

Fig. 3.3. (a) Local density of states for an s atom chemisorbed on s-band cubium (solid); and the change in the local density of states for the substrate atom directly beneath (dashed), next nearest (dot-dot-dashed), and second nearest (dot-dashed) neighbor in the surface plane. The latter two curves are four times the individual site contributions since each has four equivalent sites. The adatom level is at energy −0.3 with respect to a zero at the band center of width 6 and the coupling matrix element $V = 1.5$. The light long dashed curve is the unperturbed local density of states in the first layer. (b) Damped change in local density of state as given by eq. (3.17). The long-dashed curve gives the undamped change while the dashed curve shows the extreme case of emission only from the adatom and top substrate layer. The solid curve pictures an intermediate case with $\lambda = 5a/3$ (a the layer spacing). The dash-dot curve indicates the contribution of just the surface complex. (Einstein 1975).

ing the coordinate origin to site $\boldsymbol{R}_n$, the matrix element can be written as

$$m_{\boldsymbol{k}} = \hat{\varepsilon} \cdot \int \mathrm{d}^3 r \phi^*_>(\boldsymbol{r}) \boldsymbol{\nabla} \phi_{j,n}(\boldsymbol{r} - \boldsymbol{R}_n) = m\,\omega \hat{\varepsilon} \cdot \int \mathrm{d}^3 r \phi^*_>(\boldsymbol{r} - \boldsymbol{R}_n)\, \boldsymbol{x} \phi_{j,n}(\boldsymbol{r}),$$

where now the scattering state is described with respect to an origin at site $\boldsymbol{R}_n$. If the hole created in the photoemission is fairly delocalized, the associated hole potential will be small enough that it does not break the initial state translational symmetry. (This is a crucial simplification.) Thus by Bloch's theorem $\phi_>(\boldsymbol{r} - \boldsymbol{R}_n) \sim \mathrm{e}^{\mathrm{i}\boldsymbol{k}\cdot\boldsymbol{R}_n} \phi_>(\boldsymbol{r})$ and

$$m_{\boldsymbol{k}} \simeq \omega \mathrm{e}^{-\mathrm{i}\boldsymbol{k}\cdot\boldsymbol{R}_n} \hat{\varepsilon} \cdot \int \mathrm{d}^3 r \phi^*_>(\boldsymbol{r}) x \phi_{j,n}(\boldsymbol{r}), \tag{3.18}$$

dropping the multiplicative electron mass term.

The partial wave expansion of $\phi_>(\boldsymbol{r})$ is

$$\phi_>(\boldsymbol{r}) = 4\pi \sum_{l,m} (\mathrm{i})^l \mathrm{e}^{-\mathrm{i}\delta_l(\varepsilon_f)} Y^*_{lm}(\hat{k})\, Y_{lm}(\hat{r}) G_{\varepsilon_f}(r),$$

where $\delta_l(\varepsilon_f)$ is the final state l'th wave phase shift from the potential at n, $\hat{r}$ the angle associated with the spatial coordinate of the electron, and

$G_{\varepsilon_f l}(r)$ a continuum radial function. If $\phi_{j,n}(r) \equiv R_{jl'}(r) Y_{l'm'}(\theta, \phi)$ is an atomic-like function with good angular momentum quantum numbers l' and m', one finds (Cooper and Zare 1968, Gadzuk 1975 a, b, c, Herbst 1977, Liebsch 1978 a)

$$m_k = \omega e^{ik \cdot R_n} \sum_l a(l, m) R_l(\varepsilon_f) \langle lm | [(\hat{\varepsilon} \cdot x)/|x|] | l'm' \rangle, \tag{3.19}$$

with

$$a(l, m) = 4\pi (i)^l e^{-i\delta_l(\varepsilon_f)} Y_{lm}(\hat{k}),$$

and

$$R_l(\varepsilon_f) = \int_0^\infty r R_{jl'}(r) G_{\varepsilon_f, l}(r) \, dr.$$

The angular matrix element of the dipole operator can easily be expressed as a small sum of tabulated Gaunt coefficients (Cooper and Zare 1968, Gadzuk 1975 a, c, Herbst 1977, Liebsch 1978 a, b, c, Condon and Shortly 1967), which vanish unless $l = l' \pm 1$, $m - m' = \pm 1, 0$. This is an expression of the atomic dipole selection rule which allows eq. (3.19) to be written in terms of partial wave matrix elements

$$m_k = e^{-ik \cdot R_n} (m_k(l'+1; j, n, \hat{\varepsilon}) + m_k(l'-1; j, n, \hat{\varepsilon})), \tag{3.20}$$

and the angle resolved energy distribution is

$$J''_{oe} \sim \rho(\varepsilon_f) \sum_{in} f(\varepsilon_{in}) \delta(\varepsilon_{in} + \omega - \varepsilon_f) \times \Bigg| \sum_{j,n} c_j(\varepsilon_{in}) Q_n \, e^{i[\delta_n - k \cdot R_n]} \times (m_k(l'+1; j, n, \hat{\varepsilon}) + m_k(l'-1; j, n, \hat{\varepsilon}) \Bigg|^2 \tag{3.21}$$

where

$$m_k(l' \pm 1; j, n, \hat{\varepsilon}) = \omega \sum_m a(l'+1, m) R_{l' \pm 1}(\varepsilon_f) \langle l' \pm 1, m | (\hat{\varepsilon} \cdot x)/|x| | l'm' \rangle.$$

Equation (3.21) is the ultimate result of this section and demonstrates that photoemission from the extended MO states can be cast into the apparent form of photoionization from a coherent array of atomic-like states. The angular and spectral features depend both upon the individual atom details through the energy dependence of $m_k(l'+1)$ and

$m_k(l'-1)$ and the interferences between the $l'-1$ and $l'+1$ channels and also on the solid state coherence factors through $\exp[i(\delta_n - \boldsymbol{k}\cdot\boldsymbol{R}_n)]$. In the extreme atomic limit the summation of j, n are limited to one term, in which case eq. (3.21) reduces to

$$J''_{\text{oe}}(\text{atom}) \sim \rho(\varepsilon_{\text{f}})\delta(\varepsilon_{\text{in}} + \omega - \varepsilon_{\text{f}}) \times \{|m_k(l'+1)|^2 + |m_k(l'-1)|^2 + (m_k(l'+1)m_k^*(l'-1) + \text{c.c.})\},$$

which necessarily is of the form $a + b\cos^2\theta$ with $|a| < |b|$ and θ the angle between $\hat{\boldsymbol{R}}$ and either $\hat{\varepsilon}$ or the photon direction, depending upon the polarization state (Cooper and Zare 1968, Gadzuk 1975c, Herbst 1977, Liebsch 1978a, Dill and Manson 1977). The contrasting solid state extreme limit is defined by setting $m_k(l'+1) = m_k(l'-1) = \text{constant} \times (\hat{\boldsymbol{k}}\cdot\hat{\varepsilon})$. Using the previous specifications for δ_n and Q_n and letting $c_j(\varepsilon_{\text{in}}) = 1$ for one j value only, the mod square is

$$\sim (\boldsymbol{k}\cdot\hat{\varepsilon})^2 \left| \sum_n e^{i(\boldsymbol{k}_{\text{in}} - \boldsymbol{k})\cdot\boldsymbol{R}_n} \right|^2 = \sum_{\boldsymbol{G}} (\boldsymbol{k}\cdot\hat{\varepsilon})^2 \delta^{(3)}(\boldsymbol{k}_{\text{in}} - \boldsymbol{k} - \boldsymbol{G}),$$

with $\boldsymbol{G}$ a reciprocal lattice vector. Thus eq. (3.21) becomes

$$J''_{\text{oe}}(\text{solid}) \sim \rho(\varepsilon_{\text{f}}) \sum_{\text{in},\boldsymbol{G}} f(\varepsilon_{\text{in}}(\boldsymbol{k}_{\text{in}}))\delta(\varepsilon_{\text{in}}(\boldsymbol{k}_{\text{in}}) + \omega - \varepsilon_{\text{f}}(\boldsymbol{k})) \times (\boldsymbol{k}\cdot\hat{\varepsilon})^2 \delta^{(3)}(\boldsymbol{k}_{\text{in}} - \boldsymbol{k} - \boldsymbol{G}),$$

which expresses the solid state $\boldsymbol{k}$ conservation law. Since the unified treatment leading to eq. (3.21) easily reduces to the correct atomic and solid state limits, it should easily be applicable to the partially localized chemical bond states at surfaces. It is for this reason that we have dwelt at some length on the formal aspects.

Typical applications

Angle resolved photoemission studies from adsorbate covered surfaces are currently being pursued successfully in many laboratories. Both the fixed orientation and shapes of the initial state atomic, molecular, or chemisorptive bonding orbitals [eqs. (3.11) and (3.12)] (Gadzuk 1974a, b, c, Meyer 1977) and the diffractive solid state effects of the final state wave functions [eq. (3.14)] (Liebsch 1974, 1976, 1977, Gadzuk 1975b, c, Tong and Van Hove 1976, Tong et al. 1977, Li and Tong 1978, Jacobi et al. 1977, Scheffler et al. 1978) build retrievable chemical (orbital identification) and structural (location of adparticles relative to substrate sites) information into the angle resolved spectrum. In order to extract the useful information, it is very desirable to have available as simple a theory as possible which still remains sufficiently correct to prevent erroneous conclusions. The simplest possible expressions follow when the final state $\phi_>$ is approximated by the plane wave term in eq. (3.14); in other words, diffractive backscattering by the substrate and final state interactions with the hole are neglected. It is then straightforward to bring eq. (3.15), integrated on energy, to the form

$$\frac{\mathrm{d}J_{\mathrm{oe}}}{\mathrm{d}\Omega} \sim \cos^2\gamma \left| \sum_{j,n} c_j Q_n \mathrm{e}^{\mathrm{i}(\delta_n - \boldsymbol{k}\cdot\boldsymbol{R}_n)} \phi_{j,n}(\boldsymbol{k}) \right|^2,$$

where the sum on n is over all centers covered by the initial state orbital, $\phi_{j,n}(\boldsymbol{k})$ is the Fourier transform of the relevant atomic orbital on site n, and γ the angle between $\boldsymbol{k}$ and $\hat{\varepsilon}$. Note that the transform preserves the angular dependence of the real space function.

It is informative to consider a few examples. In the atomic limit, ($c_j = c\delta_{j,0} Q_n = Q\delta_{n,0}$) the angle resolved spectrum is just

$$\frac{\mathrm{d}J_{\mathrm{oe}}^{\mathrm{atom}}}{\mathrm{d}\pi} \sim \cos^2\gamma |\phi(\boldsymbol{k})|^2.$$

Evidence has recently been presented by Larson and co-workers (1978) which is consistent with this picture. They obtained photoelectron spectra from Cl on Si(111) using p polarized light at 45° angle of incidence which displayed lobes with well defined maxima; one in the azimuthal range $\theta = 20°–25°$ (in the $\Gamma \mathrm{M} \Gamma$ plane of incidence), believed to be emission from a Cl p_z orbital and the other near 70°, maybe from the p_x orbital. For this particular geometry, the atomic formula just written predicts maxima at $\theta = 22.5°$ and 67.5°, which is in good agreement with the data. Another example which illustrates the simplicity and ease of utilization of a plane wave theory is that of emission from an orbital in an oriented molecule (Cooper and Zare 1968, Gadzuk 1975b, c, Herbst 1977, Liebsch 1978a, Dill 1976, Dill et al. 1976, Wallace et al. 1978, Davenport 1976, 1978) such as the π_z orbital of benzene, shown in the inset of fig. 3.4. For the totally symmetric orbital composed of $\mathrm{p}_z \sim \cos\theta$ atomic orbitals on each site, with normally incident, unpolarized light, the angular distribution written above becomes

$$\frac{\mathrm{d}J_{\mathrm{oe}}}{\mathrm{d}\Omega}(\pi_z, \text{benzene}) \sim 4\sin^2 2\theta[\cos(ak_y) + 2\cos(\tfrac{1}{2}ak_y)\cos(\tfrac{1}{2}\sqrt{3}ak_x)]$$

with a the internuclear separation. Both polar and azimuthal plots of this expression, for final state energies $\varepsilon \sim k^2 = 100$ eV are shown in fig. 3.4, where the richness of structure due to the interferences in emission from

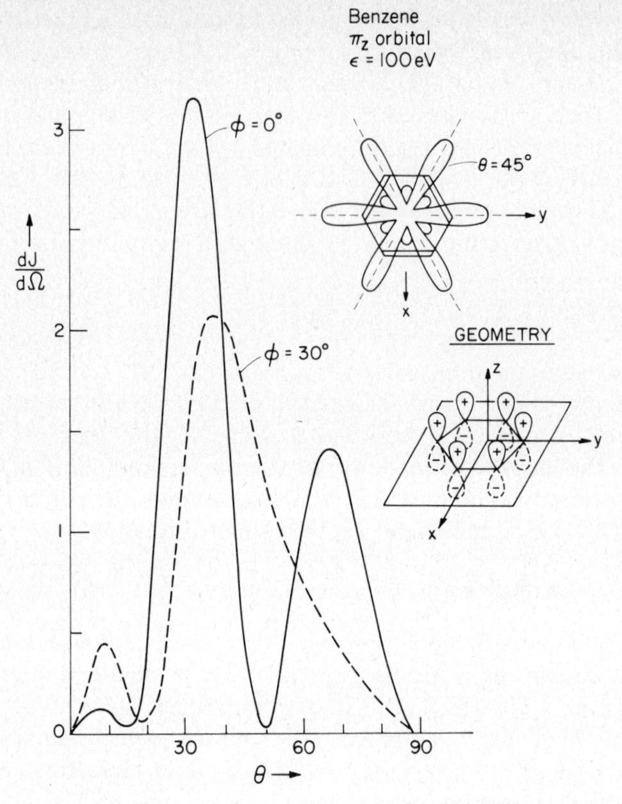

Fig. 3.4. Angle resolved photoemission spectra, within the plane wave final state approximation, from the symmetric π_z molecular orbital of a benzene molecule, oriented on the surface as depicted. Kinetic energy of the ejected electron is taken to be 100 eV. Polar variations are shown for two azimuthal angles, whereas the azimuthal variation is given only for $\theta = 45°$.

the various source atoms is apparent. It is the analysis of this type of structure which should make possible orbital and geometrical identification.

Going beyond analytically tractable model calculations, Dill and Dehmer (1978) and also Davenport (1976, 1978) have computed angle resolved spectra as a function of photon energy for fixed orientation N_2 and CO, with special emphasis on final state energies which show a resonance behavior in the total yield. In these works, wave functions are calculated with the "self-consistent-field scattered wave X-α" scheme in which final state interactions between the photoelectron and the ion are presumably included.

The importance of LEED-like effects in the final state wave functions has been discussed first by Liebsch (1974, 1976, 1977) and subsequently by others (Gadzuk 1975b, c, Tong and Van Hove 1976, Tong et al. 1977, Li and Tong 1978, Jacobi et al. 1977, Scheffler et al. 1978). Scheffler, et al. (1978) have performed calculations of the angle resolved photoemission spectra from p(2 × 2) and c(2 × 2) oxygen chemisorbed on Ni(100) (1977). The initial state was an LCAO consisting of an O atomic orbital and Ni orbitals on the 4 nearest neighbors which were chosen to replicate X-α cluster results. A LEED multiple scattering final-state was chosen. Some typical results are shown in fig. 3.5 for the two overlayer structures at various polarizations and angles. If indeed experimental angle resolved spectra are as sensitive to the matrix element variations as these calculations suggest, then a vast wealth of detailed information on the surface chemical bond should be obtainable from a correct analysis of the data.

Auger emission

Fortunately for the reader, much of what was presented in the photoemission section can readily be adapted to the theory of AES. Although Auger emission is due to an inherently two-electron mechanism, it is enlightening to draw whatever analogies with photoemission that are possible. Consider the 5-point AES diagram in fig. 3.2. The left-hand line describes the propagation of the prepared hole state with energy ε_0 until

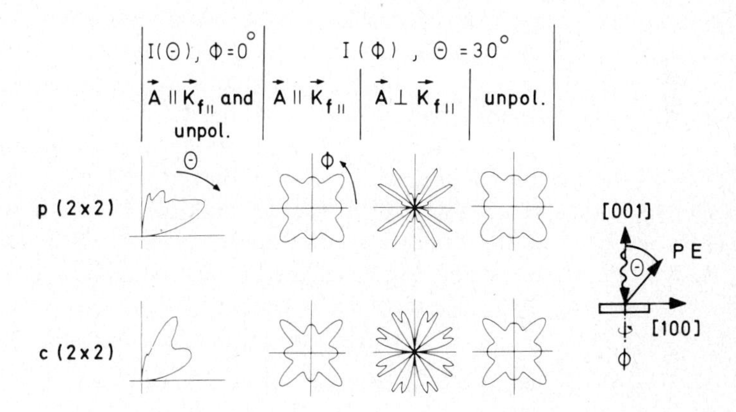

Fig. 3.5. Angular distribution from 0 $2p_z$ state of O on Ni(001) into a multiple scattered final state for two different adsorbate superstructures. Excitation with He II (final state energy: 30.6 eV), for different polarizations with the vector potential A parallel to the surface. $k_{f\parallel}$ is the projection of the wave vector of the photoelectron on the surface. Curves are normalized to maximum. (Scheffler et al. 1978.)

it arrives at x_1, at which point it makes a transition to a lower energy state ε_1. This information is communicated to the rest of the system via the dashed line, which might be thought of as representing a virtual photon (Chattarji 1976) created at x_1. From the point of view of the electron–hole pair generated at x_1', a photon of energy $\hbar\omega = \varepsilon_0 - \varepsilon_1$ is absorbed at this vertex and the subsequent time evolution is similar to that of the 3-point photoemission diagram. There are some pitfalls in this argument though (Chattarji 1976) so one may be safer with the view that the initially prepared ε_0 hole arrives in the x_1, x_1' domain. Two electrons in lower binding energy states ε_1 and ε_2 interact via their unscreened Coulomb potential one filling the ε_0 hole and one being excited to a continuum state at ε_f. Conservation of energy requires $\varepsilon_{in} = \varepsilon_{fin}$ or $\varepsilon_0 = \varepsilon_1 + \varepsilon_2 + \varepsilon_f$. Since the Coulomb potential is an electromagnetic field (or photon) propagator, the two points of view can be forced into some harmony. In either case the "one" electron transition probability is

$$\dot{N}_{oe} = 2\pi \sum_{1,2,f} \delta(\varepsilon_0 - \varepsilon_1 - \varepsilon_2 - \varepsilon_f)\, |\langle \mathrm{in}|\tau|\mathrm{fin}\rangle|^2, \tag{3.22}$$

where now the sum on final states includes all possible combinations of hole states 1 and 2 and excited electron states f which are coupled by the interaction in the Auger matrix element and which conserve energy.

Within a particle–hole interpretation, the initial one-hole state is represented as $|\mathrm{in}\rangle = \phi_0^*(\boldsymbol{r}_1 - \boldsymbol{R}_m)$ where ϕ_0 is a deep core wave function localized on one atom at site $\boldsymbol{R}_m$. Similarly the final two-hole–one-electron state is $|\mathrm{fin}\rangle = \psi_f(\boldsymbol{r}')\psi_2^*(\boldsymbol{r}')\psi_1^*(\boldsymbol{r})$, with ψ_f a continuum wave function and ψ_1 and ψ_2 either localized core or delocalized conduction band wave functions, depending upon the particular Auger transition in question. In either case they can be written in the LCAO form of eqs. (3.11) and (3.12). Note that the hole wave functions are just complex conjugates of the electron functions, at least in the no relaxation effects limit. We also neglect considerations due to multiplets, which while important, are beyond the scope of this chapter. Some of the many authoritative references on this aspect of the problem should be consulted (Feibelman et al. 1978, Chattarji 1976, Carlson 1975). The Coulomb interaction line in fig. 3.2 is $\tau = e^2/|\boldsymbol{r}_1 - \boldsymbol{r}_1'|$. Consequently the Auger matrix element can be written as

$$\boldsymbol{M}(1,2;0,f;m) = \int \mathrm{d}^3 r \psi_f^*(\boldsymbol{r}) \times \left\{ \int \mathrm{d}^3 r' \phi_{0,m}^*(\boldsymbol{r}' - \boldsymbol{R}_m)(e^2/|\boldsymbol{r} - \boldsymbol{r}'|)\psi_2(\boldsymbol{r}') \right\} \psi_1(\boldsymbol{r}), \tag{3.23}$$

where the subscript on r has been dropped. Strictly speaking, an exchange matrix element in which indices 1 and 2 are interchanged should also be included.

Consider first the bracketed term in eq. (3.23) with ψ_2 of the form given by eqs. (3.11) and (3.12)

$$\{\ \} \equiv V_a(r, R_m) = \sum_{j',n} c_{j'}(\varepsilon_2) Q_n \mathrm{e}^{\mathrm{i}\delta_n} \times \int \mathrm{d}^3r' \phi^*_{0,m}(\boldsymbol{r}' - \boldsymbol{R}_m)(e^2/|\boldsymbol{r} - \boldsymbol{r}'|)\phi_{j'n}(\boldsymbol{r}' - \boldsymbol{R}_n).$$

Due to the localization of $\phi_{0,m}$ on $\boldsymbol{R}_{m'}$ the integral will be large only when $\boldsymbol{R}_n = \boldsymbol{R}_m \equiv 0$ so

$$V_a(\boldsymbol{r}) \simeq \sum_{j'} c_{j'}(\varepsilon_2) Q_m \, \mathrm{e}^{\mathrm{i}\delta_m} \int \mathrm{d}^3r' \, \phi^*_{0,m}(\boldsymbol{r}')(e^2/|\boldsymbol{r} - \boldsymbol{r}'|)\phi_{j'm}(\boldsymbol{r}'). \tag{3.24}$$

If the Coulomb potential is written as a spherical harmonic expansion

$$\frac{1}{|\boldsymbol{r} - \boldsymbol{r}'|} = 4\pi \sum_{\kappa=0}^{\infty} \sum_{\beta=-\kappa}^{\kappa} \frac{1}{(2\kappa + 1)} \frac{r_<^{\kappa}}{r_>^{\kappa+1}} Y^*_{\kappa\beta}(\Omega_{\hat{r}}) Y_{\kappa\beta}(\Omega_{\hat{r}'}),$$

then

$$V_a(\boldsymbol{r}) = \sum_{j'} c_{j'}(\varepsilon_2) Q_m \mathrm{e}^{\mathrm{i}\delta_m} \sum_{\kappa=0}^{\infty} \sum_{\beta=-\kappa}^{\kappa} f_{\kappa,\beta,j',m}(r) Y^*_{\kappa\beta}(\Omega_{\hat{r}}), \tag{3.25}$$

where $f_{\kappa\beta j'm}(r)$ is some complicated function of the magnitude of $\boldsymbol{r}$, peaked at a value of r within the ion core of the atom at site $R_m = 0$, which depends upon the details of the specific system and Auger transition in question. No simple generalization as to the most likely values for κ is possible. The point though is that the action of the $\boldsymbol{r}'$ electron dropping from $\phi_{j'\kappa}(\boldsymbol{r}')$ to $\phi_{0,m}(\boldsymbol{r}')$ has the same effect on the $\boldsymbol{r}$ electron as would some operator or static potential with arbitrary components of angular momentum. This is in contrast to photoexcitation in which the only non-vanishing terms would be those with $\kappa = 1$. For this reason, selection rules do not provide such great intuitive insight into Auger emission characteristics as they do in photoemission.

The final state wave function of the Auger electron is formally identical with that of photoemission, eq. (3.14), with the modification that $V(\boldsymbol{r}')$ should contain the fact that two, rather than one, holes are to be included in its determination. The inclusion of electron attenuation on the excited electron is identical. Thus in analogy to eq. (3.15), the one electron expression for the Auger emission spectrum, which follows from eqs. (3.11), (3.12), (3.14), and (3.22–3.25) is

$$J''_{oe}(\text{Auger}) \sim \rho(\varepsilon_f) \sum_{1,2,m} f(\varepsilon_1) f(\varepsilon_2) \delta(\varepsilon_0 - \varepsilon_1 - \varepsilon_2 - \varepsilon_f) \times \left| \sum_{j,p} c_j(\varepsilon_1) Q_p \, e^{i\delta_p} \right.$$

$$\left. \times \int d^3 r e^{(z-z_0)/2\lambda} \phi^*_>(\boldsymbol{r}; \boldsymbol{R}; \varepsilon_f) V_a(\boldsymbol{r}; \boldsymbol{R}_m) \phi_{j,p}(\boldsymbol{r} - \boldsymbol{R}_p) \right|^2, \quad (3.26)$$

which shall now be checked in some limiting cases. Note that an incoherent sum on possible source atom locations has been introduced to obtain the total current expression.

Density of states limit

As with photoemission, if the matrix elements are set constant and thus the selection rules turned off, the density of states limit for the Auger energy distribution, from eq. (3.26), follows as

$$\frac{dJ_{oe}}{d\varepsilon_f} \sim \rho(\varepsilon_f) \int d\varepsilon_1 \, d\varepsilon_2 \rho_1(\varepsilon_1) \rho_2(\varepsilon_2) \delta(\varepsilon_0 - \varepsilon_1 - \varepsilon_2 - \varepsilon_f)$$
$$= \rho(\varepsilon_f) \int d\varepsilon_1 \rho_1(\varepsilon_1) \rho_2(\varepsilon_0 - \varepsilon_1 - \varepsilon_f), \quad (3.27)$$

which is the convolution of the density of states 1 with those of 2. For CCC transitions in which both ρ_1 and ρ_2 are delta-functions (or a series of them due to multiplets), the spectrum is simply

$$\frac{dJ_{oe}(\text{CCC})}{d\varepsilon_f} \sim \rho(\varepsilon_f) \delta(\varepsilon_0 - \varepsilon_{b1} - \varepsilon_{b2} - \varepsilon_f),$$

with ε_b the electron binding energies of states 1 and 2. A CCV transition gives rise to

$$\frac{dJ_{oe}(\text{CCV})}{d\varepsilon_f} \sim \rho(\varepsilon_f) \rho_v(\varepsilon_0 - \varepsilon_b - \varepsilon_f),$$

which replicates the valence electron density of states. Lastly, CVV transitions, with $\rho_1(\varepsilon) = \rho_2(\varepsilon) \equiv \rho_v(\varepsilon)$ in eq. (3.27), yield spectra proportional to the self-convolution of the band density of states. Considerable evidence has been obtained which suggests that the density of states limit is not actually realized in transitions involving band electrons (Powell 1973, Gadzuk 1974a, Jackson et al. 1975, Madden and Houston 1977, Rojo and Baro 1976, Cini 1976, 1978, Sawatzky 1977).

The next level of improvement is the introduction of various *local* density of states effects. Note that since $V_a(\boldsymbol{r}; \boldsymbol{R}_m)$, eq. (3.24), is larger when $\boldsymbol{r}$ is small, the matrix element in eq. (3.26) will be maximized when

$\phi_{j,p}(\boldsymbol{r}-\boldsymbol{R}_p = \phi_{j,m}(\boldsymbol{r}-\boldsymbol{R}_m)$. If we define $\bar{\phi}_{\text{in}}(\boldsymbol{r};\boldsymbol{R}_m;\varepsilon_1) = Q_m \exp(\mathrm{i}\delta_m)\Sigma_j$ $c_j(\varepsilon_1)\phi_{jm}(\boldsymbol{R}_m)$, and invoke matrix element approximations in the spirit of those leading to eq. (3.16), that is

$$(1,2;m) \simeq \mathrm{e}^{-(z_m-z_0)/\lambda}|\bar{\phi}_{\text{in}}(\boldsymbol{R}_m;\varepsilon_1)|^2 \times |\bar{\phi}_{\text{in}}(\boldsymbol{R}_m;\varepsilon_2)|^2, \tag{3.28}$$

then eqs. (3.23), (3.26), (3.27), and (3.28) yield the result that

$$\begin{aligned}\frac{\mathrm{d}J_{\text{oe}}}{\mathrm{d}\varepsilon_{\text{f}}} \sim \rho(\varepsilon_{\text{f}}) \sum_{m_z,m_\parallel} \mathrm{e}^{(z_m-z_0)/\lambda} \int \mathrm{d}\varepsilon_1\rho_1(\varepsilon_1)|\bar{\phi}_{\text{in}}(\boldsymbol{R}_m;\varepsilon_1)|^2 \\ \times \rho_2(\varepsilon_0-\varepsilon_1-\varepsilon_{\text{f}})|\bar{\phi}_{\text{in}}(\boldsymbol{R}_m;\varepsilon_0-\varepsilon_1-\varepsilon_{\text{f}})|^2,\end{aligned} \tag{3.29}$$

which is the convolution of the local density of states of 1 and 2, measured at the site of the initial core hole. The intensity from transitions involving holes located a distance z_m into the solid are exponentially reduced due to the electron attenuation.

The potentially useful aspect of CVV transitions in chemisorption studies is that one can look at processes known to originate at a specific atom in the chemisorption complex (if all atoms of the same species are equivalent) by chosing ε_0 correctly. Since the resulting energy distribution is to some degree a reflection of the valence electron density of states at the given center, one has a localized probe of electronic wave functions involved in the chemical bond. Caution still must be exercised against putting too much faith in the local density of states expressions if detailed understanding of Auger lineshapes and angle-resolved spectra are required.

Matrix elements

Proper calculation of the Auger matrix element is a *tour de force* in angular momentum algebra, which presents no problem in principle but is rather tedious in practice. Since we have reduced the algebraic form of the solid state Auger spectrum to that of a coherent sum of atomic-like processes [eq. (3.26)], the detailed atomic calculations outlined elsewhere (Feibelman et al. 1978, Chattarji 1976) can immediately be applied to our problem. Feibelman et al (1978) have shown that within the one center LCAO formulation for an Auger transition, the following selection rules must be satisfied

$$\begin{aligned}&l_0 + l_{j'} + \kappa = \text{even integer}, \qquad l_{\text{f}} + l_j + \kappa = \text{even integer},\\ &\beta = m_{j'} - m_0 = m_{\text{f}} - m_j,\\ &|l_0 - \kappa| \leq l_{j'} \leq l_{j'} + \kappa, \qquad |l_j - \kappa| < l_{\text{f}} < l_j + \kappa,\end{aligned} \tag{3.30a–e}$$

where (l_0, m_0), $(l_{j'}, m_{j'})$, (l_j, m_j) are the angular momentum quantum numbers associated with $\phi_{0,m'}\phi_{j',m'}$ and $\phi_{j,m}$ respectively (since they are of the form $R(r)Y_{lm}(\theta, \phi)$), (l_f, m_f) those of the particular partial wave in the expansion of $\phi_>$, eq. (3.18) and (κ, β) those of the particular component of the Coulomb potential expansion, eq. (3.25). There are many different outgoing channels (l_f, m_f) which satisfy eqs. (3.30a–e). Estimation of their relative intensities requires numerical calculations of radial matrix elements which are of the Slater integral form. Nonetheless, the matrix element derived selection rules are still useful on their own. For instance it seems reasonable to imagine that when looking at angle resolved CVV Auger energy distributions in which the initial core hole is localized on an adatom and the two valence electrons could be in bonding or non-bonding orbitals, it should be possible to rule out certain feasible bonding schemes; this being done on the basis of a comparison of the gross features of the observed spectrum with the selection-rule-allowed or forbidden requirements.

Typical applications

A few attempts have been made at theories for angle-resolved Auger emission from chemisorption complexes in which initial cluster and/or final LEED states have been included in a manner similar to that in photoemission theory. In general, the Auger results are less enlightening, perhaps due to the complications of the two-electron process, and hence will not be expanded upon here.

Electron energy loss

Within the limitations implicit in the left hand diagram of fig. 3.2, EELS is the least complicated of the spectroscopies considered here theoretically. This is in large part due to the fact that no secondary particles are generated and thus the initial and final state effective Hamiltonians are identical for low energy events in which electronic excitation is unlikely or forbidden. Thus eq. (3.3) (with $\omega = 0$) provides a useful starting point. The initial state must describe substrate and adsorbate electrons, the scattering electron, and phonons or localized vibrational modes of the system. Since electronic excitation is not being considered, the substrate and adsorbate electrons play a role only to the extent that they contribute to the potential that scatters the incident electron. Thus any dynamical processes can be described in terms of changes of state of the incident electron, $|\text{in; el}\rangle$ and the vibrational ground state $|0; \text{v}\rangle$. For a Hamiltonian of the form $H_{\text{tot}} = H_{\text{el}} + H_{\text{v}} + \tau(\text{e; v})$, with obvious notation, the initial state is a simple product, $|\text{IN}\rangle = |\text{in; el}\rangle|0; \text{v}\rangle$ and eq. (3.3) becomes

$$\dot{N} = \int_{-\infty}^{\infty} \mathrm{d}t \langle \mathrm{in}; \mathrm{el}|\mathrm{e}^{\mathrm{i}H_{\mathrm{el}}t}\langle 0; \mathrm{v}|\mathrm{e}^{\mathrm{i}H_{\mathrm{v}}t}\tau^{\dagger}(\mathrm{e}; \mathrm{v}) \times \mathrm{e}^{-\mathrm{i}(H_{\mathrm{el}}+H_{\mathrm{v}})t}\tau(\mathrm{e}; \mathrm{v})|\mathrm{in}; \mathrm{el}\rangle|0; \mathrm{v}\rangle. \tag{3.31}$$

If we further assume a separable interaction $\tau(\mathrm{e}; \mathrm{v}) = \tau_{\mathrm{e}}\tau_{\mathrm{v}}$ (to be discussed shortly), and insert complete sets of electron and vibrational eigenstates $(1 = \Sigma_{\mathrm{f}}|\mathrm{f}; \mathrm{el}\rangle\langle\mathrm{f}; \mathrm{el}| = \Sigma_{\mathrm{x}}|\mathrm{x}; \mathrm{v}\rangle\langle\mathrm{x}; \mathrm{v}|)$, eq. (3.31) immediately reduces to

$$\dot{N} \equiv J = \sum_{f,x} \delta(\varepsilon_{\mathrm{el}}^{\mathrm{i}} - \varepsilon_{\mathrm{el}}^{\mathrm{f}} - \varepsilon_{\mathrm{x}})|\langle \mathrm{in}; \mathrm{el}|\tau_{\mathrm{e}}|\mathrm{f}; \mathrm{el}\rangle|^2 \times |\langle 0; \mathrm{v}|\tau_{\mathrm{v}}|\mathrm{x}; \mathrm{v}\rangle|^2,$$

with ε_{x} the vibrational energy.

The angle-resolved energy distribution follows as

$$\frac{\mathrm{d}^2 J}{\mathrm{d}\varepsilon_{\mathrm{f}}\,\mathrm{d}\Omega} \equiv J' = \sum_{\mathrm{x}} \delta(\varepsilon_{\mathrm{i}} - \varepsilon_{\mathrm{f}} - \varepsilon_{\mathrm{x}})|M_{\mathrm{el}}|^2 \times |M_{\mathrm{v}}|^2, \tag{3.32}$$

where the el subscript has been dropped, summation over all vibrational states is to be done, and shorthand labeling for the electronic and vibrational matrix elements has been adopted. Equation (3.32) says that the energy distribution is just a series of delta-functions positioned at the vibrational excitation energies below the energy of the primary beam. Intensities and directions of the various peaks depend upon the matrix elements and the "selection rules" implicit within them. To go further, specific details about the form of the interaction must be given.

In order for the inelastically scattered incident electron to be observed at the detector, both energy loss and large momentum transfer are necessary. A number of possible scattering sequences are shown in fig. 3.6, in which both the oriented molecule and the substrate provided scattering mechanisms (Šokčević 1977, Persson 1977). Figure 3.6a displays the surface version of the gas phase scattering where the molecule alone is responsible for both the energy loss and momentum transfer. Unless the incident electron energy is just right for inelastic resonance scattering, as discussed by Davenport et al. (1978), large angle scattering sufficient to redirect the incident beam to the detector is unlikely. Thus we will not consider this scattering mode. The most probable mechanisms are shown in figs. 3.6b and 3.6c where (approximately) forward inelastic scattering from the molecule is preceded or followed by large angle "elastic" scattering from the substrate. Strictly speaking, referring to this as elastic is misleading since the attenuation processes within the substrate are still operative, reducing the elastic

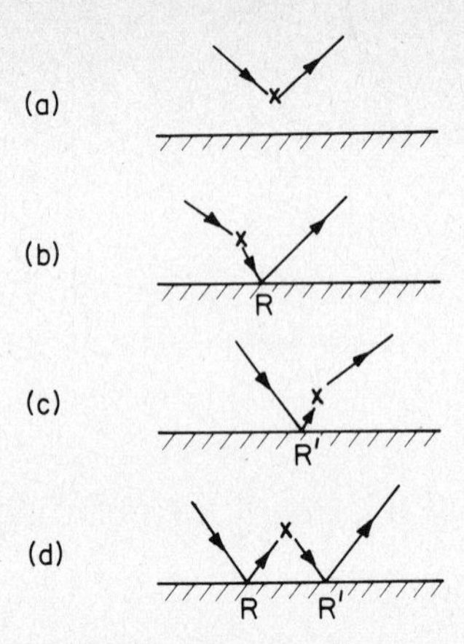

Fig. 3.6. The four contributions to the scattering amplitude when reflected plane waves are used. Only process (a) is possible in the gas phase and process (d) is significantly less probable than (b) or (c).

reflection coefficient significantly below unity (Powell 1974). In the spirit of LEED theory, this effect is allowed for by treating the solid in terms of a complex optical potential, which amounts to the inclusion of the phenomenological damping term in electronic matrix elements, discussed previously. Within the present context, inelastic scattering refers to the molecular events. The probability of double surface and intermediate angle molecular scattering (fig. 3.6d) is very low, due to the smallness of the reflection coefficient as well as the molecular scattering power. The combined elastic surface and inelastic molecular scattering can be treated theoretically within the framework of "two-potential scattering theory" (Goldberger and Watson 1964) in which the surface potential defines a Hamiltonian whose eigenstates are coupled by the vibrational excitation process. For a perfect surface with no potential variation in the transverse direction these states are of the form

$$\psi_k(\boldsymbol{r}) = \eta\, \mathrm{e}^{\mathrm{i}\boldsymbol{k}_\parallel\cdot\boldsymbol{\rho}}(\mathrm{e}^{-\mathrm{i}k_z z} + R(k_z, \boldsymbol{k}_\parallel)\mathrm{e}^{\mathrm{i}k_z z}), \tag{3.33}$$

where η is an appropriate ($\boldsymbol{k}$ dependent) normalization, and $R(k_z, \boldsymbol{k}_\parallel)$ the complex reflection amplitude which specularly scatters the incident wave $\sim\exp(-\mathrm{i}k_z z)$. The electronic matrix element in eq. (3.32) is

$$M_{\mathrm{el}} = \int \mathrm{d}^3 r \psi^*_{k_\mathrm{f}}(\boldsymbol{r})\tau_\mathrm{e}(\boldsymbol{r})\psi_{k_0}(\boldsymbol{r}), \tag{3.34}$$

with $\tau_\mathrm{e}(\boldsymbol{r})$ still to be determined. Using the wave functions, eq. (3.33), the matrix element, eq. (3.34), is the sum of 4 terms, each one corresponding to a scattering sequence shown in fig. 3.6.

The interaction must be of Coulombic origin. Since the free space adsorbate molecule is neutral, a dipole electron–molecule interaction is the lowest multipole interaction possible. Even in the case of adsorption involving charge transfer to or from the molecule, the combined entity of molecule plus substrate screening charge is neutral, thus requiring dipole interactions, which in the far field are

$$\tau(\mathrm{e};\mathrm{v}) = e\boldsymbol{\mu}\cdot\boldsymbol{r}/r^3, \qquad (3.35)$$

with $\boldsymbol{\mu}$ the total dipole moment (permanent plus oscillating) of the chemisorption complex and $\boldsymbol{r}$ the scattering electron coordinate with respect to an origin at the dipole center. Expanding eq. (3.35) about the ground state configuration

$$\tau(\mathrm{e};\mathrm{v}) = (e\boldsymbol{\mu}_0\cdot\boldsymbol{r}/r^3) + \sum_j u_j\cdot\boldsymbol{\nabla}(e\boldsymbol{\mu}\cdot\boldsymbol{r}/r^3)_{\mu=\mu_0} + \ldots, \qquad (3.36)$$

where n_j is the normal coordinate of the j'th mode. The first term in eq. (3.36) contributes only to elastic scattering and will henceforth be dropped. It can easily be shown that the coherent amplitude sum on modes j, when modulus squared, leave only those terms corresponding to an incoherent intensity sum. Thus we can drop the summation, leaving the inelastic scattering interaction with mode j as

$$\tau^j_{\mathrm{inel}}(\mathrm{e};\mathrm{v}) \simeq u_j\left(\frac{\mathrm{d}\mu}{\mathrm{d}u_j}\right)\left(\frac{e\cos\theta_{r,u_j}}{r^2}\right) = \tau^j_{\mathrm{v}}(u)\tau^j_{\mathrm{e}}(\boldsymbol{r}), \qquad (3.37)$$

with θ the polar angle between the electron and the normal coordinate. Note that eq. (3.37) is of the desired separable product form. Writing the normal mode displacement in terms of the usual second quantized oscillator operators

$$u_j = \left(\frac{\hbar}{2M_j\omega_j}\right)^{1/2}(a + a^+),$$

with M_j the reduced mass of the mode, the vibrational matrix element for single excitations is

$$|M^j_{\mathrm{v}}|^2 = \frac{\hbar}{2M_j\omega_j}\left(\frac{\mathrm{d}\mu}{\mathrm{d}u_j}\right)^2(1 + 2n(\omega_j)), \qquad (3.38)$$

where $n(\omega_j)$ is a thermal Bose–Einstein function. The resulting angle-resolved energy distribution, from eqs. (3.32), (3.37), and (3.38) is

$$J'' = \sum_j \frac{\hbar}{2M\omega_j}\left(\frac{\mathrm{d}\mu}{\mathrm{d}u_j}\right)^2 \coth\left(\frac{\hbar\omega_j}{2kT}\right)|M_{\mathrm{el}}(\boldsymbol{k}_0,\boldsymbol{k}_{\mathrm{f}})|^2\delta(\varepsilon_{\mathrm{in}} - \varepsilon_{\mathrm{fin}} - \varepsilon_j), \qquad (3.39)$$

replacing the sum on excited states by a modal sum. The electronic matrix elements can be evaluated using eqs. (3.33), (3.34), and (3.37), or with more detailed LEED wave functions if necessary (Šokčević 1977).

Much ado has been made about surface selection rules. Assuming plane-wave initial and final states and using $\tau_e^j(\boldsymbol{r})$ from eq. (3.37), the electron matrix element is

$$|M_{el}(PW)|^2 = \left(\frac{k_0}{\kappa}\right)^2 \cos^2(\theta_{\kappa,\mu}), \tag{3.40}$$

where $\boldsymbol{\kappa} = \boldsymbol{k}_f - \boldsymbol{k}_0$ is the momentum transfer and $\theta_{\kappa,\mu}$ the angle between $\boldsymbol{\kappa}$ and the oscillating dipole. Equation (3.40) implies that vibrations perpendicular to the momentum transfer cannot be dipole excited. However this is a gas phase result, which is considerably modified when surface wave functions such as eq. (3.33) are used. Although simple generalizations cannot be made, certainly those geometries in which $\boldsymbol{\mu}$ and $\boldsymbol{\kappa}$ are aligned will produce more intense inelastic loss peaks. The reader is referred to Šokčević (1977) for further specifics.

Finally, mention should be made of orientational effects for molecuues adsorbed on metal surfaces. Screening of the dipoles associated with the chemisorbed species can be represented by induced image charges. In the limits in which the dipole $\boldsymbol{\mu}$ is aligned either parallel or perpendicular to the surface, the induced dipoles are

$$\mu_{\parallel}^{in} = -\mu_{\parallel}, \qquad \mu_{\perp}^{in} = \mu_{\perp}.$$

Since the incident electron drives the total system, molecular plus induced, (ideally) $d\mu_{\parallel}^{tot}/d\mu_j = 0$ whereas $d\mu_{\perp}^{tot}/du_j = 2d\mu_{\perp}/du_j$ and it is these quantities which are to be used in eq. (3.38). This specificity for perpendicular orientations is both useful and annoying.

2.3. Many-body aspects

In actual fact, both the ion cores and the valence electrons can exhibit a dynamic response to the act of measurement. It is thus possible that not only the ground state energy of the final state differs from a value inferred through single electron theory, but also the population of excited states does. Depending upon the specific systems, the relevant excitations can include phonons, intramolecular vibrational modes, substrate or chemisorption cluster electron–hole pairs, plasmons, or surface plasmons. We will assume that everything is (in principle) known about the spectroscopic processes involving the gas phase atoms or molecules (intra-atomic or molecular) and focus our attention on those added effects due to the presence of the substrate (extra-atomic or molecular). As it turns out, the absorption spectroscopies are much less susceptible to the many-body modifications than are the emission spectroscopies. This is illustrated in fig. 3.7. Consider a chemisorption complex with an initial net charge Z and energy ε_0, which is represented by the circles. The half space cross-hatched area shows that the complex is embedded in a substrate. In an EELS experiment, an incident current, j_{in}, is scattered into the outgoing beam, j_{out}, possibly leaving the complex in an excited state ε^* but without inducing a monopole charge fluctuation within the complex. Consequently, the Coulomb coupling between the complex and the substrate is, to first order, identical in the initial and final state and thus the act of measurement does not trigger a many-particle substrate response. The transitions involved can be thought of in terms of the bottom left potential energy diagram showing ground to vibrational excitations while the system remains in its electronic ground state. Second order vibrational decay via dipole interactions with substrate electron–hole pairs will broaden the vibrational lines, but to a

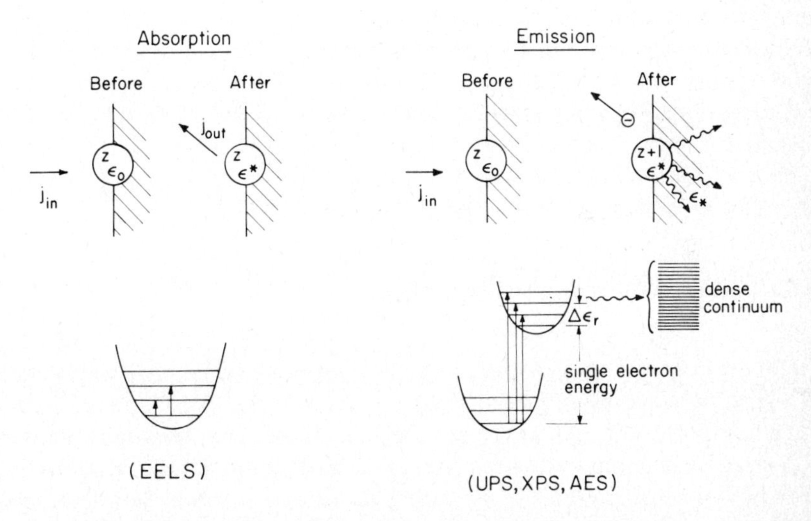

Fig. 3.7. Schematic illustration of the distinctions between absorption and emission processes, both in terms of "real" space (top) and potential energy diagrams (bottom). The circles represent localized surface complexes and the symbols within, the state of the complex before or after the process. Absorption is characterized by transitions within the electronic ground state energy surface whereas emission requires transitions to an excited state surface. The excited state can be coupled to a dense continuum of other states, reached by shake-up not included in the potential energy curves, radiative decay, or other mechanisms. The coupling is represented by the wavy pointing line.

degree which is probably small relative to the level spacing. Things are different (and more interesting) with regard to the emission spectroscopies in which a secondary charged particle is generated, as also depicted in fig. 3.7. An incident current (electrons or photons for example) impinges on the chemisorption complex. Subsequent emission of an electron alters the charge of the complex by one, if the hole produced remains localized within the sphere for a "reasonable" time span. Reasonable here means longer than other characteristic response times of the substrate. This condition is usually well met for core and chemisorption resonance holes. In the case of photoemission $Z=0$ whereas in AES, $Z=1$ resulting in a final two-hole state. Due to the sudden switch on of the monopole within the "sphere", a time varying field with Fourier components at all frequencies crosses the boundary and enters the substrate where it can be absorbed, leaving the substrate–adsorbate complex excited. This is the type of process represented by the interaction with the dense continuum of excited states. In addition, the mobile electrons of the substrate can readjust to screen the field of the ion, possibly hopping into some unfilled orbitals within the sphere or alternatively setting up a surface charge.

Both the excitation and related screening processes modify the electron spectra in non-trivial ways, due to requirements set by energy conservation. Consider first the consequences of an adiabatic relaxation in which the system is always in the ground state of the instantaneous potential. Due to the attractive Coulomb interaction between $\rho_0(\boldsymbol{r}')$ the hole charge and $\rho_{\text{in}}(\boldsymbol{r})$, the induced charge, written as

$$\Delta\varepsilon_{\text{r}} \equiv \tfrac{1}{2}\int \mathrm{d}^3\boldsymbol{r}\, \mathrm{d}^3\boldsymbol{r}'\, \rho_{\text{in}}(\boldsymbol{r})(e^2/|\boldsymbol{r}-\boldsymbol{r}'|)\, \rho_0(\boldsymbol{r}'), \tag{3.41}$$

the lowest energy state of the relaxed system will be less, by $\Delta\varepsilon_{\text{r}}$, than in the single electron Koopmans limit. Consequently, the ejected electron must arrive at the detector upshifted in energy by the same amount above the Koopmans theorem frozen orbital expectations, to conserve energy overall. In actuality, the hole potential is not switched on adiabatically and thus the ionized system stands a good chance of being left in an excited eigenstate of the Hamiltonian which includes the hole potential. When this occurs, the ejected electron must arrive at the detector with an energy which is precisely the excitation energy lower than the adiabatically relaxed energy. If the excited states are reasonably discrete in energy, the lower kinetic energy peaks in the spectrum are usually referred to as shake-up satellites. Low energy continuum excitations such as phonons or conduction band electron–hole pairs can be thought of as producing satellites, although in practice these "satellites" appear as modifications to the lineshape of the main relaxed peak, owing to the fact that the most likely excitation energies are comparable with neutral linewidths and apparatus resolution. The role of many-body theory is to provide the basis for relating relaxation and shake-up energies with intensities in each peak, thus enabling proper extraction of mean field chemical information from a complex experimental spectrum.

Absorption spectroscopies

The mathematical treatment of the many-body aspects of absorption spectroscopies such as EELS is trivial. Since EELS is a process involving transitions between eigenstates of the same Hamiltonian, $\Delta V = 0$ in eq. 3.5. Thus $U(t) = 1$ and $D(\omega') = \delta(\omega')$, which with eq. 3.7 yields $J = N_{el}(\omega)$, the single electron result given by eq. 3.32.

Some theory has been done on many-body substrate modifications to the incident electron–vibrational interaction in which the dynamical aspects of the image response to the fluctuating molecular dipole have been considered (Delanaye 1978). Such effects are only significant when the vibrational and surface plasmon energies are comparable. As this is rarely the case, vibrational energies usually being lower by one or two orders of magnitude, we will drop the subject here.

Emission spectroscopies

The characteristic feature of the emission spectroscopies that gives rise to the observable many-body effects is a (usually but not always) localized charge fluctuation within the atomic or chemisorption sphere as a consequence of the emission process. This leads to a non-vanishing ΔV in eq. 3.5, which, upon insertion of a complete set of many-body eigenstates of the final Hamiltonian $\equiv H_0 + \Delta V$, can be written

$$D(\omega) = \sum_{x} \delta(\varepsilon_0 + \omega - \varepsilon_x)|\langle 0|x\rangle|^2. \qquad (3.42)$$

This spectral function is just a series of delta-functions positioned at the relaxation shifted ground state and at the ionic excitation energies below. The relative intensities are determined by the overlap of the initial ground state wave function with the particular excited final state wave function, whose value depends upon the specific model.

One of the most popular relaxation models takes full advantage of the fact that plasmon and phonon excitations can be described as harmonic oscillators and thus boson fields. Since the quantum mechanics of forced oscillators is well established (Pryce 1966, Gross 1967, Lucas and Šunjić 1969, 1972a, b), the task at hand is to map the dynamical relaxation problem onto that of forced oscillators. A second quantized model Hamiltonian which includes the relevant physics, namely the excitation spectrum of the responding system (electron gas or phonons), the form of coupling between the hole and the excitations, and a time dependence for the hole switching on and which is a generalization of one suggested by Lundqvist (1967, 1969) is

$$H = H_{\mathrm{p}} + H_{\mathrm{el}}, \tag{3.43}$$

where

$$H_{\mathrm{p}} = \varepsilon_0 c^{\dagger} c + \sum_q \omega_q b_q^{\dagger} b_q + \sum_q c c^{\dagger} g_q (b_q + b_q^{\dagger}),$$

$$H_{\mathrm{el}} = \sum_{\boldsymbol{k},\sigma} \varepsilon_{\boldsymbol{k}} c_{\boldsymbol{k}\sigma}^{\dagger} c_{\boldsymbol{k}\sigma} + \sum_{\boldsymbol{k},\boldsymbol{k}',\sigma} V_{k,k'} c_{k\sigma}^{\dagger} c_{k'\sigma} c c^{\dagger},$$

and with the following interpretation. The first and second terms of H_{p} refer to the (assumed) structureless core or localized state and to the boson system respectively. The key features of the model are contained in the third term which describes the (again assumed) linear coupling of the hole potential with the surface plasmons or phonons. The operator $cc^{\dagger} = 1 - n_{\mathrm{c}}$ accounts for the sudden switching and proper time ordering since it equals zero (unity) when the core state is occupied, $n_{\mathrm{c}} = 1$ (vacant, $n_{\mathrm{c}} = 0$). In time space, it just says that the hole–boson interaction cannot occur until after the hole is created. The strength of the interaction is set by the coupling parameter g_q which we will return to later. The basic assumptions of this model, often called the independent boson or displaced oscillator model are that the coupling is linear in boson operators (or coordinates) and the excitation of a given mode goes on independently of all other excitations. H_{p} describes the model proposed and solved perturbatively by Lundqvist (1967, 1969) and exactly solved by Langreth (1970). The beauty of this model arises from the assumed linear coupling of the hole to the plasmons. By a straightforward unitary transformation (roughly equivalent to completing the squares) H_{p} can be diagonalized. Alternatively, the problem is that of a displaced harmonic oscillator and is thus reduced to finding the projection of the ground state oscillator wave function onto those of the displaced oscillator. This is a straightforward exercise which has been detailed in great length elsewhere (Langreth 1970, Harris 1975, Gadzuk 1978, Gumhalter and Newns 1976, Gumhalter 1977a, b, c). The end result is that eqs. (3.42) and (3.43) can be brought to the form of a Poisson distribution

$$D_B(\omega) = e^{-\beta} \sum_{n=0}^{\infty} (\beta^n/n!)\, \delta(\omega + \beta\omega_0 - n\omega_0) \tag{3.44}$$

for dispersionless bosons of frequency ω_0. The term $\beta\omega_0$ is to be identified with the screening shift $= \Delta\varepsilon_r$ given by eq. (3.41), or more generally

$$\Delta\varepsilon_r = \sum_q g_q^2/\omega_q = \beta\omega_{q=0} \tag{3.45}$$

when dispersion is included. The delta-function in eq. (3.44) carries the information that the hole state is lowered in energy by $\Delta\varepsilon_r$, but can be left in the n'th excited oscillator state with energy $n\omega_0$, the probability of this being $\beta^n/n!$. For a hole localized on an adsorbed entity, surface plasmons are the relevant field quanta. On purely classical grounds, the interaction between a point charge and an induced surface charge can be represented by that of an image charge, in which case $\Delta\varepsilon_r \simeq V_{\text{image}} = e^2/4s$ where s is the hole–image plane separation (Harris 1975, Gadzuk 1978, Gumhalter and Newns 1976, Gumhalter 1977a, b, Hewson and Newns 1974, Laramore and Camp 1974, Gadzuk 1976a, b, Šunjić et al. 1972, Harris and Jones 1973, 1974, Chang and Langreth 1973, Heinrichs 1973, Gadzuk 1977a, Gumhalter 1977c). Thus $\beta \simeq V_{\text{image}}/\omega_s \lesssim 0.2$ for experimentally realizable systems. This suggests that the screening modifications to an observed spectrum consist of a ~1–3 eV relaxation shift plus a weak, at best, surface plasmon satellite, within the present model. Parenthetically note that for optical phonon fields, $\omega_0 \sim 0.01$ eV, in which case $\beta \sim 40$. In this limit the Poisson distribution is well represented by a continuous Gaussian whose peak is displaced $\sim\beta\omega_0$ below the zero-phonon line.

Consider now a specific adsorbate–substrate model which provides a reasonable estimate of the numerical consequences of the displaced surface plasmon model. Within the linear response approximation, the $q_\parallel$'th Fourier component of the induced surface charge $\rho_{\text{in}}(r)$ is related to the hole charge through the relation $\rho_{\text{in}}(q_\parallel) = R_s(q_\parallel, \omega = 0)\, \rho_0(q_\parallel)$, where $R_s(q_\parallel, \omega = 0)$ is the static density–density response function and $q_\parallel$ is the wavenumber parallel to the surface. An approximate form for R_s, equivalent to an image charge within a non-perfect conductor (i.e. non-vanishing screening length) has been shown to be

$$R_S(q_\parallel, \omega) = \frac{\varepsilon_{\text{ads}} - \varepsilon_s(q_\parallel, \omega)}{\varepsilon_{\text{ads}} + \varepsilon_s(q_\parallel, \omega)}, \tag{3.46}$$

where $\varepsilon_s(q_\parallel, \omega)$ is the substrate surface dielectric function (Gadzuk

1977a, Datta and Newns, 1976) and ε_{ads} the dielectric constant of the adsorbate half space. The two-dimensional Fourier transfer of the hole Coulomb potential is

$$v(\boldsymbol{q}_\parallel) = \frac{2e^2}{\varepsilon_{\text{ads}} q_\parallel} \exp(-2q_\parallel s), \tag{3.47}$$

referenced to a zero at the image plane. Within the Fermi–Thomas approximation, the surface dielectric function is $\varepsilon_s^{\text{FT}}(q_\parallel, 0) = (1 + \kappa^2/q_\parallel^2)^{1/2}$ with $\kappa = 2.95/r_s^{1/2}$ Å^{-1}, the screening length. Fourier transforming eq. (3.41), for a point hole charge distribution, and combining with eqs. (3.46) and (3.47), the screening energy is then given by

$$\Delta\varepsilon_r = \frac{e^2}{2\varepsilon_{\text{ads}}} \int_0^\infty \mathrm{d}q_\parallel \left(\frac{\varepsilon_{\text{ads}} - \varepsilon_s(q_\parallel, 0)}{\varepsilon_{\text{ads}} + \varepsilon_s(q_\parallel, 0)} \right) \exp(-2q_\parallel s). \tag{3.48}$$

Equation (3.48) has been evaluated using $\varepsilon_s^{\text{FT}}$ and treating ε_{ads} parametrically for a substrate with $r_s = 2$(Al). The ratio $\Delta\varepsilon_r/\varepsilon_{\text{im}}$ (with $\varepsilon_{\text{im}} = e^2/4\varepsilon_{\text{ads}} s$) as a function of s is shown in fig. 3.8. For example, at $s = 1$ Å and $\varepsilon_{\text{ads}} = 1$, $\Delta\varepsilon_r \simeq 2.35$ eV and $\beta = \Delta\varepsilon_r/\omega \simeq 0.2$ (with $\omega_s = 10.7$ eV as for Al). Thus the strength of the substrate surface plasmon satellites is expected to be small, even though significant screening shifts are experienced (Datta and Newns 1976).

The possibility of conduction band electron–hole pair shake-up is

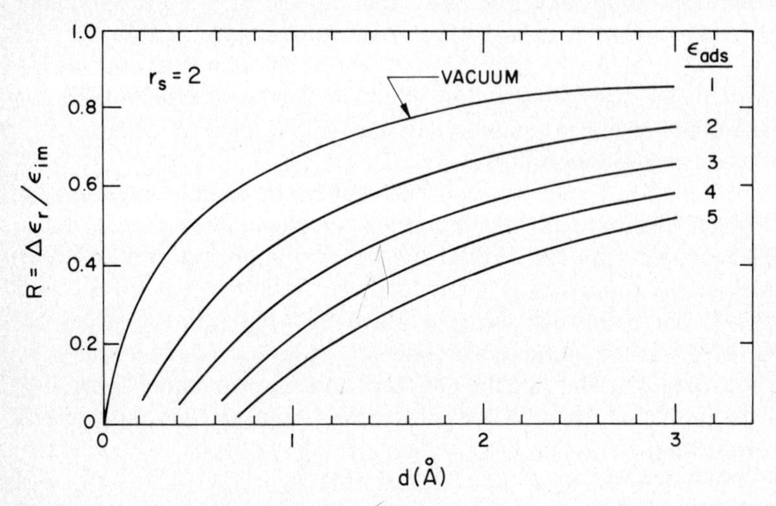

Fig. 3.8. Ratio of the surface screening energy given by eq. (3.48) to the classical image potential, $\varepsilon_{\text{im}} = e^2/\varepsilon_{\text{ads}} s$ as a function of the hole-screening plane separation treating ε_{ads} parametrically. Here $r_s = 2$.

covered by H_{el} in eq. 3.43. Here the screened hole potential $V_{kk'}$ excites electrons from within the Fermi sphere $|k'| < k_F$ to unoccupied states outside $|k| > k_F$. Assuming that H_{el} commutes with H_p(i.e. the plasmon response is independent of the long time pair response (Noziéres and deDominicis 1969)), the total hole spectral function is a convolution

$$D_{tot}(\omega) = \int d\omega' \, D_B(\omega - \omega') B_+(\omega'),$$

where $B_+(\omega')$ is the spectral function associated with H_{el}. In the case of a hole switched on within an infinite solid, $B_+(\omega') = \omega'^{\alpha-1} (\omega' > 0)$ where $\alpha = \Sigma_{l,m} (\delta_{l,m}/\pi)^2$ and $\delta_{l,m}$ are the partial wave phase shifts for Fermi level electrons scattered by $V_{kk'}$. Typically, $\alpha \sim 0.1$–0.25. The discovery and theory of this effect, often called the X-ray edge singularity, was carried out by Mahan (1974) and Noziéres and deDominicis (1969) (MND), and has frequently been commented on by Dow and coworkers (1977).

Gumhalter and Newns (GN) worked out a zero order theory for the adsorbate analog of the MND edge and found that $B_+^{ads}(\omega') \sim \exp(-\omega'/\omega_m)$ ω'^{f-1} where f (usually $\leqslant 0.1$) and $\omega_m (\gtrsim 1\,\text{eV})$ are functions of s (Gumhalter and Newns 1975, Gumhalter 1977a, b, Gadzuk and Doniach 1978). The experimental consequences in electron spectroscopies of both the MND- and GN-edges were first realized by Doniach and Šunjić (1970). Accounting for core hole lifetimes by convolving $B_+(\omega)$ with a Lorentzian, then demonstrated that the observable lineshapes should be asymmetric with broadening and long range tails on the low kinetic energy side. The degree of asymmetry depends upon the strength of the interaction and hence magnitude of α or f. Experiments have confirmed the basic soundness of the ideas, both for bulk (Citrin et al. 1977) and adsorbate (Fuggle et al. 1978) systems.

Yet another possible screening mechanism involves an atomistic process, first detailed by Kotani and Toyazawa (1974) for bulk XPS, applied by Gumhalter and Newns to the problem of X-ray absorption in adsorbed atoms (Gumhalter 1977), and adopted by Schönhammer and Gunnarsson (SG) to XPS of adsorbates (Schönhammer and Gunnarsson 1977). The idea is that upon creation of a core hole, the substrate supplies an electron which hops into the lowest unoccupied orbital of the ionized adparticle, thus neutralizing or screening the hole charge. In this picture the properties of the substrate are of minor significance compared to those of the adparticle. The specific new feature of this model is provision for an atomic orbital with initial orbital energy ε_a sufficiently above the substrate Fermi level to assure its vacancy. Upon ionization, the Coulomb attraction U_{ac} between the positive core hole and an electron which could occupy this orbital in effect pulls down the final state orbital energy to a value $\tilde{\varepsilon}_a = \varepsilon_a - U_{ac}$, which could lie below the substrate Fermi level. If a coupling between the renormalized adparticle orbital and the substrate exists, two effects result. Firstly, new system eigenstates are formed which can be represented as linear combinations of substrate and ad-particle orbitals. The form of the local density of final states in the vicinity of the adatom depends upon the coupling strength. Weak interaction merely converts the discrete state at $\tilde{\varepsilon}_a$ into a broadened resonance, whereas strong coupling could produce pairs of states split off from the substrate band which correspond to bonding and anti-bonding combinations. Secondly, electrons can be transferred from the substrate into any of these final states. If the transfer is into the lowest energy coupled state, then the photoelectron emerges with maximum kinetic energy; the totally relaxed energy. On the other hand when the transfer is into an excited state, the photoelectron must be emitted at a kinetic energy which is the excitation energy below the relaxed peak. These electrons appear as shake-up satellites. A model Hamiltonian which includes the just described atomic relaxation process is

$$H = \varepsilon_0 c^{\dagger} c + (\varepsilon_a - U_{ac}(1 - n_c))n_a + \sum_k \varepsilon_k c_k^{\dagger} c_k + \sum_k (V_{ak} c_k^{\dagger} c_a + \text{h.c.}), \tag{3.49}$$

where the notation of eq. (3.43) has been adopted. Note that initially $n_c = 1$, hence the orbital term is $H_a^{\text{in}} = \varepsilon_a n_a$, whereas finally $n_c = 0$ and $H_a^{\text{fin}} = (\varepsilon_a - U_{ac})n_a = \tilde{\varepsilon}_a n_a$. The last term in eq. (3.49) is responsible for the substrate–adparticle mixing.

It is informative to first consider a simple limiting case of eq. (3.49) to gain a feel for the qualitative properties of the spectrum (Gadzuk and Doniach 1978, Gadzuk 1978). Consider the (strong coupling) extreme in which the substrate bandwidth goes to zero, thus allowing the replacement $\Sigma_k \varepsilon_k c_k^{\dagger} c_k \to \varepsilon_m c_m^{\dagger} c_m$ where ε_m is the "band" energy; and similarly for the hopping term. Then solution of the final state Hamiltonian of eq. (3.49) is reduced to that of a two-state ("diatomic molecule") problem. Setting $\varepsilon_m = 0$ (choice of arbitrary energy zero), the eigenvalues are easily determined to be

$$\varepsilon_{\pm}/V = p \pm (1 + p^2)^{1/2}, \qquad p \equiv \tilde{\varepsilon}_a/2V, \tag{3.50}$$

which are simply bonding (−) and anti-bonding (+) combinations of the

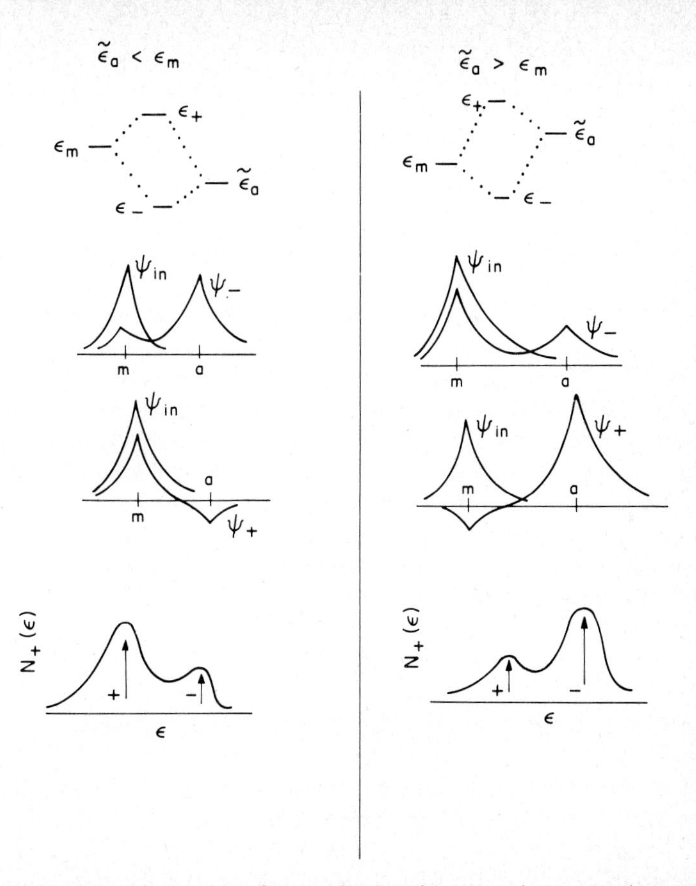

Fig. 3.9. Molecular orbital analog of the unfilled orbital relaxation model, illustrating the differences which arise depending upon the sign of $\tilde{\varepsilon}_a - \varepsilon_m$, negative for the left-hand column and positive for the right. The first row shows the final state eigenvalues ε_+ and ε_-. The second and third rows depict wave functions of the initial state, ψ_{in}, and the possible final states, either bonding (ψ_-) or antibonding (ψ_+), from which it is possible to see the origins of the relative weights in the satellite vs. main peaks in the spectra, shown in the bottom row. For esthetic reasons, the delta-functions have also been smoothed out.

atom and substrate orbitals. From eq. (3.42), the spectral function is

$$D_{ao}(\omega) = \delta(\omega - \varepsilon_+)|\langle\psi_{in}|\psi_+\rangle|^2 + \delta(\omega - \varepsilon_-)|\langle\psi_{in}|\psi_-\rangle|^2, \qquad (3.51)$$

where the overlap of the initial state ψ_{in} (localized within the substrate) with either of the final eigenstates ($\psi_\pm = \alpha_\pm\psi_a + \beta_\pm\psi_m$) is

$$|\langle\psi_{in}|\psi_\pm\rangle|^2 = |\beta_\pm|^2 = [1 + (\sigma_\pm/V)^2]^{-1}. \qquad (3.52)$$

The resulting photoelectron spectrum is "two-peaked" with the high energy relaxed peak that one in which the screening electron occupies the bonding state (ε_-). The satellite follows when the screening electron occupies the excited anti-bonding state (ε_+). The relative intensities of the satellite to adiabatic relaxed peak are determined by the sign of p, or equivalently the position of $\tilde{\varepsilon}_a$ with respect to ε_m as illustrated in fig. 3.9. When $p < 0$, the bonding adatom-substrate screening orbital $\psi_-(p<0) = \alpha_-\psi_a + \beta_-\psi_m$ is more strongly localized on the adatom (i.e. $|\alpha_-|^2 > |\beta_-|^2$) and thus the intensity in the leading peak is less than that in the satellite. As $\tilde{\varepsilon}_a$ exceeds ε_m, either due to a larger value of ε_a or smaller U_{ac}, then the majority intensity $\sim |\langle\psi_{in}|\psi_-(p>0)>|^2$ is found in the adiabatic peak, as can be understood from the right hand column in fig. 3.9. Those examples should point out the manner in which a spectrum can undergo a smooth transition from mostly "main" to mostly satellite intensity.

Schönhammer and Gunnarsson (1977) have provided a detailed and comprehensive theory for the spectral function of a core hole propagating in time according to the full Hamiltonian of eq. (3.49). In essence, eq. (3.5) was solved with

$$H_0 = \varepsilon_0 + \varepsilon_a c_a^\dagger c_a + \sum_k \varepsilon_k c_k^\dagger c_k + \sum_k (V_{ak} c_k^\dagger c_a + \text{h.c.}),$$

$$H_0 + V = (\varepsilon_a - U_{ac}) c_a^\dagger c_a + \sum_k \varepsilon_k c_k^\dagger c_k + \sum_k (V_{ak} c_k^\dagger c_a + \text{h.c.}),$$

adding to and modifying some techniques introduced by Noziéres and deDominicis (1969) in their solution to the long time response problem. Some characteristic spectral functions obtained by SG are shown in fig. 3.10 for a "reasonable" set of parameters. Finite core hole lifetimes have been allowed for by taking ε_0 to be complex. Emphasis here is on the evolution of the spectrum with increasing value of $V_{ak} \equiv V$. The inset shows the local density of states of the screening orbital, with respect to a zero at the Fermi level, before ($\varepsilon = \varepsilon_a$, solid curve) and after ($\varepsilon = \varepsilon_a - U_{ac}$, dashed curve) core hole creation. It is apparent that as V and hence both the strength and rate of relaxation increases, intensity within the upper totally relaxed peak grows.

Some effort has been devoted to model studies in which both surface plasmon [eq. (3.43)] and atomic orbital [eq. (3.49)] screening are accounted for within a single formalism. Depending upon the relative magnitudes of ω_s, λ, V_{ak}, and U_{ac}, one or the other mechanism may dominate, but rarely can the lesser one be totally neglected. For instance in the limit in which only zero or one surface plasmons and the bonding or anti-bonding orbitals are allowed, the spectrum will show a relaxed

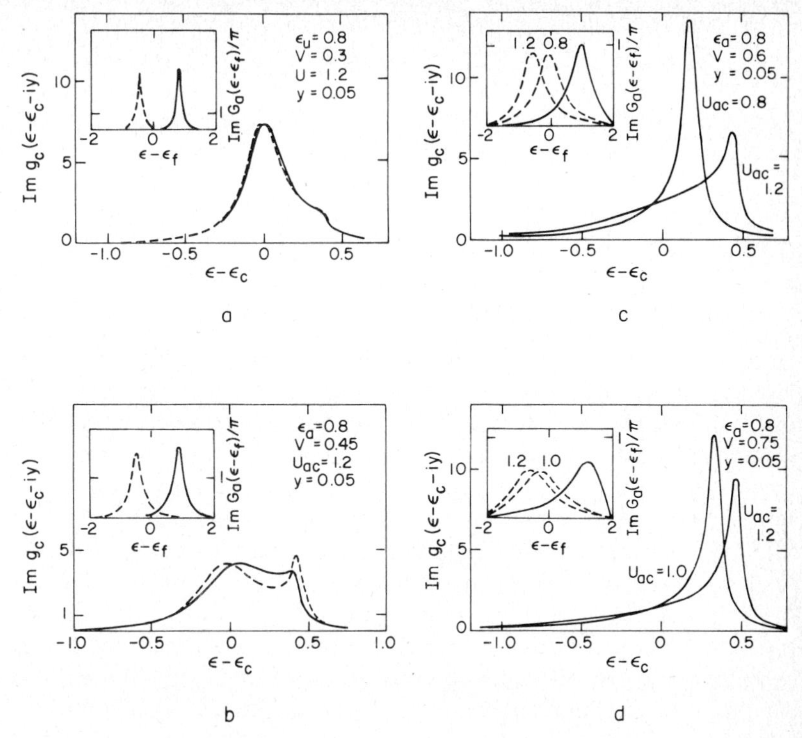

Fig. 3.10. Exact numerical results for the core hole spectral function $\equiv$Im $g_c(\varepsilon)$. The inserted figure shows the adsorbate spectral function in the initial (full curve) and the final (broken curve) states. A small imaginary part has been added to the core hole energy to account for finite lifetimes (Schönhammer and Gunnarsson 1977).

peak, shifted by a screening charge which is a mixed orbital-induced surface charge and 3 hybrid shake-up satellites, whose positions and intensities are functions of the characteristic system parameters (Gadzuk and Doniach 1978, Gadzuk 1978, Hussain and Newns 1978).

Finally we note that justification and regions of validity for the unfilled orbital mechanism has been provided through the "first principles – core hole in (or on) jellium" theory and calculations due to Lang and Williams (1977) and Williams and Lang (1978). In brief they find that when the valence shell of the adsorbate, in the initial state, is not fully occupied, then the screening charge is well approximated by the charge of the unfilled orbital, as is the case for adsorbed Na. In contrast for Cl adsorbed onto jellium ($r_s = 2$), large charge transfer into the 3p shell occurs solely due to bond formation. Consequently with a sparse distribution of vacancies on the adatom, the screening charge is forced to reside within the substrate. These two extremes point out the possible pitfalls inherent in any sweeping assertions on the nature of the many-body relaxation response in core level spectroscopies and demonstrates the users' need for understanding the fundamentals of the relaxation process before one can obtain reliable results.

3. Experimental principles

3.1. General considerations

The recent development of electron spectroscopic methods for surface analysis has provided a variety of microscopic measurements which are capable of providing a detailed basis for making critical interpretations of chemical bonding at surfaces. They can be described for convenience in terms of four general groups; surface crystallography, electron excitation phenomena, vibrational spectroscopy measurements, and energetics of adsorption*. The importance of LEED for surface structural chemistry has been widely recognized (Ertl and Küppers 1974) since Germers and Farnsworth's pioneering LEED work. Its significant contribution to the subject of surface bonding is treated separately in this volume by Van Hove (chapter 4). Another significant aspect, that of the energetics of chemisorptions on metals, is likewise discussed by Ertl (chapter 5).

The contributions to the subject of chemical bonding through the methods of electron spectroscopy, principally photoelectron emission, Auger and electron energy loss phenomena are the specific topics to be considered here. Different bonding properties are measured in each case; discussion therefore is focussed accordingly on the following specific aspects; internuclear distances and symmetries, electron charge transfer and molecular orbitals, and electron or atom binding energies. Since these studies are still very much in a stage of development, the picture of surface chemical bonding is far from completely defined on a microscopic level or capable of complete theoretical interpretation. Recent advances in the field of electron spectroscopy, nevertheless, have opened up promising new approaches to the subject (Ibach 1977a, b, Willis et al. 1976, Siegbahn et al. 1976, Shirley 1972). The general features of the different approaches and their experimental and theoretical characteristics set working limits on their degree of ap-

* Definitions relevant to applications of electron spectroscopy to the surfaces of metals are described in appendix B.

plicability to solid state surface chemistry. It is likely that substantial advances in the area of electron spectroscopic interpretation of surface chemical bonding can be expected in the future.

The organization is to present the main experimental features of the specific electron spectroscopic methods based on photoemission, Auger and electron loss approaches respectively. The relevance of these approaches to evaluation of surface chemical bonding is introduced followed by conditions of applicability, ranges of typical applications and a general critique of pertinent experimental procedures and implications. Discussion of applications of the methods in a subsequent section will be directed to interpretation of specific solid surface–adsorbate interactions systems.

3.2. Photoemission spectroscopy

Introduction

Interpretation is based to a first approximation on a simple one-electron excitation process in which electron emission within the electron emission skin dpeth of the solid surface is stimulated by photon radiation (Carlson 1975). Since the energy of the photon can be varied over a considerable range from just a few eV to many keV, the magnitudes of the electron binding energies which can be studied vary broadly. In addition, since the intensity, energy and momentum distribution of the photoemitted electrons can be effectively measured, photoemission spectroscopy provides a powerful approach to the characterization of electron orbitals associated with both stable and perturbed surfaces of solids as well as the perturbation of the molecules interacting with those surfaces (Eastman 1972a, b, Liebsch 1978a). Major efforts pioneered by Siegbahn and others have helped to emphasize the promising capabilities of photon stimulated emission of photo- and Auger-electrons. Subsequently, this approach has been broadly applied to the study of structure and bonding in solids, surfaces, liquids and free molecules (Siegbahn 1976).

The binding energy of electron, E_b, is simply related by the photon energy, $\hbar\omega$, and the work function, ϕ_e to the kinetic energy of the photo-electron, E_k, by the relationship

$$E_b = \hbar\omega - \phi_e - E_k. \tag{3.53}$$

The energy of emitted electrons in Auger or autoionization processes is independent of the mode of excitation (Carlson 1975). The different processes can be most simply categorized by the diagrams shown in fig. 3.11. The X-ray emission process is included for completeness. The simple Bohr model is shown on the left, the onc-electron energy level diagrams in the middle, and the many-particle pictures on the right. A many-body complication in photoemission results from the simultaneous excitation of other electrons in the target, referred to as shake-up phenomena. These can be important in the interpretation of the spectra but, as previously indicated, are not readily dealt with at this stage of theoretical understanding. This process leads to photoelectron emission with a reduced kinetic energy and the production of an excited residual ion as indicated.

An impetus in the application of photoemission spectroscopy to the study of clean and chemisorbed surfaces of both metals and semiconductors resulted from the work of Shirley (1972), Eastman (1972a, b) and others (Koch et al. 1974, Plummer 1975a, b). The unique capability for this purpose of a differentially pumped photon source in the vacuum ultraviolet combined with an ultra-high vacuum chamber suitable for preparing clean surfaces and an appropriate electron detector is generally recognized. It stimulated to a large extent the present major effort in this area. Since then, techniques used for sample preparation and characterization, for the generation and monochromization of the photon beam, for the combined characterization by supplementary surface probes, for the enhanced sensitivity and capability in the electron detectors, and in the recording, storage display, and processing of the data have progressed substantially (Lee 1976, Czanderna 1975, Hannay 1976).

In the elucidation of surface structure and bonding by photoelectron spectroscopy, the following factors can provide chemical bonding information in XPS-spectra: core-level chemical shifts, multiplet splitting, shake-up and shake-off satellites, and line-broadening (Brundle 1975a, b). Bonding information from UPS-spectra usually reflect the band structure of the metal with the superimposed broadened orbitals of the adsorbate. The assignment of orbital energy levels and orbital character from such spectra provides a major contribution to the interpretation of molecular and atomic bonding at surfaces.

It has also become evident in the past few years that substantial benefits may be derived in photoelectron spectroscopy by the use of synchrotron radiation sources generally associated with electron storage ring facilities. The potential use of synchrotron radiation for studies of adsorption, reflectivity, fluorescence spectroscopy and particularly photoelectron spectroscopy to examine solids, surfaces, adsorption, atom

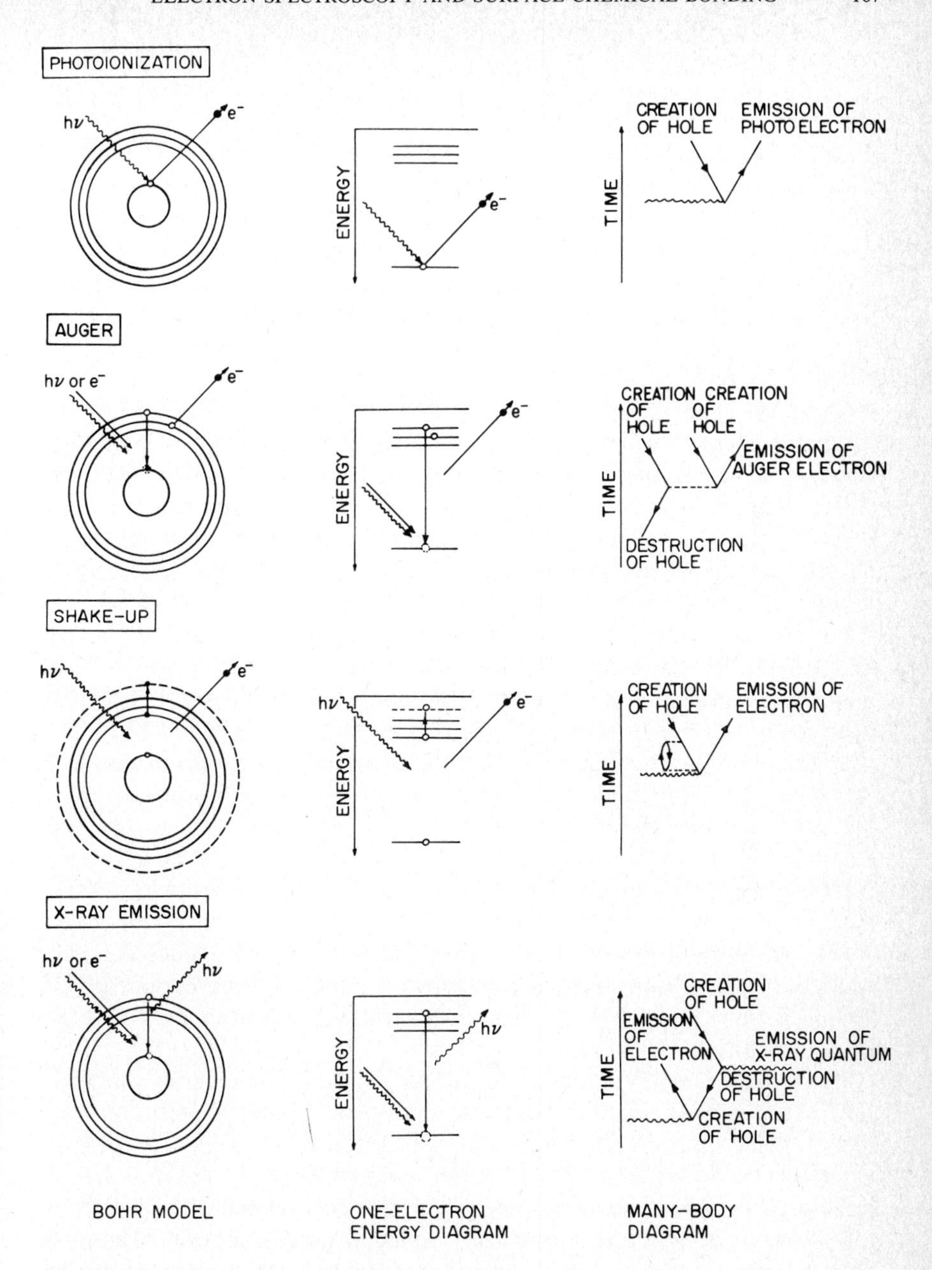

Fig. 3.11. Excitation of electron spectra and electron emission. (Siegbahn 1976). This is a simpler version repeated here for the convenience of the reader of the same processes depicted in fig. 3.2 for independent particles.

and molecular systems is well documented (Koch et al. 1974, Castex et al. 1977a). These advantages accrue largely from the availability of a frequency-tunable, collimated, polarized, relatively intense photon beam variable over a range of energies from 5 eV to at least 10 keV. The ability to vary the frequency continuously over a wide energy range is an essential feature. This usefulness depends greatly on the range and efficiency of the available vacuum monochromators coupled to it. Applicability of advances in synchrotron radiation photoelectron spectroscopy depend in large measure on the availability of sources with higher energy and intensity and of new and versatile monochromators for this purpose (Hodgson et al. 1976b).

A promising development in photoelectron spectroscopy from the viewpoint of chemical bonding is the examination of the angular distribution of the emitted photoelectrons (Liebsch 1978a). The incident and emitted beam directions are defined relative to the symmetry planes of the crystal, the normal to the crystal, and to the angle of incidence of the photon relative to the crystal. The important angle for unpolarized photons is complementary to the angle between the incident photon beam and the emitted electrons. For the plane polarized photons available from a synchrotron radiation source, an important angle is that between the electron field vector of the photons and the emitted photoelectrons.

Applications of photoelectron spectroscopy to chemisorption are well documented in the literature. Some recent reviews of particular relevance to interpretations relating to surface bonding on metals are available (Feuerbacher and Fitton 1977b, Eastman 1972, Plummer 1975a). Reference can also be made to the recent Proceedings of the Noordwijk Symposium (Willis et al. 1976), of the NATO Advanced Study Institute (Ghent 1977), and of the 3rd International Conference on Solid Surfaces (Dobromzemsky et al. 1977) and to Feuerbacher et al. (1978). Typical examples of important types of measurements will be described in the next section.

Much experimental information has been produced recently through the use of angular resolution energy analyzers combined with continuously tunable and polarized light sources. Schemes have been developed to interpret differential photoionization cross sections for different metal surface systems including clean surfaces, physically adsorbed complexes, chemisorbed atoms and molecules and reaction products. The examples to be discussed will show how the characteristics of the photocurrent can be used to obtain important information on the one-electron properties of the clean surface and on the perturbations in it as well as in the interacting atom or molecule caused by molecule–surface interactions. The development of angle-resolved photoemission and partial yield spectroscopy also provides sophisticated information on the two-dimensional aspects of the band structure at the surface and the characteristics of unoccupied surface states.

A simple but useful application of photoemission spectra relates to its use as a "chemical finger-print" of the chemisorbed surface (Brundle 1975b, Brundle and Baker 1977, Plummer 1975a). Spectra for chemisorbed systems are often complex and may or may not be clearly related to the spectra for the isolated molecule in the gas phase. Nevertheless, each such spectrum is uniquely characteristic of a specific surface electron configuration and, to the extent that it can be resolved from the effects of background as well as those of spectra associated with the presence of other molecules, provides a powerful and useful indication of the electron structure characteristic of a given state of bonding. An interesting extension of this method has been suggested which exploits a concept common to coordination chemistry. The relative bonding strengths and photoemission spectra of a series of related compounds to a surface is determined by conducting a series of displacement reactions in a series starting with the less strongly bound and proceeding systematically to the most strongly bond adsorbate for a given adsorbent (Muetterties et al. 1971).

There are a series of stages by which the photoemission can be usefully applied to chemisorption studies. The finger-print stage permits one to distinguish among various surface phases. The second stage consists of identifying the chemical nature of an adsorbate by comparison to the gas-phase spectra. The third and most difficult stage is to apply theoretical analysis of both the photoemission and the surface chemistry to provide quantitative information of the energies associated with valence orbitals, molecular interaction effects and chemical bond formation.

Many of the schemes for interpreting the differential cross sections for ionization from chemisorbed molecules, fixed on the surface of the adsorbent, are derived from similar schemes depicting the isolated, randomly oriented molecule (Dill and Dehmer 1974, 1975). Since the substrate surface tends to orient preferentially chemisorbed molecules along a preferred direction, applications of angle-resolved photoemission are promising for the determination of the orientation of molecules at surfaces. The scattered wave method has been applied to this end thoeretically by Davenport (1976). Predictions have been experimentally confirmed for nickel by Allyn et al. (1977), Smith et al. (1976), for platinum by Apai et al. (1976), and for iridium by Rhodin et al. (1977a, 1978) mainly for chemisorbed carbon monoxide on single crystals of the corresponding metals.

Dynamical calculations of angle-resolved ultraviolet photoemission from simple ordered atomic chemisorption is an important step on the way to an interpretation of the electron interaction of a chemisorbed molecule with a reactive metal surface. The important factors are the dependence of photoemission intensity on the angle of the collection, the plane of collection, the photon incidence and the photon polarization. The purpose of such a study is to separate and assign contributions from surface orbitals with different symmetries that participate in the process of bonding at the surface. This will be discussed later with specific reference to the atomic adsorption of gases on transition metals.

Dynamical calculations of surface crystallography using low-energy electron-diffraction (LEED) have been successful in particular cases where a series of special requirements have been satisfactorily fulfilled [see Van Hove (chapter 4)]. They include availability of a realistic surface model, of an effective scattering potential and of adequate corrections for surface force interactions such as thermal vibrations. It is of value to have an independent method for the determination of surface structure to which the LEED results may be compared and evaluated. Extraction of surface structural information from angle-resolved ultraviolet photo-emission spectroscopy provides such an alternative source of crystallographic information (Liebsch 1978a). The degree to which this has been effectively achieved, including a comparison with similar information obtained from LEED analysis, can be considered as an alternative approach to the analysis of surface crystallography.

The photoemission approach to chemical bonding through crystallography structure can also be pursued through the study of molecular metal clusters. In this case emphasis is placed on the analogy between the molecular structure characteristic of ligand-stabilized polymetallic clusters and that of ordered chemisorption layers containing the same ligand molecule bonded to an extended metal surface. The usefulness of this analogy between cluster chemistry and catalysis has been developed by Muetterties (1977) and others (Ozin 1978). It postulates that variations in the fine structure of the photoemission or Auger spectra could reflect alterations in the nature of the bonding of a molecule to a surface and thus might show up in the spectra of the corresponding metal cluster. Exploratory studies of this hypothesis will be considered later.

Experimental factors

The importance of photoemission measurements for the determination of features of chemical bonding at surfaces is determined largely by the present experimental state of the art as well as by constraints intrinsic to the characteristics of the emission process. Details relating to measurements and design have recently been treated in detail (Eastman 1972b, Larabee 1974, Hercules and Hercules 1974, Carlson 1975, Rhodin and Brucker 1976, Feuerbacher and Fitton 1977b, Feuerbacher et al. 1978). It is useful to consider critically the particular experimental aspects which define or limit effort to draw useful conclusions pertinent to bonding at solid surfaces using these methods. Five features which deserve specific consideration are; excitation source, target surface, measurement chamber, energy analyser, supplementary surface probes, and data acquisition and processing.

Three very useful types of excitation sources are soft X-ray generators, gas discharge lamps and synchrotron radiation (SR). Their design and operating characteristics have been discussed (Carlson 1975, Sampson 1967). The two most important X-ray sources are the characteristic K_αX-rays of Mg and Al since they yield the narrowest K_αX-ray lines that can be obtained practically. For example the line width of the K_αMg line is 0.8 eV at 1254 eV and that of K_αAl is 0.9 eV at 1487 eV. Nevertheless, the 80–90% emitted at the above energy is mixed in addition with residual photon emission at somewhat different energies which must be removed with a monochromator to make more precise energy shift measurements. Although other useable "soft" X-ray sources including the Cu-K_α line at 8048 eV are available, the present tendency is to shift over to the use of a synchrotron radiation sources for energies in the soft X-ray region.

Characteristics of inert gas vacuum ultraviolet wavelength sources most widely used are summarized in table 3.1 (Sampson 1967). They are

TABLE 3.1

Source	Energy, eV	λ, Å
He II	40.8	303.8
He I	21.22	584.8
Ne I	16.85	735.9
	16.65	743.7
Ar I	11.83	1048.2
	11.62	1066.7
H (Ly α)	10.20	1215.7

readily adapted to UHV operations and are generally used without monochromators because one or two wavelengths are produced to the exclusion more or less of the others.

Synchrotron radiation sources provide a continuous wavelength beam of photons over an energy range starting at about 10 eV and going up easily to 1000 eV (Hodgson et al. 1976a, National Research Council 1976). The practical upper limit is determined in part by the luminosity–energy characteristics of the electron accelerator and in part by the interfacing of appropriate vacuum monochromators to the storage ring. Since the photons emitted internally in storage rings associated with accelerators now in the operational (or developmental) stages can go up as much as several hundred keV (Batterman and Ashcroft 1978), there is a large unexplored region of wavelengths waiting for the availability of appropriate storage ring facilities to probe new regimes of excitation modes and bonding on surfaces. This is indicated in fig. 3.12 which shows in a comparative scheme (a) experimental values of electron inelastic mean free path lengths, (b) the SR-continuous energy spectrum from the H-Ly α emission at 10.20 eV to the Al-K_α line at 1487 eV and (c) a proposed division into characteristic electron structure regimes, particularly the surface effects regime starting at about 20–30 eV. The wavelength region between 20–100 eV is of particular interest because it is within the energy range of many chemical bond-making or bond-breaking processes.

It should be pointed out that surface science research successfully carried out using electron storage ring facilities requires considerable effort and special instrumentation. Hence this kind of research is best limited to those measurements which fully exploit the unique features of this type of photon source (National Research Council 1976). Experimental objectives towards which synchrotron radiation measurements are most effectively applied will be discussed.

The nature of the target defines the special techniques most appropriate for study by methods of electron spectroscopy (Carlson 1975, Siegbahn 1976, Rhodin and Adams 1975a, Ertl and Küppers 1972). Since we are concerned mainly with solid crystalline surfaces, consideration is given to the problem of preparing and characterizing clean metal surfaces. The usual UHV procedures of ion bombardment, gas reactions at elevated temperatures, high temperature vaporization or cleavage are all applicable (Kane and Larabee 1974, Czanderna 1975). Problems of combining effectively the surface probes of Auger, LEED, SIMS, etc. with high resolution angle-resolved photoemission measurements especially in conjunction with synchrotron radiation require special designs

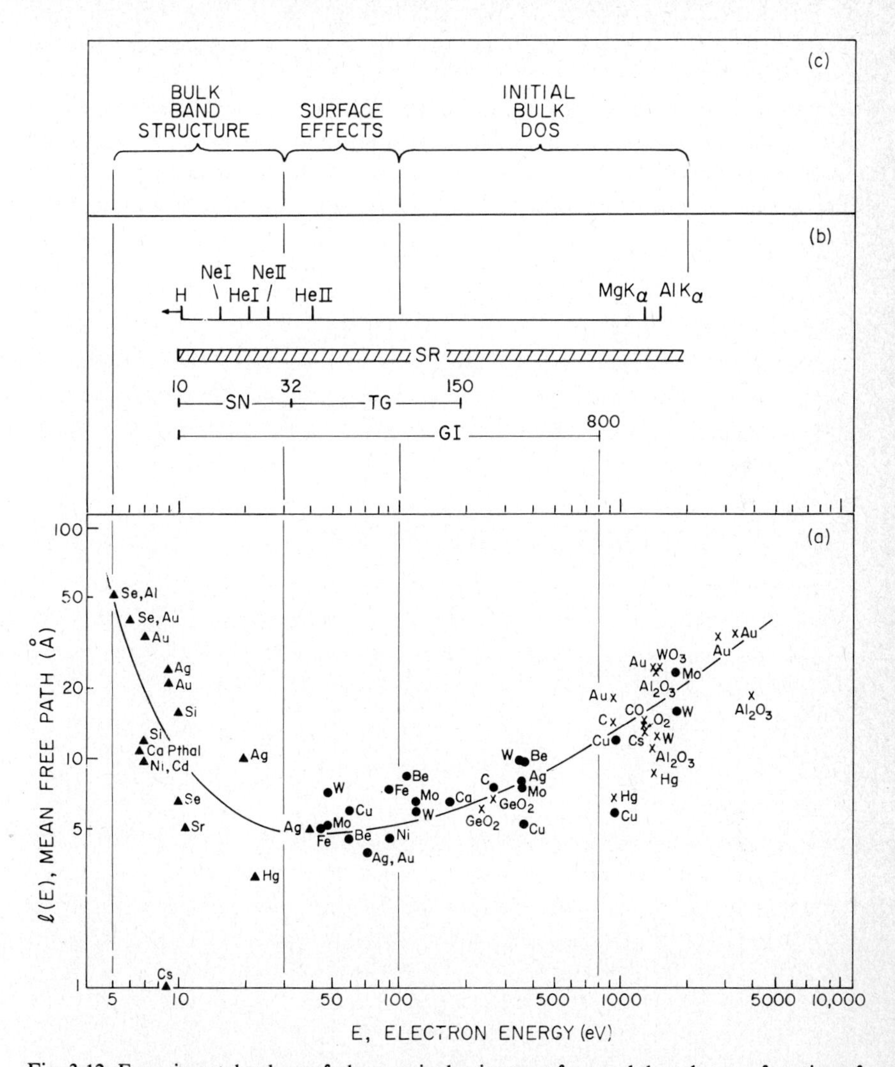

Fig. 3.12. Experimental values of electron inelastic mean free path lengths as a function of electron energy from UPS, XPS, and AES studies. Schematic comparison of the synchrotron radiation (SR) continuum with VUV and X-ray line sources: (≦10.2 eV), Ne_I (16.7–16.8 eV), He_I (21.2 eV), Ne_{II} (26.8 eV) He_{II} (40.8 eV), MgK_α (1253.6 eV) and AlK_α (1496.6 eV). Existing monochromators include the normal incidence Seya–Namoika (SN) and grazing incidence (GI) "grass hopper". Toroidal grating (TG) monochromators show promise of improved performance over the entire range $10 \leq h\nu \leq 2000$ eV. (Rhodin and Brucker 1977.)

of the probes and the manipulator (Koch et al. 1974, McGowan and Rowe 1976, Rhodin and Brucker 1977). It is also essential that all residual gas pressures be maintained sufficiently low to preserve a well defined surface clean of contaminants for the longer counting times often required for resolving relatively small spectral effects. One such scheme is shown in fig. 3.13 where the surface characterization is achieved at one level and the photoemitted electrons at a second level in

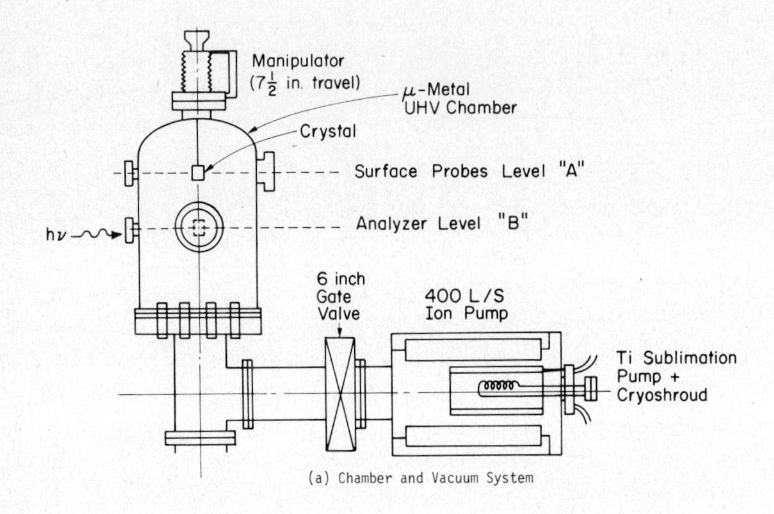

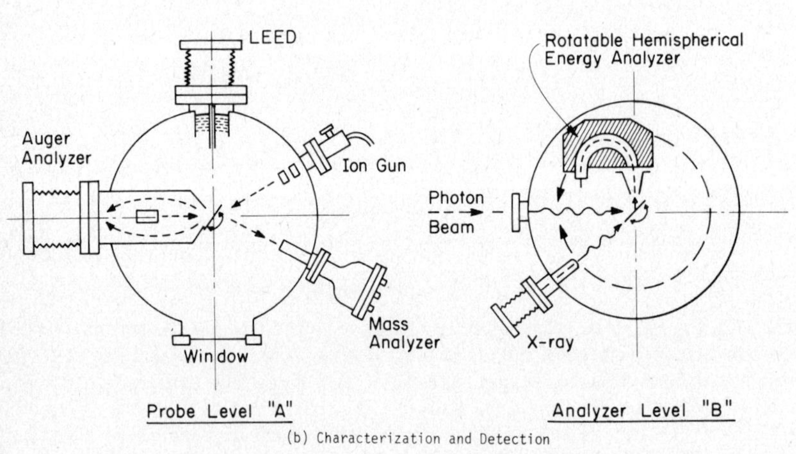

Fig. 3.13. Schematic of over-all system, (a) Vacuum chamber pumps (elevation) and (b) Characterization level "A", electron detection level "B" (Rhodin et al. 1978).

the measurement chamber. Provision for translational motion and spatial clearance of the probes and the manipulator is essential in this design. The target is positioned to obtain a predetermined incident angle of the photons and the electron detector is swung independently in both the azimuthal and polar planes relative to some chosen plane of incidence and to some symmetry direction in the target. It is, in addition, highly desirable to cool or to heat the target over wide temperature ranges, if possible without impairing the angular adjustability of the detector.

The gas pumping system on the chamber must be well shielded if it is sputter ion-pumped and should be well trapped if it is diffusion-pumped. Surface charging, especially for non-conducting solids, can contribute to one of the most serious interpretational hazards in photoelectron spectroscopy of solids, that of line broadening. Line broadening also results from the natural width of the photon source for soft X-ray generation and from instrumental effects in monochromators as well as from the intrinsic width of the subshell from which photoemission occurs. Charging problems for insulators can be minimized but can not always be completely eliminated by various procedures. Sharpness of the soft X-ray lines can be improved with monochromatization. Lifetime broadening associated with the vacancy created in the photoemission process decreases with an increase in binding energy. Hence it can be minimized by working at higher energies. In addition, the valence band in solids, particularly transition metals, is generally intrinsically broad. The net effect is that photoelectron peaks generally have a normal width of 0.5–1.0 eV regardless of the response function of the detection system.

The two most critical features of the measurement chamber other than its UHV capability are in providing adequate shielding of the electron detector (residual fields of less than 10 milligauss) and a geometry favorable to the optimum manipulation of the target and the electron detector. Mu-metal chambers provide the advantage of minimizing the degree of additional shielding required inside or outside the chamber and lead to a relatively simplified arrangement.

The characteristics of energy analyzers based either on the retarding potential or dispersion principles are well reviewed in the literature (Roy and Carette 1977). Although magnetic and electrostatic dispersing systems have comparable resolution and performance, the latter are usually preferred for cost and convenience. A convenient hemispherical analyzer is a 150 degree sector type with a 50–100 mm radius incorporating Herzog electrodes for correct field termination and interchangeable exit/entrance apertures to optimize intensity vs. resolution.

Factors to be optimized include the intensity of the emitted electrons accepted by the detector, the maximum resolution and the compatibility of positional adjustability over a wide range of angles in two dimensions inside the UHV chamber. One possible scheme is indicated in fig. 3.13. Special additional shielding of the analyzer is usually also desirable even in a mu-metal vacuum chamber. The electron detection devices in the analyzer best suited to UHV measurements and with adequate sensitivity and dynamic range are the channeltron or channel electron multipliers, devices based on glasses impregnated with metal.

The importance of supplementary probes for surface preparation and characterization has already been noted with reference to the properties of the target. It is essential that characterization of surface structure and composition be systematically monitored in conjunction with the photoelectron measurements to insure the most reliable interpretation of the spectra. Combined probes of this kind can be complex to install and cumbersome to operate. On the other hand, compromising them can lead to serious confusion and uncertainity in the interpretation of the surface phenomena. This is particularly pertinent to the study of rather small or subtle spectral effects on very reactive metal surfaces.

Emphasis is placed on the data acquisition system because with it one can monitor precise read-out of rather small difference effects which can be very important in photoemission studies of adsorption. In other cases, great quantities of data can be amassed rapidly such as in angle-resolved studies. Normalization and calibration of the data particularly in synchrotron radiation work is usually critical. Display of the spectral curves, their comparison, and statistical analysis are often essential to interpretation. A multiplicity of control or measurement data may result from combining the surface characterization studies with the photoelectron measurements. For all these reasons and others, interfacing and programming a small computer with the digital read-out of the electron detection device in the analyzer are desirable. A larger initial effort is required than for implementation with a multichannel analyzer, but the final effectiveness is significantly enhanced in scope and versatility.

3.3. Auger spectroscopy

Introduction

Auger electron emission involves basically a non-radiative transition to an inner shell by having one electron from a less tightly bound orbital fill a vacant hole while a second electron is ejected into the continuum with an energy equal to the difference in total energies of the initial and final states. It is essentially a two-electron Coulombic readjustment to the initial hole. Theoretical interpretation has already been considered. There are also accounts of the theory of the Auger process (Carlson 1975) with reference to its applications to compositional and chemical analysis. Although the Auger process is essentially an atomic phenomena, it is of particular interest to chemical bonding at solid surfaces because of its surface compositional sensitivity as well as its sensitivity to chemical environment. Although Auger spectroscopy is widely used for surface compositional analysis the Auger spectra are also highly characteristic of the state of chemical bonding. We will be primarily concerned with it here as a probe of local chemical environment. It appears that increasing applications of Auger spectral chemical shifts to chemical bonding will be made as more high resolution electron spectroscopy is applied to Auger spectra.

Chemical bonding effects are reflected in both the shape and intensity of the Auger peak as well as shifts in its energy. Because of limitations at present in our ability to fully interpret the Auger process, peak shapes are used as a sensitive but essentially qualitative indication of the chemical environment of the atom. Distinctive Auger shapes for the derivative plots are particularly characteristic of changes in the molecular nature of combination during the transition, for example, of a clean metal surface through various states of chemisorption to the formation of a surface metal compound. Chemical shifts tend to be significant but more complex for Auger than for XPS-photoelectron emission. Hence, the former is capable in principle of yielding chemical information equivalent in significance to that of photoelelctron spectroscopy. It requires that instrumental resolution be emphasized over signal-to-noise ratio. The latter is more generally stressed now in most AES set-ups used for compositional analysis.

The effects of chemical environment involve two important contributions to the chemical shifts in the Auger spectrum. The extra-atomic or dynamic relaxation, which represents the response of the localized atomic environment in the solid metal to the positive charge on the ionized atom, contributes to the fine structure in the spectra to the extent that it is not screened out by the conduction electrons. A second complication introduced by the metal environment is its influence on the lifetime of the core hole states which contribute to the linewidth. This lifetime depends on the valence electron density in the atom as well as on its local environment. Adsorbate line shapes and multiple lines in both XPS and Auger spectra are sensitive to the screening of the core hole by the metal electrons (Fuggle et al. 1978).

The chemical shifts in Auger spectra can reflect transitions involving both levels in the valence bond as well as core levels. In this case it has a wider implication than photoemission in either the vacuum ultraviolet (UPS) or soft X-ray (XPS) regions which tend to emphasize either valence or core level excitations, respectively. When Auger chemical shifts reflect primarily transitions involving valence levels, a rather complicated modification in the final state of the doubly charged molecular ion may occur. In summary, the more sophisticated but complex chemical shifts from Auger spectra are likely to include valuable core level effects as well as those from the valence band level. On the other hand, the UPS photoelectron spectra contain chemical shift information for chemisorbed layers, particularly for the nature of the chemical bond in these layers.

There are some additional problems associated with the observation of Auger electron emission as a technique for the interpretation of surface electron structure. The problems of radiation damage to the surface and back-scattering from the surface can lead to uncertainities in interpretation of the spectra. The degree to which this is important varies considerably with impact energy and with the angle of incidence as well as with whether electron or X-ray impact is used to excite the radiation. High energy satellite lines can originate from initial multiple ionization or resonance absorption in the gas phase. In addition, electrons emitted by excitation from a solid can suffer characteristic energy losses. These can be rather difficult to resolve since they are often complex and spread out over a rather large range of energy. Previous reference was made to the procedural problem associated with surface charging. It is common to electron spectroscopy from non-conducting surfaces in general, but is particularly severe for Auger emission because of the larger electron current normally involved.

Interpretations of the chemical shifts arising out of Auger transitions are not simple, as pointed out. Complications arise from the double ionization process involved as well as from combinations of valence and core level excitations. In addition, extra-atomic relaxation effects occur which can vary considerably depending on the metallic or dielectric nature of the target. Chemical shifts can be very informative on the other hand, when these complications are absent or accounted for. Shifts associated with electron negativity can be closely correlated, for example, with Auger transitions associated with core level excitations, Auger transitions associated primarily with transitions in the valence band can, at least in principle, be directly related to the density of states.

In summary, although the observation of chemical shifts is common to both photoelectron and Auger spectroscopy, more progress has resulted in the former case due mainly to the relatively greater ease in measurement and interpretation. There is, however, a great deal of useful information on chemical bonding present in Auger spectra which will become more useful as high resolution Auger electron analysis advances and interpretation of the complex variations of the Auger spectra become more clearly understood. A combination of UPS- and XPS-photoelectron spectroscopy with Auger peak shapes and positions has been applied to the study of the binding of carbon monoxide in metal carbonyl clusters in a comparison to chemisorption on extended metal surfaces (Plummer et al. 1978).

Experimental factors

Auger spectroscopy is usually an electron-excited core level spectroscopy where the emitted electrons are detected from solid surfaces under ultra-high vacuum conditions in a manner similar to that used for the photoemission measurements. It should be noted, however, that the background in the Auger spectra is large, e.g. the signal-to-noise ratio is poor, due to the electron excitation. It must be either suppressed or subtracted. Although the stimulation involves high energy electrons (1–50 keV) rather than photons, there is a strong similarity in target preparation, in the beam-sample geometry and in the nature of the electron emission process itself.

Dispersive detectors are superior especially if the acceptance angle is not very large. A high resolution spectrometer with preretardation of the emitted electrons is often used in conjunction with a cylinderical mirror analyzer (CMA). Resolution is enhanced at higher electron energies by keeping the primary energy and the pass energy constant and varying the retarding potential on the analyzer.

Characteristics and design of electron detectors are well covered in the literature (Roy and Carette 1977). In addition to the CMA, a very useful type of band pass detector is the hemispherical deflection analyzer (HDA). It is characterized by a superior limiting energy resolution (up to 10 meV) which is particularly appropriate for the precise measurements of chemical shifts. There exist definite advantages in the observation of fine structure in the Auger peaks to be gained by the use of a band-pass type of detector since they avoid the need to utilize differentiation-enhanced signal detection. Useful comparisons between Auger (AES) and photoelectron techniques (XPS) for the observation of chemical bonding can be made as indicated. They both examine a region of the surface limited by the mean free path of the emitted electron. Although initial signal intensities are larger for Auger emission, the superior signal-to-noise ratio for the photoelectron results in comparable sensitivity for both. Both methods can be combined effectively with other types of surface probes but not without some complications associated with beam focussing and spatial constraints associated with the detectors themselves. There is no doubt that chemical bonding effects are easier to interpret from the peak shifts in the photoelectron rather than the Auger spectra. This could change with the development of a more sophisticated theory of the Auger emission process from solids and the more effective use of directional-sensitive high energy resolution analyzers.

3.4. Electron energy loss spectroscopy

Introduction

The study of phonon interactions at clean and chemisorbed surfaces has long been recognized as critical to the understanding of bonding and chemical reactivity at metal surfaces (Rhodin and Adams 1975b general introduction, Ibach 1977b, Froitzheim 1977 detailed review). With a few notable exceptions, e.g. the early work of Propst and Piper (1966) and the subsequent significant advance by Ibach (1970), as well as contributions by Andersson (1977a, b) and others (Bertolini et al. 1976, Backx et al. 1976), progress in experimental investigation using high resolution energy loss spectroscopy of surfaces and adsorbed layers has been used less broadly than other methods in applications to solid surfaces. Surface phonon effects, originally studied using infrared techniques (Greenler 1966, Pritchard and Sims 1970, Shigeishi and King 1976) or inelastic electron scattering experiments (Plummer and Bell 1972, Lambe and Jaklevic 1966) are not in general readily applicable to UHV studies of well defined surfaces because of limitations in execution and interpretation.

Recently significant advances have occurred in development, both in the theory of collective and localized surface vibrational modes (Wallis 1968) and in the measurement of very small but discrete phonon losses on well defined surfaces (Ibach 1977b). Comparisons between theory and experiment are no longer restricted to thermodynamic average properties but are related in detail to microscopic surface parameters and models of chemical bonding and of surface reactivity. It is, therefore, relevant to a discussion of surface chemical bonding to consider in detail some recent advances associated with the application of high energy resolution electron energy loss spectroscopy of solids applied specifically to measurement of vibrational modes associated with atoms and molecules at solid surfaces.

A complete and general theoretical description of the elastic and inelastic losses in energy and momenta suffered by an electron upon passage through a metal is a difficult problem that is not yet completely developed. Two theoretical treatments which have proved to be useful for the interpretation of electron loss spectroscopy studies for the characterization of metal surface are related to a classical description of the dielectric theory, reviewed by Raether (1965) and by Geiger (1968) and a full quantum mechanical description of the dielectric theory presented by Evans and Mills (1973) and by Mills (1975).

The classical description strictly within the dielectric theory approach treats the moving electron as a point charge under conditions which can be shown to apply quantitatively only for transmission experiments. It is significant to note, however, that a semiclassical description within the dielectric theory framework developed by Lucas and Šunjić (1972a, b) describes quite well the early results of Ibach (1972). In this case, the electron is dealt with as a particle, the crystal surface as an infinitely repulsive barrier, and the effects of elastic scattering on the scattering efficiency are neglected. The wave vector conservation is introduced kinematically into the treatment and only single excitations of a collective plasmon or phonon mode by a relatively fast moving electron are considered.

A full quantum mechanical treatment within the dielectric theory framework in which the quantization of both the electron and the lattice restrictions on the nature of the elementary excitations are treated in full is now available (Evans and Mills 1973, Mills 1975). The nature of this treatment regarding the interpretation of experimental data on surface vibrational losses is discussed (Ibach 1977a) in terms of contributions from the excitation of optical surface phonons in two general types of matter, infrared-active and infrared-inactive. Other contributions from the excitation of plasma waves and from electronic surface transitions are also considered. Comparisons between theory and experiment have been fruitful but additional refinements of theory will be required as experimental studies increase in sophistication. Some of these points will be considered later with reference to specific experimental applications.

Experimental factors

To perform electron loss studies effectively on well defined solid surfaces requires the use of spectrometers with an overall resolution of at least 10 meV and with design parameters which permit stable operation including angular rotation over a reasonable range within the UHV system. It is also required that the sample surface be readily prepared and characterized in situ. It is often desirable to have the capability of making other supplementary measurements such as LEED, Auger, SIMS, etc. on the same surface. The effective sensitivity and energy resolution are critical parameters. Angular resolution can also be important where momentum effects are involved. The large effort devoted to the design of high performance electron monochromators and analyzers for the purpose of surface analysis has been reviewed (Roy and Carette 1977). Some of the more important features will be noted here to put the technique and its performance into perspective. Two spectrometer designs which combine performance with relative ease of construction and operation in an optimum manner for surface studies are based on either the cylinderical deflection analyzer (CDA) or on the spherical deflection analyzer (SDA).

A typical spectrometer (Froitzheim 1977e) such as that used by Propst and Piper (1966) and by Froitzheim and Ibach (1974) for high resolution of low energy electron losses from solid surface in UHV, is composed of a monochromator and an analyzer stage in tandem, based on the cylinderical deflection principle. It involves an electron gun, an electron monochromator, a chamber containing a sample which can be heated, cooled and positioned, a rotatable energy analyzer, a high response electron multiplier and appropriate electron optics for optimizing the trajectories of the electrons into, through, and out of the spectrometer. The fundamental design used for the monochromator and analyzer is of the cylinderical deflection type utilizing gridded electrodes and low operating energy. Effective mu-metal shielding to minimize interacting magnetic fields is essential. Current measurements are carried out with a resolution of 10 meV and a useable electron current of about 5×10^{-9} A. An effective and somewhat simpler version has been effectively used by Andersson (1976, 1977b) in combination with LEED studies. It operates at primary electron energies in the range 1–2 eV with a resolution of 7–14 meV. The absolute energy calibration is accurate to 0.5% for a sweep in the energy loss range of 250–500 meV.

Although spectrometers based on the spherical deflection principle possess superior focussing and energy-dispersing properties to the cylinderical deflection ones they tend to be more complicated in design and operation. Comparisons of these analyzers are available (Read et al. 1974). A well developed prototype of the spherical deflection principle is the Simpson–Kuyatt design (Simpson 1964) which has been extended in various versions to vibrational spectroscopy as well as to applications in field-emission (Kuyatt and Plummer 1972) and photoemission spectroscopy (Allyn et al. 1978). An effective spectrometer for studying surface phonons and vibrational losses of adsorbed molecules can be effectively based on either design. It will possess in general an overall resolution of 7–30 meV relatively insensitive to the primary energy which in turn may be varied from 1–10 eV. The angular resolution is about 1.5° in the horizontal and vertical direction with respect to the plane of incidence. The angle of incidence can be varied by rotating the sample between 45 and 90° and the angle of emission by rotating the analyzer within the same limits. It should be stressed that although reliable performance at the levels of the above parameters is essential to effective measurements of surface phonons effects, high resolution electron energy analysis is not so readily attained. There are a limited number of such spectrometers which are operating effectively. The factors of electron optical alignment, of magnetic shielding, of angular rotation, and of sample preparation under stringent UHV conditions (1×10^{-10} torr or less) require optimization. Resolution can also be limited by space charge effects in the monochromator and non-uniform work function effects in both the monochromator and the analyzer.

In summary, EELS studies of chemisorbed metal surfaces can yield unique information on the microscopic structure of the adsorbate, the symmetry and strength of the surface bond and the nature of the binding potential at the surface. Accurate changes in work function are measurable and the measurements are all made in a non-destructive manner. EELS results can be particularly effective when combined with other surface probes to achieve more refined interpretations of the microscopic nature of bonding modes and symmetries at surfaces.

4. Typical experimental applications

4.1. Introduction

This section deals with the nature of surface chemical bonding in terms of application for different specific metal–gas systems starting with consideration of the clean surfaces of nearly-free-electron and transition-metals. It proceeds to consideration of both atomic and molecular interpretations of metals in a sequence from the nearly-free-electron metals to the 3d-, 4d-, and 5d-transition metals. Discussion is focussed on observations and conclusions relevant to the nature of surface chemical bonding in terms of specific metal–gas systems using pertinent electron spectroscopy data.

4.2. Surface structure of clean metals

The principal use of UPS-spectroscopy for free molecules in the gas phase has been the investigation of molecular orbitals, whereas the principal application to clean metals relates mainly to the determination of band structure. Its relevance to surface chemical bonding involves a combination of both. Much of the earlier work, stimulated by the efforts of Shirley (1972), Eastman (1972a), Spicer (1970) and others, was initially limited to conduction band studies by the nature and maximum energy cut-off of the available excitation sources. Since then with the availability of more efficient gas discharge and X-ray tubes and, particularly of continuously tunable plane-polarized radiation up to much higher energies from synchrotron radiation sources, both core level and valence level excitations are now being effectively employed to evaluate the electron structure of both clean and chemisorbed metal surfaces. Reference to the literature is recommended for details (Koch et al. 1974, Carlson 1975, Siegbahn 1976, Joyner 1977).

Photoemission spectroscopy contains much information on the energy and momentum distribution of the excited electrons that escape into the vacuum from both the conduction and valence bands*. It is sensitive to the influence of adsorbates at the surface whose presence is reflected not only in a redistribution of the metallic electrons in the bands but also in the presence of perturbed features of the adsorbate molecular orbitals. Since some of the most significant photoemission results for chemical overlayers relate directly to the surface orientation and bonding of molecules, some very useful comparisons can be made to both vibrational data (EELS) and surface crystallography data (LEED). Although the Auger spectra contain, in principal important chemical information

* Originally, photoemission results distinguished between valence level spectra associated with excitation energies up to about 40 eV and core level spectra associated with excitation energies above about 1400 eV. The sharp distinction between UPS- and XPS-spectroscopy disappears with the availability of continuously tuneable radiation over the complete energy range.

on binding of the core electrons, it has not yet been widely applied for this purpose.

The nearly-free-electron metals such as aluminum, zinc, cadmium, and the free electron metals such as the alkali and alkaline earth metals have simpler electronic structures than the transition metals. On the other hand, their surface-related electron spectra have not been extensively studied by electron spectroscopy methods partly because their electron structure is less sharply delineated than for the transition metals, partly because their chemisorption effects are less striking and partly because study of their surface properties can be experimentally difficult. Selected physical electronic properties such as electron–electron scattering and collective electron effects have been explored, but very little of this has been related directly to chemical bonding. Some UPS-photoemission studies up to 1970 of the valence band of alkali metals and of aluminum (Eastman 1972a) were done in a photon energy range limited to the lithium chloride window cut-off of 11.6 eV. With the availability of higher photon energy sources and sophisticated UHV surface preparation and characterization methods, chemisorption studies of these metals using both valence level and core level spectroscopies is proving to be fruitful. Core level photoemission spectroscopy of the groups IA and IIA elements as well as of Al has been devoted mainly to chemical shift studies of their compounds as a function of degree of ionicity (Carlson 1975).

There is extensive literature on the UPS-photoemission and on the electron properties of solid metals (Shirley 1972, Eastman 1972a, b, Koch et al. 1974), particularly with reference to the polycrystalline surfaces of noble and transition metals. Although related to elucidation of surface chemical bonding, much of the earlier work is not readily interpreted in terms of microscopic parameters because it is not based on surfaces with well characterized properties. More recent photoemission studies in UHV systems have been pursued on a variety of well defined single crystal surfaces of simple metals, noble metals and transition metals from the 3d-series, 4d-series and the 5d-series (Castex et al. 1977, Willis et al. 1977, Kumagai and Toya 1974, Dobrozemsky et al. 1977). Of these only a limited number of clean single crystal metal surface structures have been systematically examined using energy- and angle-resolved photoemission spectroscopy, e.g. copper, nickel, molybdenum, silver, iridium, and gold (Feuerbacher et al. 1978, Dobrozemsky et al. 1977, Liebsch 1978b). Energy- and angle-resolved photoemission spectra have been done systematically on atomic adsorbates (oxygen and sulfur) and on molecular adsorbates (carbon monoxide) for nickel, iridium, platinum and tungsten single crystal surfaces (Willis 1977, Dobrozemsky et al. 1977, Liebsch 1978a). It is likely that angle- and energy-resolved studies will be pursued on a wide number of clean and chemisorbed metals in the future over a broad energy range.

Reference to the preceding section on theoretical formulation and interpretation is recommended for a discussion of the formulation required for determining the differential cross section for emission from a clean surface. The essential objective is to develop a description of the frequency-, polarization-, and angular-dependence of the photoemission in terms of a simplified band structure description of the metal surface usually within the one-electron theory (Liebsch 1978a, Schaich and Ashcroft 1971, Feibelman and Eastman 1974). The first step is then to extend the description of the effects of photoemission to interpret the electron interactions between atoms, molecules and overlayers and metal substrates (Liebsch 1978a, Dill and Dehmer 1974, Davenport 1976).

It should be noted that the role of the photon excitation energy can be important in discriminating between the relative contributions of the initial and final states. The former corresponds approximately to the initial photoionization process and the latter to the scattering and diffraction of the photoelectron on the way out of the metal, respectively. The initial state effects tend to become more emphasized, for example, on copper at photon energies above about 130 eV. The surface effects are most noticeable at lower photon energies of between 30 and 100 eV. The mean free path of the electron is most favorable to escape from the surface region in this range, as indicated in fig. 3.12. The uniqueness of synchrotron radiation provides the capability to tune the photon energy to emphasize, selectively, the bulk band structure effects, the surface effects or the initial state effects as desired. The significance of these results for chemisorption refers to the usual procedure where one subtracts away the clean surface contribution in order to isolate and emphasize the adsorbate-induced spectral features to a first approximation. In the case of wide d-band metals like tungsten, platinum and iridium, complications arise in the interpretation of the spectra because both band structure and surface effects can be important at photon energies as low as 10 eV.

Since from the viewpoint of surface bonding we are particularly interested in how atoms and molecules perturb the surface component of photoemission, the relative magnitudes of the bulk and surface contributions to the amplitude of the photoemission is important. Measurements of total photoemission yield from nearly-free-electron metals like aluminum and magnesium (Flodström and Endriz 1975, Flodström et al. 1975) indicate that the surface contribution becomes dominant at low photon energies for p-polarized light at high angles of incidence. Photoemission studies of the valence band region for the (100), (110) and (111) faces of high purity aluminum (Bachrach et al. 1977) show deviations from a simple free-electron structure associated with intrinsic surface resonances. These states are particularly sensitive to observation by angle-resolved measurements (Gartlend and Slagsvold 1977). Development of significant additional fine structure in both the core levels and in the valence band of oxygen is indicated by the work of Flodström et al. (1977a, b). Experimental and theoretical effort on the nearly-free-electron metals like aluminum (Pendry 1977) has been much less than that for noble metals such as copper (Nilsson and Ilver 1976, Wehner et al. 1977).

Extensive comparisons between theory and experiment are available for single crystal copper surfaces (Lloyd et al. 1976, Ilver and Nilsson 1976, Pendry 1976, Pendry and Titterington 1977) in terms of angle-resolved plane-polarized, frequency-variable parameters for Cu(001) and Cu(111). Much of the spectral structure can be interpreted using existing one-electron calculations corresponding to coherent superpositions of many transitions in the d-band. The relative amplitudes of these various transitions can be correlated with polar and azimuthal angles of emission on which they show a strong sensitivity. A comparison of calculated spectra by Pendry (1976), Feuerbacher et al. (1978) with various experimental data in terms of the polar and azimuthal angles of emission is indicated in fig. 3.14 for the clean Cu(100) surface. Good agreement, generally obtained between measured and calculated spectral results for Cu(100) and Cu(111), suggests that analysis of peak shapes and intensities can indeed provide information on the surface electron structure for a "typical" noble metal.

Ultraviolet photoemission is particularly well suited for probing the behavior of electron states unique to the surface region, due to its high surface sensitivity. In addition, the existence of surface states has been strongly indicated for the Cu(111) surface by both experimental (Lloyd 1976, Ilver 1976) and theoretical (Pendry (1976, Feuerbacher et al. 1978) UPS-photoemission results. It is likely that angle-resolved photoemission will be particularly fruitful in studying the momentum-dependence of electron emission associated with electron states sensitive to the presence of the surface. Recent theoretical (Kar and Soven 1976) and experimental (Plummer 1975a, Billington and Rhodin 1978) studies using both field electron emission and UPS-photoemission point strongly to

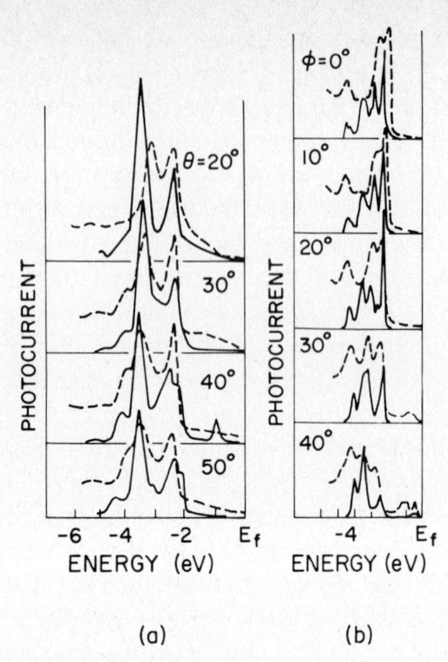

Fig. 3.14. Calculated (solid curves) and measured (dashed curves) photoemission spectra for Cu(001) taken at a photon energy of (a) 21.2 eV and (b) 16.8 eV. Unpolarized light is normally incident. In (a), the polar emission angle is varied, the azimuth is in the (110) direction; in (b), the polar angle is held at 45° and the azimuth is varied. (Nilsson and Ilver 1976, Pendry 1977.)

the presence of the surface state at about 0.4 eV below the Fermi level for Cu(111) (Heinmann et al. 1976b). Good evidence also exists for the existence of surface states in the gaps of the (111) surface of silver and gold. Reasonable potential models lead to approximately the correct energy and dispersion with momentum parallel to the surface, as indicated by angle resolved studies on copper (Fadley et al. 1976). It should be noted in these comparisons that the polarization of the photon field can have a strong influence on the angular dependence of the cross sections for photoionization. Hence measurements using photon beams with well defined polarization are useful (Sass et al. 1976, Feibelman 1975a, b). The use of photon excitations of well defined polarization over a range of energies not only provides a good measure of the surface localization of the electron emission but can be effectively used to define both the photoemission process itself as well as the orientation of chemisorbed molecules (Liebsch 1978a, Feuerbacher et al. 1978).

Since the earlier measurements (Eastman 1972a, b), many additional UPS-photoemission spectra have also been made on both bulk and surface emission for nickel (Williams and Norris 1976, Heimann and Neddermeyer 1976). Energy band calculations lead to the result that the bulk and surface electron structure of nickel is probably one of the better understood of the ferromagnetic transition metals. It is not surprising that many of the more useful surface chemical studies (to be discussed later) using UPS-photoemission spectra have also been done on nickel. Many similarities exist in the d-band structures for the different transition metals. Characteristics of the d-band for example, obtained by UPS, XPS, and band calculations are summarized in table 3.2 (Eastman 1972c) for selected transition metals from the 3d-, 4d-, and 5d-groups of the periodic table. Studies of nickel surfaces have been extended in depth and are being continuously expanded using angle-resolved photoemission with synchrotron radiation to higher photon energies (Eastman 1972d). Some interesting results, for instance, have been obtained on the ferromagnetic exchange splitting of peaks just below the Fermi level (Williams and Norris 1976) and on the role of initial state emission (Heinmann and Neddermeyer 1976) from the Ni(111) surface. This indicates the future possibilities of extending

TABLE 3.2
Summary of d-band widths obtained by UPS, XPS, and band calculations (Eastman 1972d)

Metal	UPS widths[a)]	XPS widths[b)]	Calculated widths[a)]
Cu	3.0	2.9	3.3
		2.7	
Ag	3.5	3.5	3.2
		3.5	
Au	5.7	5.4	5.2
		5.7	
Ni	3.3	2.7	5.0 eV
		3.0	4. eV
Pd	4.6	4.5	4.5
		4.1	
Cr	4.2	—	4.2
Ti	1.7	—	1.6

a) Widths for UPS measurements and calculated DOS curves are taken as the full width at ~10–15% of maximum intensity (after subtracting secondary emission from the emission spectra).
b) XPS widths are defined as full widths at half maximum (FWHM); this definition is reasonable in view of the ~1 eV XPS resolution.

characterization of exchange splitting in the clean surface to the influence of adsorbates.

A great number of UPS- and XPS-photoemission spectral studies of single crystal gold have been made with the main interest being in evaluating the photoemission process and relating the emission to band structure caclulations. Angle-dependent studies have shown core and valence level band structure effects in good agreement with calculations (Fadley et al. 1976). Surface chemical effects, however, on the noble metals, (copper, silver, and gold) have been pursued to a lesser degree using mainly photoemission because the associated chemisorption effects are less spectacular than for the transition metals. Molybdenum is an interesting 4d-transition metal of interest since the bulk contribution of photoemission tends to dominate over surface effects, as might be predicted (Gartland 1976). Moreover, surface states are predicted just below the Fermi level for the (100) and (110) orientations. Angle-resolved photoemission studies of surface electron structure using synchrotron radiation indicate the occurrence of some surface resonances, but are not definitive so far as evidence for the existence of surface states (Guillot et al. 1976). This makes systematic angle-resolved photoemission studies of surface chemical effects on molybdenum single crystals as a function of photon energy of particular interest.

Among the 5d-transition elements, the clean surface electron structures of iridium, platinum and gold particularly have been studied using angle-resolved and/or synchrotron radiation (Castex et al. 1977b). Since their band structures are rather broad (6 to 8 eV) and complex it is important to define clearly the relative effects of bulk and surface band structure contributions from the clean surfaces. This is particularly important for iridium and platinum, the surface chemistry of which are of particular practical interest. A typical band structure regime from angle-averaged spectra for the iridium 5d-bands of a (111) surface is illustrated in fig. 3.29a. There are many features in the spectra which change in position and shape as a function of photon energy in conformance with the selection rules governing the conservation of crystal momentum. Certain spectral behavior is reminiscent of direct transition effects observed for gold and copper. The existence of considerable variation of spectral structure with surface crystallography indicates the likelihood of enhanced surface effects, particularly for the reconstructed Ir(100) (5×1) surface as well as volume effects in the energy range up to 40 eV. Similar results with some attenuated effects of fine structure in the spectra have also been reported for platinum (Apai et al. 1976). They clearly establish the need for additional systematic angle-resolved spectral studies for clean iridium and platinum over a wide range of excitation energies as a foundation for interpretation of chemisorption on the surface band structure.

It is interesting to note that XPS-spectroscopy can also be used to study angular dependencies at high energy resolution in the energy distribution curves of photoelectrons to elucidate electron structure. This was previously discussed with reference to Cu(001) (Fadley et al. 1976). A direct transition model gives a good first order description of the angle- and energy-dependent spectral features (Wehner et al. 1977). The effects of angular studies on surface analysis using XPS-spectroscopy have been discussed in detail by Fadley (1974). Some high resolution Auger spectroscopy studies of surface electron structure using X-ray excitation have also been explored for magnesium, tungsten and ruthenium (Fuggle and Menzel 1977). It is not clear, however, how generally useful this approach will be for the study of surface chemistry. The relative effectiveness of chemical shift observations using Auger shifts over XPS-chemical shift studies have been compared (Fuggle et al. 1976a, b).

The spectroscopy of inner shells is well covered by Carlson (1975). Chemical shifts associated with core level excitations relate to chemical states of the metal and are best considered with reference to specific chemisorption or chemical reaction effects on metal surfaces. The alkali metals, having only a single oxidation state, are somewhat less interesting from this viewpoint. The transition metals have received more attention because of the wealth of oxidation states and bonding configurations characteristic of formation of a wide variety of surface- and organometallic-complexes. The multicomponent structure in the spectra associated with multiplet splitting and shake-up effects can provide useful information on chemical bonding. Interpretations of the trends in chemical shifts with electron structure as well as the multicomponent structure in the spectra can, however, be difficult. Theoretical understanding of their nature is only now becoming established. It is important to note that since core binding energies depend on the crystal potential through the influence of both adsorbed atoms as well as conditions of stress and dimensionality; such measurements are sensitive to the state and composition of the surface. The factors of line width, of surface charging (for insulators), of multiple chemical coordination (e.g. terminal vs. bridge bonding) and of resonances can all play a role in the interpretation of the core-excitation spectra.

Finally, characteristic energy losses associated with bulk or surface plasmon losses are important. In that sense XPS spectra of solid metals are to a large extent a reflection of the surface region, the depth of which is directly related to the limited energy-sensitive mean free path of the photoelectrons (see fig. 3.12). Fadley and Bergstrom (1971) have studied the effects of angular measurements on surface analysis. The effect of channeling along preferred crystallographic directions, as reflected in the polar angle dependence of the emitted intensity, is well established (Fadley and Bergstrom 1971). Brundle (1975b) and others have emphasized the importance of combined data on both core and valence level excitations, e.g. concurrent XPS- and UPS-photoemission measurements as well as the importance of other surface characterization parameters, to optimize interpretation of the relative contributions of these different surface and bulk contributions to the spectra.

In summary, it is clear that useful information on surface band structure and surface states can be obtained from angle-resolved UPS- and XPS-photoemission studies of clean metal surfaces. It is of special importance for chemisorption and chemical reaction studies that the clean metal surface structure be well characterized to facilitate reliably the interpretation of molecular bonding on the surface.

4.3. Atomic bonding on metal surfaces

It is important in the development of a general theory of surface chemical bonding to understand in detail typical systems in terms of their electronic, compositional, and crystallographic microscopic parameters. Although this remains to be fully accomplished, it is informative to review specific cases where significant progress has been made towards this end. In all cases to be considered, at least one, and usually several of the electron spectroscopy techniques previously described have been effectively employed. Of the many gas–metal systems to be discussed from this viewpoint, one of general importance pertains to the dissociative chemisorption of electronegative diatomic gases on transition metals. The gases most widely studied to date are hydrogen, oxygen, nitric oxide, hydrogen sulfide, hydrogen selenide and carbon monoxide. The metal surfaces most extensively studied are nickel, tungsten and copper. Whereas different aspects of bonding are probed by the various techniques described, in no single case are all the data needed to support a given model available. Certain gas–metal systems have been chosen to illustrate the specific types of information available. There is no intention of presenting a complete review of the extensive information on the subject. Reference to the literature is recommended for details (Kumagi and Toya 1974, Gomer 1975a, Dobrozemsky et al. 1977, Hannay 1976).

Oxygen on aluminum

Aluminum single crystal surfaces offer an ideal example for the study of the surface chemistry of a nearly-free-electron metal. Until recently, relatively little definitive work was done on this system mainly because of experimental difficulties. Surface resonances have been measured using photoemission (Flodström et al. 1977a, b, Fuggle et al. 1976a, b). LEED studies of low index single crystal surfaces provide no evidence for the existence of ordered overlayers of oxygen at room temperature. The electronic structure of chemisorbed oxygen on aluminum has been calculated using a tight-binding method by Batra and Ciraci (1977a, b) and found to be in agreement with UPS-spectral data. Photoemission studies (Flodström et al. 1977a, b) over the appropriate photon energy range ($\hbar\omega = 100$ eV) have provided significant new information on the existence of multiple oxidation states in terms of substrate core level shifts. The oxygen adsorption process has been followed in terms of the Al-2p core level shift and of the nature of the oxygen-2p valance level structure (see fig. 3.15). An intermediate oxidation state has been reported for Al(111) and bulk-like amorphous oxides for the (100) and (110) surfaces (Flodström et al. 1977a, b).

These observations are important in defining the nature of intermediate chemisorption states using core level shifts to study the atomic interaction of a reactive gas for well defined nearly-free-electron metal

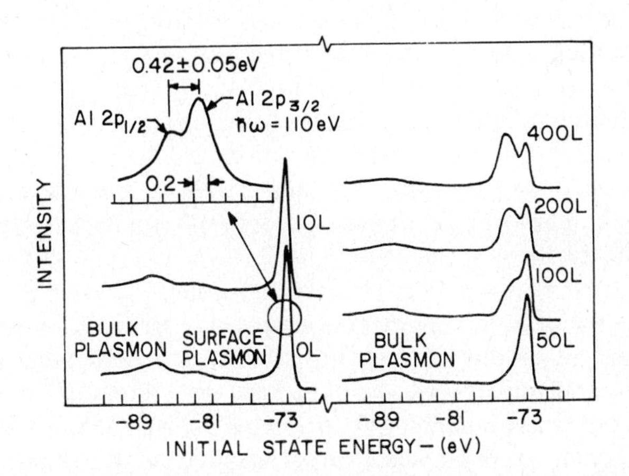

Fig. 3.15. Photoelectron spectra of the Al-2p region for clean and oxygen-exposed aluminum films at a photon energy of 170 eV. The oxygen exposures are given in Langmuirs ($1\,L = 10^{-6}$ Torr·sec). The expansion shows the resolved Al 2p doublet obtained at a photon energy of 110 eV (Flodström et al. 1976).

surface. Angle-resolved studies at selected photon energies of oxygen overlayers on Al(100) reveal further interesting features of the electronic structure, such as splitting and broadening of the O-2p levels. Adsorbate–substrate interactions in the initial oxidation of single crystal faces of aluminium have also been studied to some advantage using Auger photoyield spectroscopy (Bianconi et al. 1977).

Oxygen and sulfur on nickel

The sensitivity of the UPS-photoemission technique for following adsorbate energy level changes with exposure was first effectively demonstrated using angle-integrated emission by Eastman and Cashion (1971) for adsorbed oxygen on evaporated nickel film as a function of exposure at room temperature. The chemisorption and dissolution of the dissociated surface oxygen corresponding to variations in exposure and temperature were readily followed in terms of the intensity of the O-2p level emission. This experiment and others like it set the stage for the very effective development of UPS-photoemission spectroscopy as a tool for tracking the electron structure of metal surfaces and of surface molecules during adsorption and reaction. Since then angle-and-energy resolved photoemission has emerged as a powerful technique in the study of surface electron properties. Proposed schemes now exist by Liebsch (1976), by Gadzuk (1974a, b, c) and by Doyen and Grimley (1976) for calculating the essential features of the surface electron structure, the spectra and the photoemission process itself. The current state of photoemission theoretical analysis, as discussed in a preceding section, involves some difficulties in defining the surface potential, relaxation effects, and the spatial dependence of the vector potential. Nevertheless, the characteristic variation in the adsorbate induced resonances can be predicted sufficiently well to identify orbital symmetries and adsorbate–substrate interactions (Liebsch 1978a, Feuerbacher et al. 1978) .

Nickel surfaces have been increasingly singled out for systematic study of chemisorption parameters for several reasons. Among these are included the facts that it is a typical 3d-transition metal, the band structure has been studied in detail, LEED structures of the chalcogens have been established, conditions for preparing well defined surfaces are well documented, many gases interact strongly with it both atomically and molecularly, and it possesses interesting heterogeneous catalytic properties. Systematic studies of adsorption parameters on this system have been discussed up to 1975 by Rhodin and Adams (1975a).

More recent studies using angle-resolved photoemission are reported

(Allen et al. 1977, Petersson and Erlandson 1977, Guillot et al. 1977). Some interesting new angle-resolved photoemission data on chemisorbed c(2 × 2)O and c(2 × 2)S structures on Ni(100) by Weeks and Plummer (1977a, b) have been analyzed using dynamical LEED calculations by Tong et al. (1977). The spectra were measured for various set angles of incidence of unpolarized light at 21.2 eV. Very different dependence on the angle of incidence was observed depending upon which components of the vector potential of the light were responsible for emission. The phase relationships between components of the vector potential parallel and perpendicular to the surface are important in determining the effective orientation and magnitude of the vector potential at the emission site. In a related study (Weeks and Plummer 1977a, b) evidence was also obtained on the angle-resolved spectra to separate out two binding configurations: one where the sulfur was on the surface and a second one where the sulfur had diffused into the substrate within the mean free path of the photoelectron. These results are consistent with two different binding sites for sulfur at the Ni(100) surface and suggest a promising method for observing bonding effects associated with bulk-surface diffusion. Other efforts to interpret angle-resolved photoemission results in terms of calculations of combined initial and final state effects for nickel atom–oxygen or sulfur atom clusters of various configurations and sizes are critically discussed by Liebsch (1978c) and by Scheffler et al. (1976, 1978).

The recent dynamical calculations of Tong et al. (1977) for the above data show encouraging agreement for the dependence of the photoemission intensity as a function of emission angle, plane of collection, photon incidence, photon polarization and azimuthal plots. It is apparent from such calculations applied in detail to experimental results, that valuable information can be obtained on the individual emission properties of different surface orbitals. This occurs because the calculations separate the contributions from orbitals with different symmetries. It is also significant that the oxygen and sulfur systems have very different emission properties. The apparent success of the relatively simple scattering model employed in the calculation suggests that such calculational methods may be even more useful in future analysis of chemical bonding from angle-resolved photoemission data. This example illustrates much potential for the characterization of clean and chemisorbed surface layers by this approach.

The important possibility that surface crystallographic data might be obtained from angle-resolved UPS-photoemission was formulated by Liebsch (1974) and by Gadzuk (1975a, b, c). The dependencies of photoionization cross sections on the incident photon energy and the polar angle of emission have now been effectively used (Li and Tong 1978) to determine the location of adsorption sites for the c(2 × 2) S–Ni(100) and CO–Ni(100) adsorption systems. It is found that, whereas the angle-resolved photoemission polar angle dependence for surface crystallography can be very sensitive to the coordination number of the adsorption site, it is rather insensitive to the overlayer–substrate interlayer spacing normal to the surface. Angle-resolved UPS-photoemission and LEED apparently probe different aspects of the surface crystallography. The former is particularly sensitive to local directions and symmetries of surface bonding, whereas the latter is a sensitive reflection of the multiple scattering associated with the long range order of atoms on the surface. For example in fig. 3.16a the sensitivity and shape of the polar emission plots corresponding to the four and one coordination sites indicate clearly best agreement for the former for emission from the S 3p-derived level. The plots also depend on the adsorbate–substrate interlayer d-spacing but are not nearly so sensitive. This is shown in fig. 3.16b for two different spacings on the four-fold site. It should also be noted that LEED is much more sensitive to interlayer spacings due to the phase relationships among the

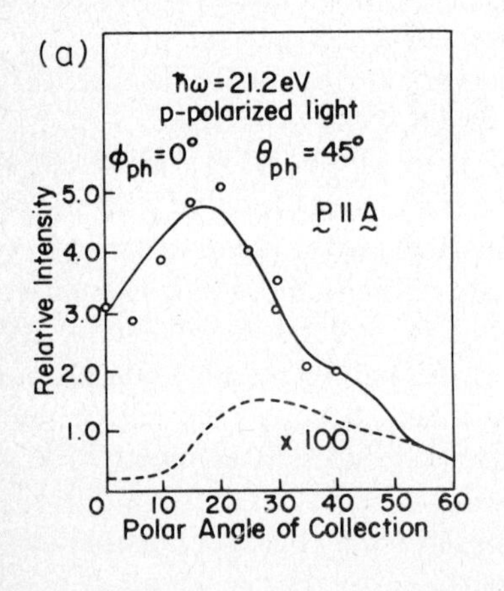

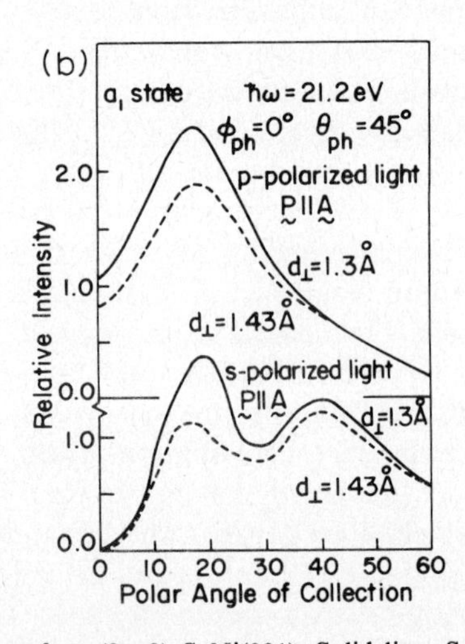

Fig. 3.16 (a) Comparison for polar intensity plots for c(2 × 2) S–Ni(001). Solid line, S at fourfold site; broken line, S at peak site; circles, experiment. The nomenclature of the photon azimuthal angle is the same as in Weeks et al. (1976). The notation $\vec{P} \parallel \vec{A}$ indicates electron-emission plane containing the photon $\vec{A}$ vector (Weeks and Plummer 1977b, Weeks et al. 1976, Tong et al. 1977). (b) Sensitivity of polar intensity plots to changes in S–Ni interlayer spacing $d_\perp$ [for c(2 × 2)S–Ni(001)]. (The S atom is at the fourfold site.) (See Weeks and Plummer 1977b, Weeks et al. 1976, Tong et al. 1977).

scattered beams. On the other hand, since LEED depends on differences in scattering factors to determine the orientation of the surface molecule, the adsorption orientation of two different molecules such as CO and NO with similar scattering powers would be more readily determined by angle-resolved photoemission. It appears to be amply clear that each of these methods has much to offer in a complementary relation for defining surface coordination and bond distances at surfaces. Best results are probably obtainable by combining data obtained by both methods on the same gas–metal system.

Vibrational excitations of oxygen in the p(2 × 2)O–Ni(100) and the c(2 × 2)O–Ni(100) surface structures have also been investigated by means of high-resolution electron loss spectroscopy (EELS) (Andersson 1976). The EELS spectra show energy loss peaks at 53.0 meV for the former and 39.5 meV for the latter. They are interpreted as due to vibrational excitations of the dissociated chemisorbed oxygen in the two structures. Only one mode is observed which is attributed to the vibrational excitation of oxygen normal to the surface plane. It is difficult to reconcile the rather large decrease of vibrational energy on going from the p(2 × 2) to the c(2 × 2) structures since electron spectroscopy measurements show only minor changes in electron structure and LEED shows no appreciable change in bond length. A plausible mechanism suggested by Andersson (1976) is that the potential barrier towards the nickel surface decreases due to dissolution of oxygen on going from the latter to the former structure. The energy shift on going from the p(2 × 2) to the c(2 × 2) structures is probably associated with a simultaneous change both in binding energy and bond length.

Oxygen on iron

There have been a great many studies (Pignocco and Pellisier 1965, Sewell et al. 1972, Melmen and Carroll 1972, Horoguichi and Nakanishi 1974, Simmons and Dwyer 1975) of atomic gas interactions and chemical bonding on iron surfaces. Relatively few studies have benefited in detail from a combination of electron spectroscopy techniques on well defined surfaces (Eastman et al. 1974, Leygraf and Eklund 1973, Yu et al. 1976a, Brucker and Rhodin 1976, Brundle 1978). Experimental difficulties are associated both with the highly reactive nature of the surface as well as with the relative complexity of the surface mechanisms and structures which occur. Photoemission spectroscopy is particularly well suited to a detailed study of chemical bonding on iron in terms of changes in the valence and core level electron structures associated with chemisorption and oxidation. It appears that initially a half-monolayer of chemisorbed oxygen forms with a $c(2\times 2)$ LEED structure characterized by a single O-2p peak in the UPS-spectra. The stages through which this progresses into a two-dimensional oxide film and then into a three-dimensional iron oxide can be systematically followed in terms of combined information from the valence level changes, LEED structures and peak-splitting in the Auger spectra (Brucker and Rhodin 1976).

The same sequence can be followed by monitoring the O-1s and Fe-2p $-\frac{3}{2}$ binding energies in the XPS-spectra (Brundle 1978, Ertl and Wandelt 1976). Neither the UPS nor the XPS-spectra alone can completely characterize the electronic nature of the reaction products. At high coverage the dominant species is Fe(III) but the Fe(II) state is also observable at lower exposures. In addition, although there exists no direct evidence for the presence of molecular oxygen on the surface, existence of a mobile precursor molecular state is suggested by the adsorption kinetics (Ertl and Wandelt 1976). At both room temperature, and lower, the major oxygen species formed is associated with the electron binding energy of O^{-2}. It is not yet clearly established that the surface layer below monolayer coverage may contain some as Fe^{II} although at monolayer and higher coverages the major species is clearly Fe^{III}. It further appears that the adsorption kinetics are consistent with the same chemical bonding at low (80 K) and high (300 K) temperatures (Brundle 1978). In a related core and valence level photoemission study of iron oxide, a combination of valence level differences and core level effects are observed which are very useful in the determination of bonding and valence states of iron and oxygen characteristic of iron oxidation at low and moderate temperatures (Brundle 1978).

In another study of the initial stages of oxidation of Fe(110) using UPS–XPS photoemission and surface potential measurements by Chesters and Rivière (1977), interesting correlations among chemical bonding, surface penetration and work function were discovered. Dissociative chemisorption producing monolayer coverage occurred followed by surface rearrangement associated with nucleation and growth of iron oxide layers in the subsurface region. The presence of iron in the Fe^{III} form in a 20 Å deep region was inferred.

It should be noted finally with reference to summarizing atomic bonding on iron (Bozso et al. 1978, Heimann and Neddermeyer 1977), only the simplest and best understood cases have been considered. There are other complex and probably more important ones as well involving chemisorption from atomic fragments split off from larger hydrocarbon or inorganic molecules by the reactive iron surface. Although these cases will be referred to briefly in a subsequent section with reference to molecular interaction, understanding of the complex and competitive bonding processes involved in gas kinetics on iron is only beginning to be unravelled.

Hydrogen and deuterium on tungsten

Many of the early pioneering surface studies initiated on clean single crystal surfaces were done on tungsten, in spite of the complexity of chemical bonding characteristic of tungsten chemistry. A great deal of useful surface science research continues on this metal. There is probably no other metal surface for which a comparable wealth of surface science data exists (Plummer et al. 1976, Plummer 1975a, Rhodin and Adams 1975a, Schmidt 1975). It is appropriate here to summarize selective examples of chemical bonding on tungsten using electron spectroscopic methods mainly to illustrate the effectiveness of combined measurements leading to the development of generalized interpretations. It is significant that, in spite of the intensive study of this particular gas–metal system, puzzling situations remain. These refer specifically to the state of association of the adsorbed species, to the electronic nature of the binding sites and to the specific features of the chemical bonding modes. It is precisely to these types of questions that electron spectroscopic techniques lend themselves most effectively. The major objective now in the study of any gas–metal prototype system such as tungsten with diatomic gases, is the correlation of kinetic mechanisms with specific aspects of chemical bonding. A detailed picture of the chemical bonds involved in the dissociative chemisorption of oxygen and hydrogen on tungsten requires more complete and detailed information on atomic geometry and electron structure than is yet available. It is instructive, however, to consider some significant advances made since 1975 (Plummer 1975a, Plummer et al. 1976) using high resolution electron loss vibrational- and angle-resolved photoemission spectroscopy in this connection.

The chemisorption of hydrogen on W(100) has been the target for study by many techniques on the assumption that tungsten is easily cleaned and that metal–hydrogen is the simplest adsorption system to understand theoretically. Understanding of this prototype system remains incomplete in spite of the interesting new information obtained from the electron spectral work. The essentials of the adsorption phenomena are as follows. The thermal desorption spectra (Menzel 1975) of hydrogen saturated W(100) at room temperature consists of two unequally sized desorption peaks at 450 K, (β_1) and 550 K, (β_2). At low coverages all of the adsorbed hydrogen desorbs in the high temperature

β_2-state with second-order kinetics, but for coverages greater than about 0.2 monolayer, the desorption spectrum also shows the low temperature β_1-state which obeys first-order kinetics. Even though the first-order kinetics is generally considered to imply molecular adsorption, both states show complete isotopic mixing. The β_2-state is characterized by a c(2 × 2) LEED structure while a new and more complex sequence of LEED structures result at higher coverages. The more conventional interpretation suggests two sequentially filled binding states (β_2 followed by β_1); the β_2-state consisting of atomic hydrogen and the β_1 state being molecular in nature (Menzel 1975). Comprehensive overviews were made by Plummer et al. (1976) and Plummer (1975a), emphasizing techniques providing data obtained directly on the energy levels, e.g. UPS-photoemission and field emission tunneling. In contrast to the sequentially-filled models, interpretation can also be based on a model which involves an adsorbate surface density-dependent conversion between two surface structures, referred to as the two-state-conversion model. Neither model explains the data satisfactorily. At this point, our understanding is fairly consistent with the original model proposed by Estrum and Anderson (1966). It proposes that hydrogen initially dissociatively adsorbs into alternate four-fold sites forming islands characterized by a c(2 × 2) LEED structure (see fig. 3.17). Structural rearrangements induced by repulsive interactions among the adsorbed hydrogen atoms occurs at coverages of 0.1–0.2. At higher coverages the surface atoms are compressed until there are two hydrogen atoms per tungsten atom, possibly located at bridge sites.

The system has been studied in additional detail using angle- and energy-resolved photoemission and electron loss spectroscopy with the following supplementary conclusions. According to Gadzuk (1974a, b, 1975a, b) the angular features of the adsorbed system should lead to direct identification of the chemisorption bond geometry neglecting the

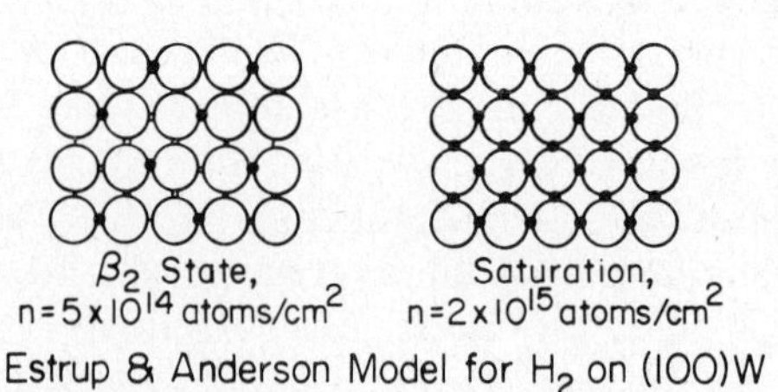

Fig. 3.17. Proposed adsorbate structure for H_2 on the planes of W. Two models have been proposed but neither simultaneously satisfies photoemission LEED and flash desorption observations (Estrum and Anderson 1966).

importance of final state effects. From angle-resolved photoemission spectroscopy one can plot the position of the resonance adsorbate peaks as a function of the *k* (parallel) vector along the two symmetry lines of the two-dimensional band structure (Feuerbacher and Fitton 1977b). The intensity of the adsorbate peaks is then plotted along the vertical axis. Adsorbate-induced peaks at 2 eV, 5 eV and 12 eV below the Fermi level are observed to disperse and split with the angle of emission in a complex but distinctive manner. The adsorbate features of the surface band structure present an array of resonant emission characteristics which are not yet completely interpreted. It is clear that angle-resolved spectra reveal much detail, requiring complete data along the chief symmetry directions to make them fully understood. (It is interesting to note that integrated spectra with less detail can sometimes present a more useful overall perspective.)

Other important features accessible through the use of synchrotron radiation are the angle-resolved polarization-dependent properties of the optical transitions, reported by LaPeyre et al. (1976) for a hydrogen saturated surface. One kind of polarization involves the electric field polarization vector parallel to the surface (s-polarization). The emission intensity depends on the direction of the polarization vector with respect to planes of symmetry in the surface. Photoemission results on the W(110)–H system are of this type. Emission in the (100) mirror plane for the four combinations of polarization parallel or normal for clean and chemisorbed W(100), show strong structural features which exhibit clear-cut four-fold symmetry under crystal rotation. Upon further analysis it is concluded that the existence of the polarization effect indicates that the energy bands formed by the adsorbate reflect the mirror plane symmetry of the substrate. Here again the interpretation is convoluted by interpretations obscured by the highly directional character of the two-dimensional band structure of the chemisorbed surface. A more detailed analysis requires a quantitative calculational analysis of directionality, energy and band character for the clean and chemisorbed systems.

The same authors have also performed an interesting angle-resolved photoemission study (Anderson and Lapeyre 1976) of the β_2-state corresponding to the c(2 × 2) structure of hydrogen on W(100) previously discussed. The UPS peak 2 eV below the Fermi level for this adsorption state has been attributed to tungsten d-orbitals originating from the specific site geometry. Angle-resolved energy distribution curves for clean and hydrogen c(2 × 2) covered tungsten show the angular properties of the hydrogen-induced structure. The chemisorbed atoms modify the photoemission process through the same elastic scattering effect which gave rise to the extra fractional order ($\frac{1}{2}\frac{1}{2}$) spots in the c(2×2) LEED pattern. This work indicates that for angle-resolved photoemission spectra especially, consideration must be given to the elastic scattering by the ordered overlayer. At this time, although the use of polarization and angular-resolved features at selected photon energies have indicated singular features in the surface electron structure, it is premature to draw definite conclusions as to the more precise nature of the chemical bonding involved in terms of either orbital properties or surface energy band structure for this particular system.

The classical adsorption system of hydrogen saturated W(100) has also been studied using electron energy loss spectroscopy by Propst and Piper (1967) and by Frotzheim et al. (1977a, b). The loss spectra for different exposures of hydrogen taken from the latter work are shown in fig. 3.18. At low coverages, ($\theta = 0.4$) only one loss at 155 meV occurs whereas at higher coverages a loss at 130 meV appears. The latter loss saturates at a coverage of $\theta = 2.0$ corresponding to two hydrogen atoms per surface tungsten atom while the loss at 155 meV disappears. It is now concluded that both adsorptions occur dissociatively, the 155 meV loss corresponds to hydrogen with a coordination of one and the 130 meV loss peak to hydrogen with a coordination of two, e.g. terminal vs. bridge-bonded modes. The two stages are neither sequentially filled not result from a simple transition between atomic and molecular

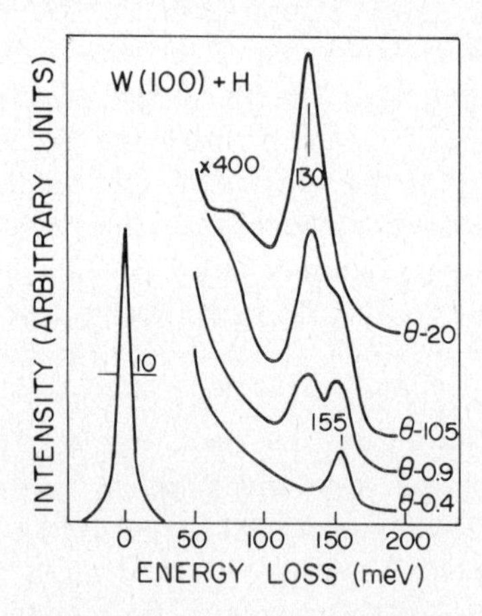

Fig. 3.18. Electron-energy loss spectra of H on W(100). The two losses at 155 and 130 meV correspond to atomic hydrogen adsorbed in on-top and bridge sites, respectively. The peak around 70 meV at higher coverages is caused by a small CO contamination ($\theta \simeq 0.01$) during hydrogen exposure (Froitzheim et al. 1977a).

adsorption but apparently result from a reversible exchange and equilibrium between the relative occupation of atomic adsorption at two different sites. The latter conclusion is in clear agreement with the relative change in intensity of the two losses with surface coverage.

The surface structure models for the various observed surface vibrational modes observed for dissociatively adsorbed hydrogen on all three low index planes of tungsten, W(100), W(111) and W(110) have been analyzed (Backx et al. 1977c). The character of the bonding indicates that the site of coordination one, e.g. on top of the tungsten atom, is preferred on all planes and that the two-coordinated (bridge-bonded site) or the four-coordinated (face-centered bonded site) form subsequently. It is significant that the EELS results on W(100) are in good agreement with coverage-dependent electron stimulated desorption (Menzel 1975) as well as photoemission studies (Backx et al. 1976, 1977a, b). It is also noteworthy that the structures and the vibrational modes are coverage-dependent, indicating that variations in the binding energy probably reflect the effect of interaction with neighboring adsorbed atoms. This is a good example of how useful the EELS results can be in developing a clearer picture of the bonding geometry when the electron loss data is carefully correlated with supporting data from thermal desorption, LEED and photoemission.

Oxygen on tungsten

In contrast to most interpretations of hydrogen and nitrogen adsorption on tungsten, for the case of oxygen adsorption, there seems to be little doubt that reconstruction and formation of surface oxides occurs under a variety of coverage and heat treatment conditions (Rhodin and Adams 1975b). The complex structures which form are attested to by the variety of LEED patterns observed (Ertl and Küppers 1974), for which there has been to date relatively little successful structural analysis based on dynamical theory (see chapter 4 by Van Hove). Some models of surface structures for oxygen on W(100) formed as a function of coverage at 300 K are shown in fig. 3.19. These structures have also been the object of analysis and interpretation using EELS combined with thermal desorption and photoemission data. Electron loss data for W(100)–O as a function of oxygen surface coverage are illustrated in fig. 3.20 from the work of Froitzheim et al. (1976a, b). An involved sequence of losses are observed which change in intensity and position with coverage in a complex manner relative to the previous case of W(100)–H. The following main interpretations nevertheless may be drawn from a consideration of the combined surface parameters.

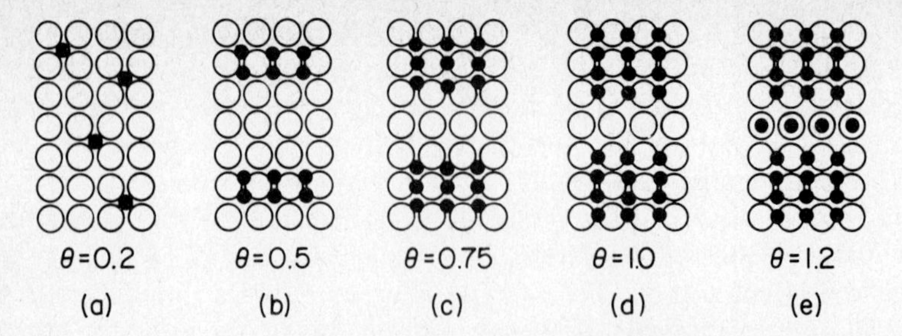

Fig. 3.19. Models for the surface structure of oxygen on W(100) adsorbed at 300 K.

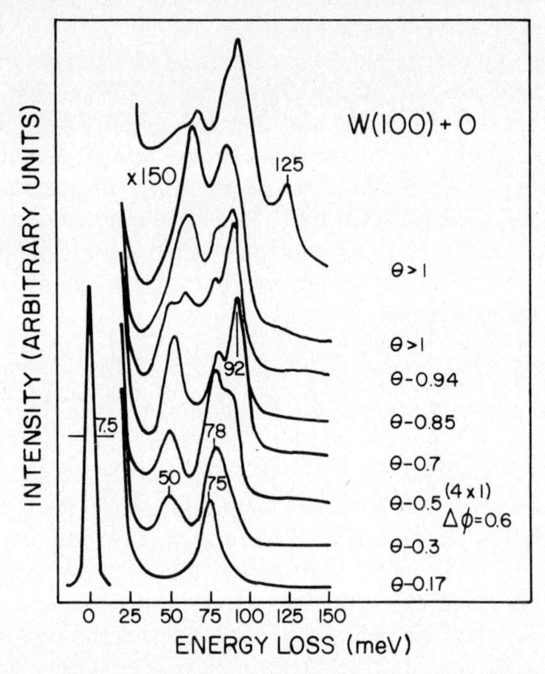

Fig. 3.20. Electron energy loss spectra of W(100) for several coverages of oxygen. The spectra are recorded at a primary energy of 5 eV with an angle of incidence $\theta_1 = 70°$ (Froitzheim et al. 1976b).

The single loss peak observed at 75 meV for very small coverage indicates a dissociative adsorption in the four-fold coordination site (fig. 3.19a).

The two domain (4 × 1) structure which forms at coverages between 0.25 and 5.0 corresponds to a shifting of the 75 meV peak to 78 meV and the emergence of a second loss peak at 50 meV. The structure consists of rows of oxygen atoms arranged in such a manner as for the surface tungsten atoms to have a valency of 4+ and a nearest neighbor configuration similar to the rutile-type structure of WO_2 (fig. 3.19b).

The loss spectra observed at somewhat higher coverages (0.5–1.0) correspond to a related multiplicity of similar oxygen rows of higher order, e.g. triple, quadruple rows etc. (figs. 3.19c, d).

Finally, for the losses corresponding to coverages greater than one it is suggested that additional oxygen atoms are forced at this higher coverage to assume adsorption sites with coordination of one, e.g. top position with terminal bonding (fig. 3.19e).

It is apparent that this hard sphere model explanation is rather over-simplified and idealized. The significance is clear, however, that a *combination* of the EELS loss spectra combined with LEED data on atomic structure and photoemission information on binding energies leads to useful and reasonable interpretations of the chemical bond geometry characteristic of a rather complex series of surface structures.

4.4. Molecular adsorption on metal surfaces

Interpretation of surface electron structure associated with the chemisorption and reaction of molecules on metals is more complicated and probably even more interesting than atomic adsorption. It is evident that both the electron spectra and the options for decomposition and reaction are more numerous in the former case. On the other hand, a wealth of interesting information is available of great interest with reference to understanding chemical bonding and reaction mechanisms of molecules on reactive transition metal surfaces (Liebsch 1977). There exists a great body of analytical effort on both associative and dissociative molecular adsorption*. The chemisorption and reaction of carbon monoxide in particular has long been used as a probe for molecular adsorption on metals because of the great variety and intensity of bonding modes with which it interacts with transition metals. Analyses of this collection of interesting results are well documented (Plummer et al. 1976, Plummer 1975b, Gomer 1974).

It is not appropriate here to attempt to reconsider in depth this particular ingenious and intricate body of interpretative adsorption chemistry. It is relevant, however, to point out some recent significant

* The concepts of associative and dissociative adsorption are used here in the surface scientists sense to indicate the surface bonding of a molecule without breaking bonds upon adsorption in one case and with partial or complete dissociation of the molecule in the other.

clarifications as well as complications of chemical bonding interpretations illustrative of molecular adsorption produced specifically from photoemission and electron loss spectroscopy studies. UPS-photoemission applied to molecular adsorbates has provided useful information not only on orbital symmetries and molecular orientations but on adsorption geometries and electronic properties of chemical bonds at surfaces (Liebsch 1976, Gadzuk 1974a, b, Rhodin and Adams 1975a).

It is in the application of molecular adsorption that these methods have perhaps made some of their major and more stimulating contributions. Although consideration focuses chiefly on carbon monoxide as a prototype adsorbate, many of the implications relate in general to the adsorption of other relatively stable molecules which chemisorb onto a wide variety of metals with no or little decomposition. The transitions between associative and dissociative adsorption depend very much on exposure, temperature and condition of the surface. Cases in which the adsorbate-molecule tends to maintain its entity upon adsorption will be considered here.

With reference to the UPS-photoelectron spectroscopic methods it should be noted that the interpretation of multiple adsorption states from photon-stimulated electron spectra of adsorbed molecules can be made difficult because of the possibility of strongly overlapping spectral features. These effects, as well as changes in valence levels, can often be large and unpredictable. In contrast, the study of core levels using XPS-photoemission is essentially atomic in nature and can give a more quantitative elemental analysis of the surface. Core-level chemical shifts provide a definable distinction between the various states of molecular adsorption. Bonding characteristics can, in some cases, be more clearly interpreted from XPS-spectra because they tend to relate more directly to individual atoms in the adsorbate. Combined emission studies (UPS, XPS, Auger, and EELS) provide information most likely to be free of ambiguities on the state of surface chemical binding of molecules. The relative merits and uncertainties of electron- and photon-stimulated electron emission for this purpose have been reviewed by Brundle (1975b).

The more significant recent advances to be considered here (Castex et al. 1977, Willis et al. 1976, 1977) are the applications of angle- and energy-resolved photoemission combined with both LEED structural data and high sensitivity electron loss measurement to the determination of the bonding configuration of molecular adsorbates, particularly carbon monoxide. Unique configurational insight on bonding (Liebsch 1978a, Feuerbacher et al. 1978) is available on both the orientation of the molecule and the redistribution of electrons in the molecule upon adsorption. There is a close coupling in understanding for both the geometric orientation and position of the molecule and the electron configuration of the bonding orbitals. Interpretation of the angle-resolved data for molecular adsorbates is complicated by the need to combine understanding of the initial excitation state with the final scattering state through a transition matrix for a complex adsorbate–surface system. Interpretation of integrated data is relatively simplified first by analogy to the free gas molecular spectra and second by the fact that many of the gas-phase orbitals are likely to be largely unperturbed upon adsorption. Considerable uncertainties can nevertheless be introduced through relaxation and polarization shifts in their peak positions and shapes. The more the gas-phase character is maintained upon adsorption, the more definitive is the interpretation of the UPS-spectra in terms of the adsorption process. Conventional procedures for the interpretation of UPS adsorption spectra on tungsten in terms of difference curves and of thermal desorption data is discussed in some detail by Plummer et al. (1976).

A more recent alternative approach to this problem is the analysis of bonding structure and reactivity of molecules on extended metal surfaces by analogy to the same parameters in metal–organic clusters. Following the earlier work of Conrad et al. (1976) on $Rh_6(CO)_6$ a study by Plummer et al. (1978) of the photoemission spectra for the transition metal carbonyl complexes indicates interesting correlations between their spectra and those of chemisorbed carbon monoxide on extended transition metal surfaces. The implications of this hypothesis have been explored in detail by Muetterties (1977), Muetterties et al. (1979) and Ozin (1978). The relevance to chemical bonding will be noted later.

It is well demonstrated that diatomic gases like carbon monoxide, nitrogen, nitric oxide and oxygen can chemisorb associatively or dissociatively on a number of transition metals. A criterion has been established based on the carbon–oxygen distance in adsorbed carbon monoxide compared to gaseous carbon monoxide by referring to the energy displacement of the non-surface bonding 4σ and 1π orbitals in the UPS-photoemission spectra. A rationalization based on a variation of the Blyholder model describing chemisorption of carbon monoxide has been proposed relating the bond stretching to dissociation at room temperature (Brodén et al. 1976a) (fig. 3.21b). Although the hypothesis is rather simplified, it is consistent with a broad base of data and can be effectively used to indicate, with reference to position in the periodic table, the propensity of a metal to promote dissociative adsorption at

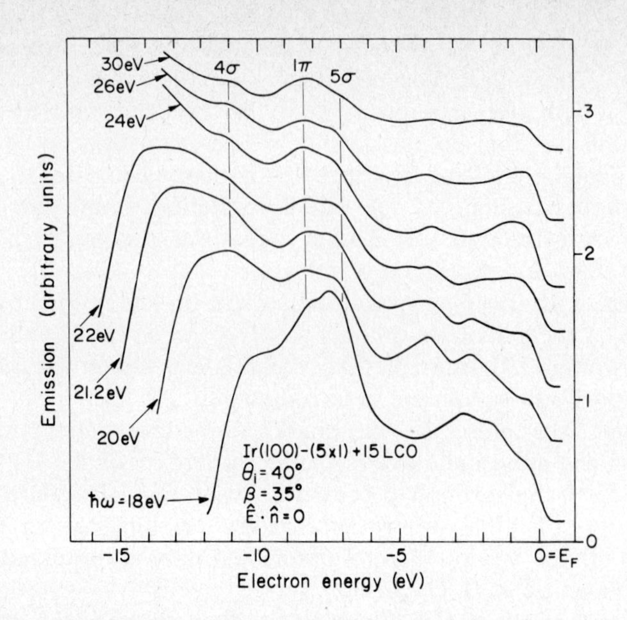

Fig. 3.21a. Photoemission curves showing chemisorption of 15 L of CO on Ir(100)–(5 × 1) for photon energies between 18 eV and 30 eV, s-polarized radiation. The curves are normalized to the incident yield. α and β defined as in the caption of fig. 3.30a (Brodén et al. 1976a).

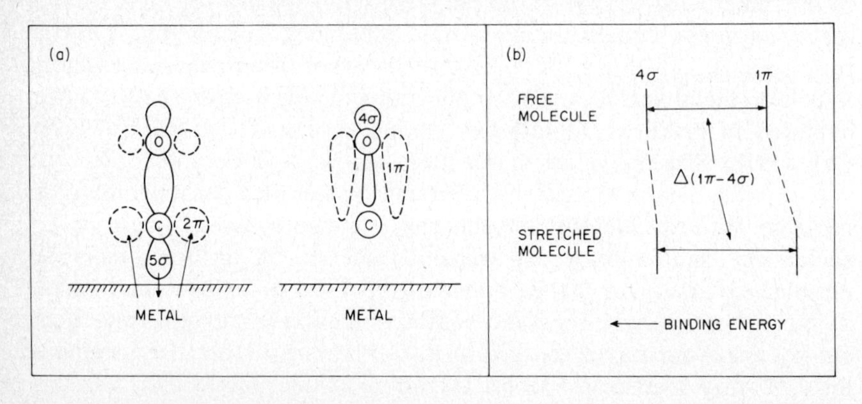

Fig. 3.21b. Blyholder model describing chemisorption of CO to a metal surface. The 2π and 5σ orbitals of CO are indicated by dashed and full lines, respectively. The CO molecule is supposed to bond with the carbon atom closest to the surface. Only the orbitals participating in the surface bonding are indicated. The approximate shape and size of the 4σ orbital (full line) and 1π orbital (dashed) of CO. These orbitals do not participate significantly in bonding to the metal surface. Shift in energy levels with adsorption shown on right (Brodén et al. 1976a).

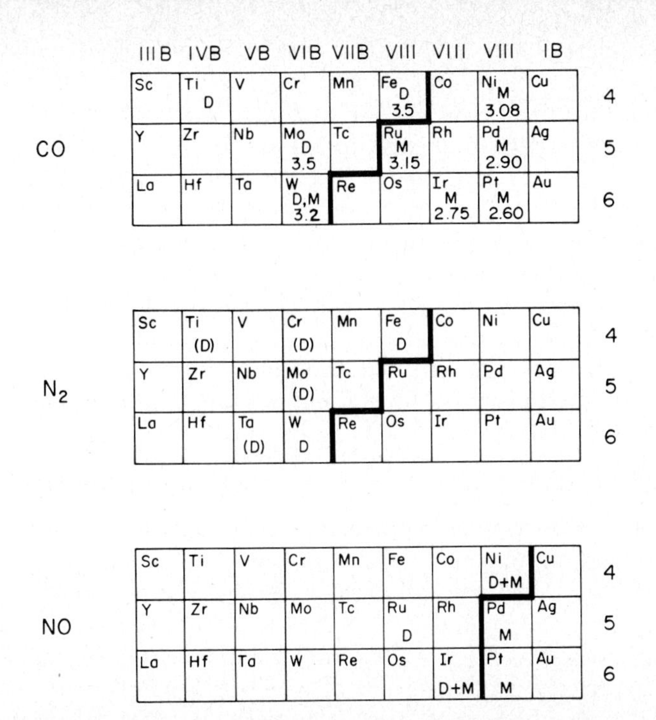

Fig. 3.21c. Section of the periodic table show the room temperature adsorption behavior of CO, N_2 and NO. M denotes molecular adsorption and D dissociative adsorption on at least one surface, single crystal or polycrystalline. (D) indicated that the adsorption is likely to be dissociative. The thick line in the periodic table gives the borderline between molecular and dissociative adsorption at room temperature. The numbers in the upper section for CO show the average value for the energy separation of the 4σ and 1π peaks $\Delta(1\pi - 4\sigma)$. (Brodén et al. 1976a.)

room temperature, as indicated in fig. 3.21c. A borderline, exists, as shown (fig. 3.21c), on the basis of available evidence at room temperature, to the right of which molecular adsorption can be expected and to the left of which dissociative adsorption is likely. In general, electron structure is the single most crucial factor except for the borderline elements like tungsten where both types of adsorption may occur at room temperature depending on the state and orientation of the crystal surface. Similar analyses for the nitrogen and nitric oxide systems indicate a corresponding basis for discrimination, but with a shift in the borderline position. It is clear that the electronic nature of the surface is dominant here also; a secondary role being played by the factors of surface crystallography and defects for the borderline metals. A more detailed understanding of these basic trends will emerge with the development of a more precise spectral measurement of the nature of metallic valence band orbitals. This approach will be discussed later with reference to dissociative chemisorption on iron.

A body of recent molecular chemisorption studies existing on the chemisorption of carbon monoxide on tungsten and nickel will be considered specifically in terms of recent interpretations of electron structure and surface molecular orientation. In addition, particular features of these effects associated with chemisorption of other molecular gases such as nitrogen and nitric oxide on other metals such as platinum, iridium and iron will be covered. Consideration will be limited to our main task of appraising the unique features inherent in surface chemical bonding.

Carbon monoxide on tungsten

The adsorption of simple, diatomic gases on tungsten is one of the most widely studied prototype adsorption systems. The routine preparation of clean surfaces using current surface probe characterization enables contact with a large body of previous work. This work has been reviewed in detail by Gomer (1974), Plummer et al. (1976), and Plummer (1975a). Hydrogen, nitrogen, and oxygen are generally but not always believed to adsorb dissociatively, especially at room temperature, but the conditions for dissociative adsorption of carbon monoxide are much more variable. The orientation of the molecule and the nature of multiple bonding states for the associatively adsorbed state have been clarified using electron spectroscopic methods. Bonding and geometry of oxidizing molecules associated with dissociative adsorption often leads to oxide formation and surface facetting and is more difficult to interpret. A detailed picture of the chemical bonds formed by hydrogen, nitrogen, oxygen, carbon monoxide and nitric oxide with tungsten surfaces requires a more quantitative understanding of surface geometry and electron orbitals than is currently available (Plummer 1975a, 1977). New significant understanding from recent electron spectroscopy studies has lately been developed.

Chemisorption of carbon monoxide on tungsten is usually classified according to the binding modes originating from thermal desorption studies as alpha-, beta-, and virgin-states (Menzel 1975) (see fig. 3.22). The alpha-one and alpha-two states are electropositive and associatively adsorbed; the higher binding α_2-state thermally desorbs above 350 K and the lower binding α_1-state thermally desorbs at 350 K and sometimes converts to α_2 during thermal desorption. The three beta-states are most

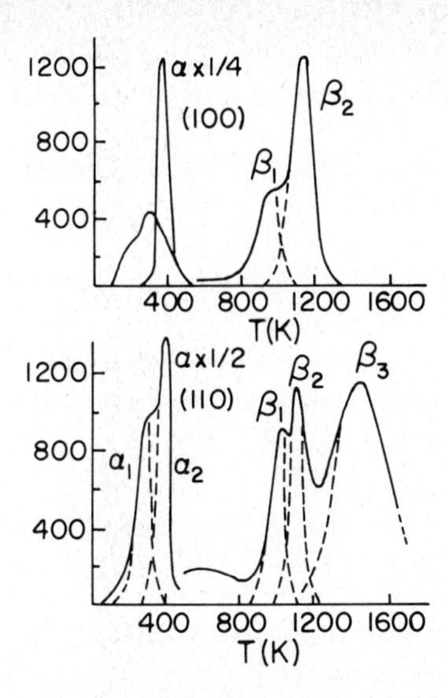

Fig. 3.22. Flash desorption spectra of CO on the (110) and (100) planes of tungsten (Schmidt 1975).

stable, until about 600 K when they desorb leaving no carbon or oxygen on the surface. They exhibit a variety of LEED structures and had been considered to be typical of associative adsorption until looked at recently in more detail by photoemission and EELS spectroscopy (Baker and Eastman 1973, Froitzheim et al. 1977a, b). Virgin carbon monoxide is considered to be molecularly adsorbed, forming directly on tungsten, and converting to β-CO upon heating at 300–600 K. Although much interesting information is available about these adsorption states, some critical questions remain with reference to the nature of the lateral interactions which lead to the observed density dependent adsorption kinetics, the role that linear and bridge bonding play in defining the different states, as well as the factors critical in the dissociative adsorption kinetics of β-CO. The following models are now proposed with reference to interpretation of the W(100)–CO and W(110)–CO adsorption systems.

Adsorption of CO on W(100)

The alpha- and virgin-states are definitely molecular and concluded to be linearly bonded by comparison of the photoemission spectra to that of the singly bonded tungsten carbonyls. A more conclusive test would be the absence of π-level splitting, although this may be difficult to establish conclusively due to line broadening and energy resolution. The more stable β-states form at higher temperatures in a variety of LEED structures consisting of carbon and oxygen surface atoms in a series of geometric arrangements which change with exposure and temperature but not in any simple sequential manner. Formation of the β_3-state at 1100 K from β_1 and β_2 at lower temperatures, for example, definitely produces a dissociated molecule characterized by a c(2×2) structure. A possible configuration has the carbon atoms in every other four-fold coordination site with the oxygen atoms filling in the other identical site. At this time precise crystallographic analyses of these and similar structures are not available.

Adsorption of CO on W(100)

The virgin state is molecularly adsorbed at room temperature. It desorbs and converts to the β-states upon gentle heating. The CO levels in the UPS-spectra attenuate and the oxygen–tungsten levels appear with this transition. The CO molecule is lying down on the surface at higher temperatures probably in all three β-states with both the oxygen and the carbon atoms forming bonds with the metal substrate. Although differences among the β_1-, β_2-, β_3-states are not completely clear, the degree of dissociation of the chemisorbed CO varies for the β-states, being essentially complete for the β_3-state and less marked for the β_1-state.

There are indications that for rather strong bonding, changes in the density of states of the adsorbed atom should be a reflection of the geometry of the binding site (Liebsch 1977). All the dissociated β-states are related to specific LEED structures which form reversibly depending on the surface density. The more weakly bound virgin- and alpha-states definitely consist of molecularly adsorbed carbon monoxide, an observation (Plummer et al. 1978) consistent with the strong similarity in photoelectron spectra for gas phase $W(CO)_6$ and adsorbed CO on both W(100) and W(110). Finally, the photoemission spectra make possible the tracking of the fingerprint of the CO molecule as it converts upon adsorption into the dissociated β-state with changes in temperature and surface coverage.

Many of the bonding aspects postulated for CO on low-index planes of single crystal tungsten from the photoemission studies for the different adsorption states have been strikingly corroborated and even extended by interpretations of the EELS spectra for CO-adsorption at 300 K at various gaseous exposures. An interesting comparison is shown in fig. 3.23 from the work of Backx et al. (1977a, b) and Froitzheim et al.

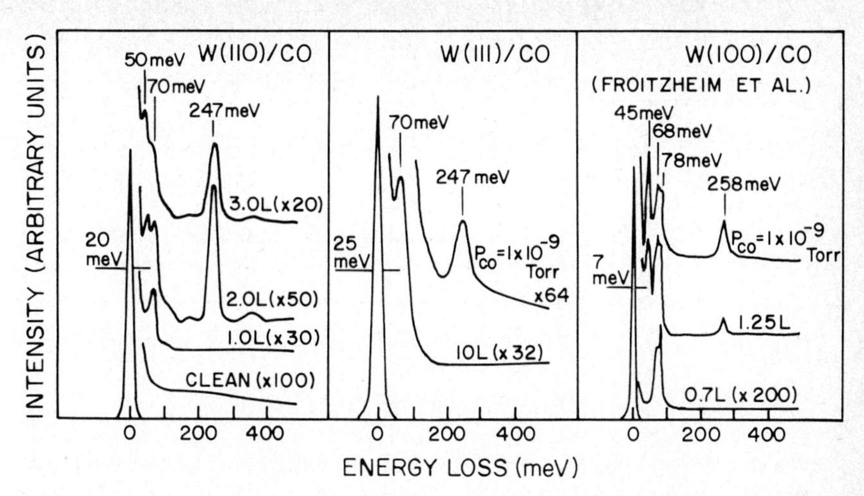

Fig. 3.23. Energy loss spectra of W(110) W(111) and W(100) surfaces exposed to 10^{-8} Torr partial pressure of carbon monoxide for various periods (Backx et al. 1977a, b). The W(100) spectra were obtained by Froitzheim Ibach and Lehwald (1977a).

(1977a, b) for all three low-index planes. The high resolution spectra $\Delta E = 7$ meV) in the data for W(100)–CO for small exposures show clearly the loss peaks at 68 meV and 78 meV due to vibrations of isolated chemisorbed carbon and oxygen atoms respectively. It is presumed that they are positioned in four-fold sites characteristic of the dissociated β-state adsorption. The two losses at 40–50 meV and 240 meV relate to the characteristic vibrations of the undissociated α-adsorption state corresponding to a molecule oriented normal to the surface. The losses at 45 meV and 258 meV for W(100) relate closely to the W–C stretching and the C–O stretching frequencies respectively reported for tungsten carbonyl, $W(CO)_6$ (Jones 1963).

The appearance of only one frequency of vibration for the molecule as a whole suggests that the molecule sits at the coordination site of one on W(100) (Froitzheim et al. 1977a) similar to that observed for CO on Ni(100) (Andersson 1977a) and Pt(111) (Froitzheim et al. 1977a). They show other losses at lower frequencies corresponding to adsorption on higher coordination sites. This suggests bonding between the carbon and more than one metal atom at a time, e.g. a bridge- or face-centered position. The interesting comparisons among these various metals and orientations are summarized in table 3.3, showing the similarities and differences between adsorption on tungsten compared to platinum and nickel surfaces. It should be noted that two related losses can result from a carbon atom in a bridging position, one associated with a low frequency

TABLE 3.3
Electron energy loss spectroscopy of CO on d-band metal surfaces

Metal	C–O Stretch	Metal–CO	Site	Reference
W(100)	258 meV	45 meV	on-top	Froitzheim et al. (1977c)
W(110)	247	50		
W(11)	247		disordered	Backx et al. (1977a, b)
Pt(111)	261	58	on-top	Froitzheim et al. (1977a)
>0.2 L	232	45	4-fold	
Ni(111)	256	60	on-top	Andersson (1976)
	240	45	out-of-registry bridge	
	249	54	disordered	

symmetric mode and one associated with a somewhat less intense higher frequency asymmetric mode. A doublet loss of this kind has been reported by Andersson (1977a) for Ni(100)–CO.

In summary, one sees that the EELS-spectra bring out additional information on the nature and orientation of the surface modes and very effectively extend the conclusions indicated by the photoemission spectra. It is also indicated by the EELS-spectra that there can exist additional modes at specific coverages not readily discernible by the UPS-spectra.

Reference was made to the hypothesis that an analysis of the intrinsic structural, sterochemical and thermodynamic features of molecules chemisorbed on extended metal surfaces can be compared to those of descrete metal–ligand clusters to present useful models for the interpretation of photoemission spectra and of chemical bonding (Muetterties 1977, Muetterties et al. 1979). This interesting idea was explored earlier for $Rh_6(CO)_6$ in an experimental photoemission study by Conrad et al. in 1976. In a similar vein, classical displacement reactions and the methods of the organometallic chemistry can serve to complement electron spectroscopic results to gain an enhanced understanding of chemisorption states and bonding on metal surfaces (Muetterties 1977). Recently an exploratory study by Plummer et al. (1978) of a comparison between the photoelectron spectra of a series of transition metal carbonyls and chemisorbed CO on the corresponding extended metal surface has produced the following pertinent observations:

(i) Referring to the chemical shifts associated with the energy levels not involved in the surface binding, four-fifths of the shifts in the one- and two-electron binding energies for CO adsorbed on a transition metal surface are present in the spectra for a single metal carbonyl.

(ii) The spectra of CO on an extended transition metal surface is almost

identical to that of the multimetal carbonyls. Apparently the increased delocalization of the metal valence levels and the 5σ-derived bonding orbital in the carbonyl is sufficient to make the carbonyl spectra and the chemisorbed metal surface spectra look alike. It appears that three to four transition metal atoms will suffice to bring the CO and metal levels essentially into agreement.
(iii) The distinction between the photoemission spectra for bridge- versus terminal-bonded CO does not appear to be sufficiently large to be a definitive indication of the type of surface bonding. In contrast to this conclusion, Brodén et al. (1977) conclude that bridge-bonded chemisorbed CO has a larger bond length than on-top CO to the extent that the photoemission data show an increase in the energy separation between the 1π and the 4σ orbitals from about 3 to 4 eV indicated for Ni(111)–CO chemisorption. This interesting hypothesis remains to be more fully explored.

There is much to be learned about the interpretation of photoemission spectra and the nature of surface chemical bonding from the analogy between classes of metal–ligand clusters and chemisorbed extended metal surfaces. The preliminary work (Conrad et al. 1976, Plummer et al. 1978, Muetterties 1977, Muetterties et al. 1979, Ozin 1977, 1978) indicates a fruitful area for experimental and theoretical exploration of surface chemical bonding.

Carbon monoxide and nitric oxide on nickel

Definitive studies on nickel surfaces using combined surface probes are attractive for a number of reasons, namely, the established well defined methods for surface preparation and characterization, systematic LEED studies on ordered overlayers, band structure typical of a 3d-metal and a wide variety of interesting chemisorption and reaction effects involving both atomic and molecular interactions. The interpretation specifically of angle-resolved photoemission measurements on chemisorbed surfaces has been particularly fruitful for CO-molecular adsorption on nickel. This is due to the availability of precise data on gas-phase photoelectron spectroscopy and the ability to draw useful analogies between spectral effects characteristic of differences in the free gas vs. chemisorbed phases. It is important to note in this regard that the partial photoionization cross sections for N_2, CO and CO_2 have been measured using continuum radiation, 14–50 eV, in the electron storage ring (Plummer et al. 1977). Systematic angle-resolved work on the surface configuration of CO and NO has been done on single cryatal surfaces of nickel (Allyn et al. 1977, Smith et al. 1976), platinum (Apai et al. 1976) and iridium (Rhodin et al. 1977a, Kanski and Rhodin 1977) in this regard.

The gas-phase theory by Davenport (1976) for the valence orbitals of CO and N_2 established for the random molecule, can be effectively used to calculate the differential photoionization cross sections as a function of energy and direction of polarization for a molecule oriented on a chemisorbed surface. These predictions have been checked for CO on Ni(111) (Allyn et al. 1977a, Smith et al. 1976) and on Pt(111) (Apai et al. 1976) as well as on Ir(111) (Rhodin et al. 1977a) on the assumption that the molecule is oriented normal to the surface with the carbon end down. The spectra for CO on Ir(111) are more complicated than for CO on Ni (111) and indicate that the molecule orientation and bonding mode may not fit the simple picture which works so well for nickel. The bonding and the orientation of the adsorbed molecule in this case appears to be sensitive to surface coverage and surface preparation (Brodén et al. 1976a, Rhodin et al. 1977a).

Actually the scattering of photons by molecules at surfaces depends in a complicated way on geometric and electronic factors (Liebsch 1978a). The geometric effects are sensitive to the degree and direction of polarization of the photon beam and its relationship to the reflection plane. The electronic effects which reflect specifically the details of chemical bonding depend on the electron structures of both the adsorbate and the substrate, interactions between them as well as with the photon beam, and on the directional aspects of excitation and emission. The latter problem has been worked out for certain conditions by Plummer (1977) but the former requires additional detailed analysis. The calculational approach to photoemission from metal molecule clusters is treated by Messmer in chapter 2.

The use of angle-resolved photoemission to determine the orientation of surface molecules does not depend on a detailed knowledge of the geometric scattering process of the light or of the electronic interactions of the photon beam with the combined orbitals of the molecule and electron bands of the metal substrate. The works of Allyn et al. (1977) and of Plummer and Gustafsson (1978) indicate that the properties of the "shape-resonances" associated with the final state of the excitation process can be used in a more direct and simple manner to define surface orientation. Such a resonance occurs when the kinetic energy of the emitted photoelectron coincides with the energy of a scattering resonance in the molecule. This bound electron state in the molecule occurs when the photoelectron has just the right energy and has a specific symmetry with reference to the geometry of the molecule. Since both the conditions for excitation of the resonance as well as the associated directional behavior of the emitted electron are defined by quantum mechanical selection rules, the photon-stimulated resonance-emission can be employed directly to define the surface orientation of the molecule when polarized light of the correct energy is used.

The "shaped resonance" approach has been used by Allyn et al. (1977, 1978) to determine the orientation of CO chemisorbed in a saturated layer at room temperature on Ni(100) and Ni(111). They showed that a shaped resonance occurs in CO at 15 eV electron kinetic energy or about 35 eV photon energy. It is associated with the 4σ orbital, the center of charge of which tends to be localized on the oxygen end of the molecule. The 4σ state can only be excited to resonance emission by an electric field whose vector is oriented parallel to the molecular axis. A plot of the measured intensity from the 4σ level against photon energy for the electric field vector normal to the surface is compared with the theoretical calculations in fig. 3.24 for three different possible surface orientations of the molecule. It is clear that the orientation normal to the surface with the carbon-end down is the correct one. The same approach can be used to determine the orientation of a molecule that is not placed normal to the surface by using a photon excitation where the electric

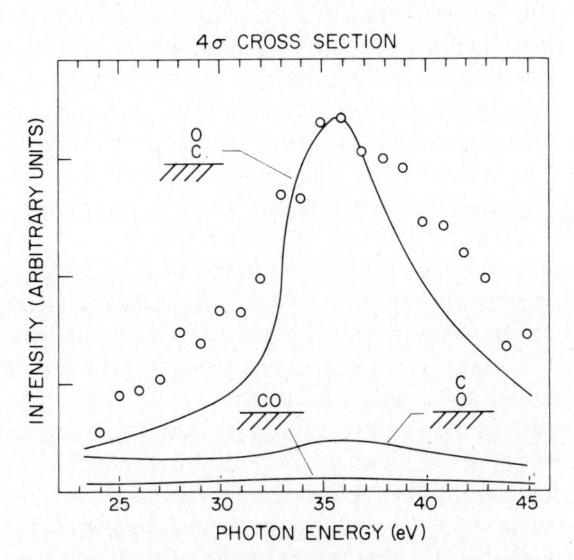

Fig. 3.24. Experimental cross section dependence (open circles of the 4σ orbital for normal emission from a CO-saturated Ni(100) surface, s/p polarization. Calculations are included for carbon-down (solid line) and oxygen down (solid line) bonding modes (Allyn et al. 1977, Gustafsson and Plummer 1977, Plummer and Gustafsson 1978).

field vector is oriented parallel to the surface (e.g. s-polarization). In this case enhanced emission can only occur with a tilted orientation. Such is the case for NO on Ni(100) (Loubriel et al. 1977) and for NO on Ir(111) (Kanski and Rhodin 1977). It is not surprising that the CO and NO molecular orientations occur in the manner they do when one examines the orientational effects in metallic clusters where the stabilizing ligands are CO or NO respectively (Cotton and Wilkinson 1972). The use of the "shaped resonance" emission to determine the orientation of molecules at surfaces in this manner is in an early stage. It requires the availability of a synchrotron resonance source to obtain the variable and polarized photon excitation. Other measurements using both angle-resolved and angle-integrated photoemission symmetry of the molecule on the surface can also be employed to define molecular geometry, but involve more detailed analysis.

The problem of the proper assignment of the energy levels in the observed photoemission spectra for CO chemisorbed on transition metals compared to the gas-phase orbitals has also been the object of considerable theoretical and experimental effort. This problem was dramatized by the photoemission spectra reported by Eastman and Cashion (1971) for the Ni(100)–CO system. Reference to the literature on this problem is recommended for an extended account of the various interpretations (Liebsch 1978a). Resolution is complicated by the difficulties of properly accounting for the weighted contributions of shifts in peak positions on chemisorption from such factors as molecular polarization, surface relaxation and chemical binding. Referring to the diagram of the molecular orbitals for gaseous carbon monoxide in fig. 3.25, one recognizes that the problem centers principally on the large energy shift of the photoemission peak for the 5σ orbital, the one principally involved in binding of the molecule to the metal surface. The proposed energy shift illustrated in fig. 3.26 from the study of Williams et al. (1976) for Ni(111)–CO is in agreement with the cluster calculations of Batra and Bagnus (1975) but is not the one generally accepted (see Liebsch 1978a, Feuerbacher et al. 1978).

On the other hand, consideration of orbital symmetry selection rules using polarized light (Allyn et al. 1977, Smith et al. 1976), as well as the photon energy dependence of the cross sections for photoionization (Gustafsson et al. 1975) favor the same orbital assignment sequence as in the gas-phase, that is, starting from the Fermi level and moving to higher binding energies, 5σ, 1π and 4σ. Although this question has yet to be conclusively settled, the weight of accumulated results appear to be supportive of the view, in general, that the sequence of orbitals observed

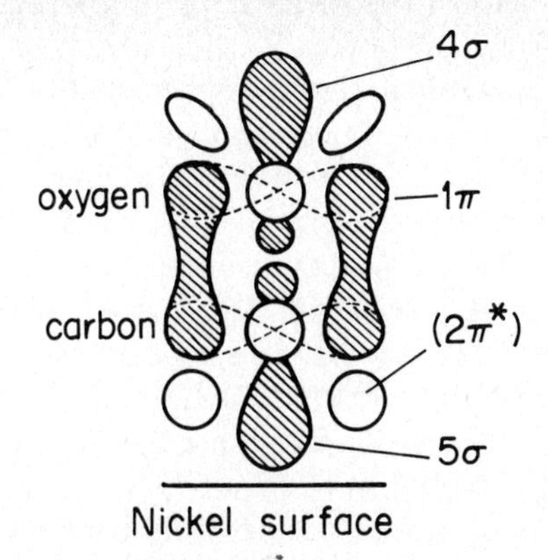

Fig. 3.25. Schematic molecular orbitals for CO adsorbed on a nickel surface in a normal orientation. The main 3σ bonding orbital (between the C and O) is not shown.

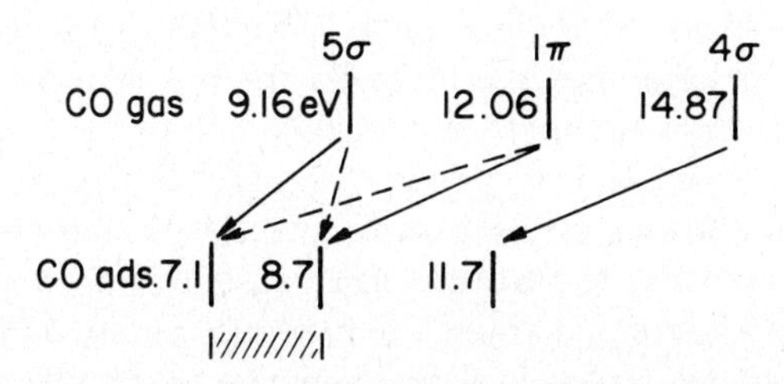

Fig. 3.26. Energy levels of molecular CO in the gas phase referred to the Fermi level (assuming work functions to be 4.85 eV) compared with binding energies of features in angle-resolved EDC's from Ni–CO adsorbate system. The proposed assignment of features is based on their polar angle dependence from which the binding energy of the 1π component was inferred (Williams et al. 1976). This peak assignment, however, is not that presently accepted for photoemission for a saturated CO-layer on Ni(001) at room temperature. The accepted assignment, based on parity and other considerations, is in the same order as for the gas phase system (see Liebsch 1977). The currently accepted arrangement is indicated by the solid arrows (see Gustafsson and Plummer 1977). The precise position of the 5σ and 1π peaks varies for different chemisorption systems; neither is the $5\sigma - 1\pi$ spacing precisely determined.

for a specific system is determined mostly by the degree to which the 5σ orbital is shifted relative to the other two upon adsorption. A precise assignment of the three peaks' positions needs to be defined for each specific adsorption system.

In addition to the photoemission studies on Ni(100), a series of studies of vibrational excitation using high-resolution electron energy loss spectroscopy combined with LEED structure measurements by Andersson (1976, 1977a) on clean and chemisorbed Ni(100), have contributed additional insight into the influence of ordered chemisorbed layers on chemical binding on this metal. Stretching vibrations of linearly bonded CO corresponding to the Ni–C and C–O bonds are observed for the Ni(100)–c(2 × 2)CO structure (Andersson 1977a). A non-ordered structure forming at low temperatures and small coverages is concluded to correspond to two bond configurations, a linear mode and a second mode which appears to indicate stretching vibrations of CO bridge-bonded to two nearest neighbor nickel atoms. Pre-exposure to hydrogen at 175 K immediately causes the bridge-bonded component to become dominant (Andersson 1977b). Hydrogen is considered to be dissociatively chemisorbed on Ni(100)–CO–H but no C–H or O–H vibrations are detected, indicating the absence of any direct bonding between chemisorbed CO and hydrogen. The sequence of Ni–C and C–O stretching vibrational energies observed is what one expects for a bridge-bonded arrangement. Whereas the clean surface shows linear as well as bridge-bonded sites, CO chemisorption on oxygenated nickel [Ni(100p(2 × 2)O and c(2 × 2)O] (Andersson 1977b) also shows principally the presence of CO molecules bridge-bonded to two nearest neighbor Ni atoms. It appears that the chemisorbed hydrogen and oxygen atoms affect the electron structure of the nickel surface atoms so as to promote the bridge-bonded over the linearly bonded modes. Vibrational spectroscopy of CO-chemisorption on well defined ordered chemical overlayers pursued in this manner provides a promising approach to probing the systematics of sequential surface bonding on clean and chemisorbed surfaces of nickel. It is premature at this stage to comment on combined vibrational studies with photoemission results in this adsorption system. Correlation of these two powerful approaches, however, applied to a well defined system of ordered chemisorbed layers can provide essential details on molecular chemisorption on a typical 3d-transition metal such as nickel.

Carbon monoxide and nitric oxide on platinum

Although the Ni–CO adsorption system has developed as a prototype for recent comparative photoemission and vibrational spectroscopy studies of surface–molecular bonding, significant angle-integrated and angle-resolved photoemission studies have also been reported for the Pt–CO, and Pt–NO systems as well as the Ir–CO and Ir–NO adsorption systems. Chemisorption studies on the noble metal surfaces provide a special insight into chemical bond formation. It should be noted that interpretation of the spectra is more complicated by the broader and more complex structure in the d-band characteristic of the clean 5d-transition metals. The problem of CO and NO adsorption on platinum and iridium is of particular importance for an understanding of the characteristic features of chemical bonding for the Fischer–Tropsch synthesis, in emission exhausts and for the ammonia-oxidation catalytic reactions.

Angle-resolved photoemission spectra of the molecular orbitals of the chemisorbed Pt(111)–CO surface shows a strong angular variation of the peak intensity ratio. Comparisons with the calculations of Davenport (1976) for the free gas molecule appear to indicate that the CO molecule is oriented normal to the surface (Apai et al. 1976) with the carbon end down, similar to that observed for Ni(100) (Allyn et al. 1977) and Ir(111) (Rhodin et al. 1977a, Rhodin and Brucker 1977b). It is also interesting that the 5d (t_{2g}) orbitals of the platinum from the UPS-spectra also appear to be strongly involved in the chemisorption bonding. The latter result is consistent with previous studies reported for chemisorbed rhodium using field electron tunneling spectroscopy (Dionne and Rhodin 1974).

Thermal desorption and photoemission were also used to study the adsorption of CO on the smooth hexagonally close packed (111) plane of platinum compared to the stepped 6(111) × (100) and the 6(111) × (111) surfaces. This work (Collins and Spicer 1977) explored the role of steps and defects on the adsorption of simple diatomic gases. It was indicated that adsorption on the steps leads to a higher binding desorption site than for the smooth surface. The CO adsorption on Pt(111) was also observed to indicate a resonance state 8 eV below the Fermi level. These interesting results were also correlated with work function changes. It is not possible, however, at this time to make any more detailed interpretations or conclusions for this system in terms of analysis of chemical interactions on these surfaces specific to molecular bonding modes.

Evidence for the existence of two different adsorption sites of CO on Pt(111) had also been obtained using highly sensitive double-beam infrared reflection spectroscopy by Krebs and Lüth (1977). Two stretching vibrations of adsorbed CO were observed at 58 and 258 eV for one site at low coverages and at 45 and 258 meV for a second site at higher coverages in agreement with the earlier work of Froitzheim et al. (1977a, b, c). The vibrational results correlated very nicely with LEED structures observed for the same two exposures by Ertl et al. (1977). In each case the losses correspond to Pt–C and CO stretching vibrations; the first for adsorption on an on-top position with a coordination of one and the second for adsorption on a bridge position. This is indicated by curves b and c in fig. 3.27 for the loss spectra and positions b and f in fig. 3.28 for the LEED structures. It appears for CO-adsorption that adsorption tends to occur in general on-top for low coverages (about $\theta - 0.3$) and

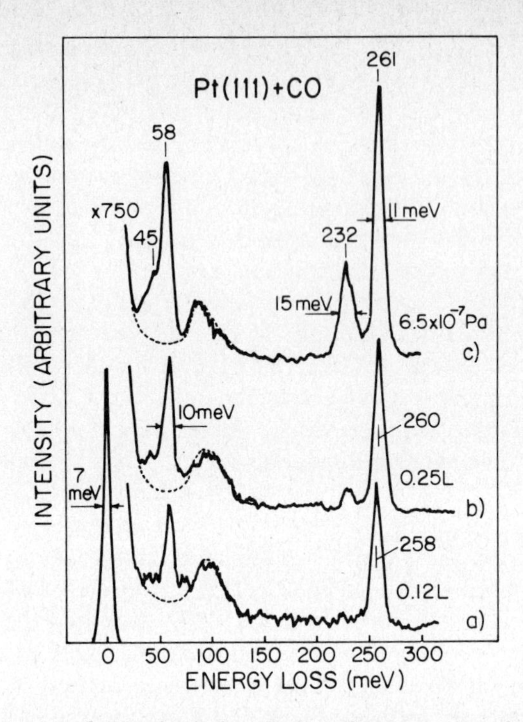

Fig. 3.27. ELS spectra of CO adsorbed on Pt(111). Spectrum (a) and (b) were obtained after exposing the crystal to 0.12 and 0.25 L CO, respectively with the crystal temperature at 320 K. Spectrum (c) was obtained with the crystal at 350 K exposed to an ambient pressure of 6.5×10^{-8} Torr. An identical spectrum (except for 20% reduction in the intensity of the 260 meV loss) was obtained after exposing the crystal to 5 L CO at 320 K. The dashed curve indicates the background obtained from a clean surface. (Krebs and Lüth 1977.)

becomes a mixture of about 1 : 1 of on-top and bridge-bonded molecules at higher coverages (about 0.5). Detailed information of this kind on the dependence of vibrational modes and surface lattice structure on surface coverage is essential to develop theoretical models and to make general selection rules for molecular interactions during adsorption.

The behavior of NO on clean Pt(100) studied by Auger, LEED and photoelectron spectroscopy (Bonzel and Pirug 1976) was also found to interact in the molecular mode but only in the lower temperature range (313–473 K). Dissociation into chemisorbed oxygen and desorbed nitrogen occurs at higher temperatures. The indications are that the NO-molecule is chemisorbed in a bent position on the surface from the observed inequality of the XPS-photoelectron electron binding energies

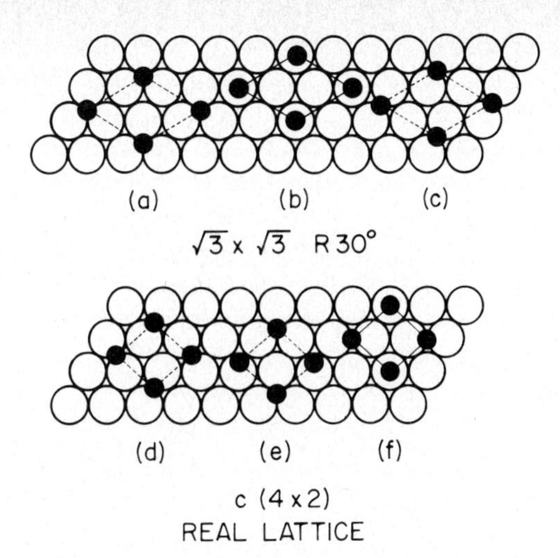

Fig. 3.28. Different positions of the observed unit meshes with respect to the substrate. The combination (b)–(f) is favored by the vibration spectra (Ertl et al. 1977).

noted for the O(1s) and N(1s) core levels of 6.8 and 4.2 eV respectively. The interesting analogies between extended metal chemisorption and metal-cluster geometries with reference to bond orientation effects for CO and NO will be considered again with reference to CO- and NO-chemisorption on iridium (Rhodin et al. 1977a, Kanski and Rhodin 1977, Rhodin and Brucker 1977b).

Carbon monoxide and nitric oxide on iridium

Studies of molecular adsorption on iridium using photoemission are of interest mainly because its clean single crystal surfaces are readily prepared; there exists an increasing body of chemisorption data from other types of surface probes (LEED, TDS, etc.), and its chemisorption behavior shows distinct and interesting differences from that of platinum. Consideration of metallic clusters based on iridium also presents interesting interpretations of bonding and atomic structure by comparison to the analogous photoemission measurements.

That the clean (100) surface exists in two different crystallographic configurations presents an opportunity to investigate effects of surface crystallography on surface chemical bonding and reactivity (Rhodin and Broden 1976). The microscopic nature in terms of electron structure should reflect the differences in atomic structure characteristic of the two forms of the Ir(100) surface. The normal Ir(100)–(1 × 1) surface structure is equivalent to an extension of the bulk lattice whereas the reconstructured Ir(100)–(5 × 1) surface structure corresponds to a close-packed more stable arrangement primarily of the surface layer. The atomic arrangements of these two structures plus the corresponding Ir(111) structure are shown in fig. 3.29. The electron density and related properties of the surface band structure of these related but different surface structures should be reflected in different chemical bonding and reactivity behavior. The use of these surfaces as a controlled structural probe for this purpose will be discussed later with reference to chemisorption of carbon monoxide, nitric oxide and unsaturated hydrocarbons on these surfaces. Chemisorption of the first two adsorbates shows a striking and not unexpected similarity of behavior between the

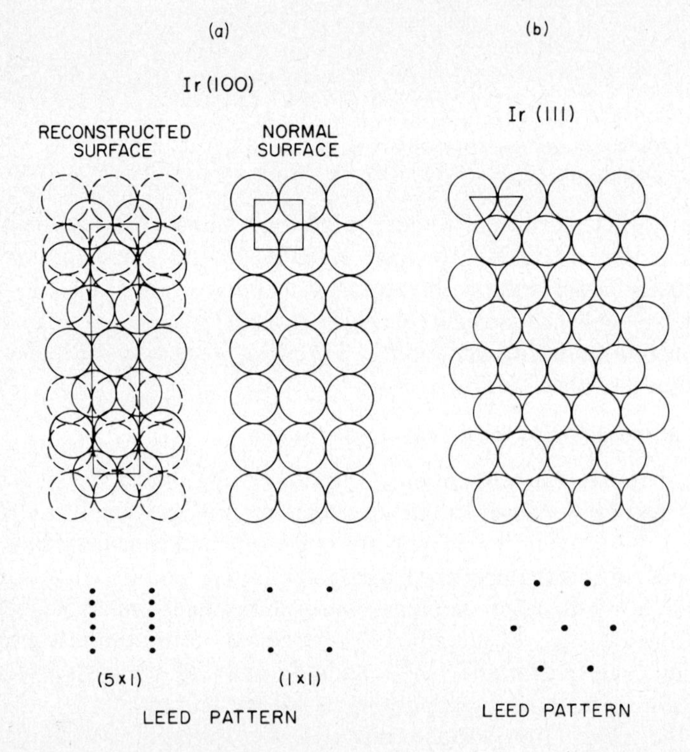

Fig. 3.29. Schematic representation of the outer atomic layers of an Ir(100) surface in its reconstructed (left) (5 × 1) structure and in its nonreconstructed (middle) structure. The outermost atomic layer is indicated by solid circles whereas the dashed circles give the positions of the atoms in the second layer for the reconstructed surface. The structure of the (111) surface is included for comparison to the (5 × 1). It should be noted that all three surfaces have been shown to be free of trace contaminants within the sensitivity limit of Auger spectroscopic analysis (see Rhodin and Brodén 1976).

close-packed Ir(100)–(5 × 1) and the Ir(111) structures (Kanski and Rhodin 1977).

That it possesses the broad and complicated d-band character typical of the third row transition metals adds an additional complication to the interpretation of the electron structure (see fig. 3.30a) for spectra typical of the clean iridium surface.

Synchrotron radiation in both the angle-resolved and angle-integrated modes has been applied to a study of the orbital photoionization cross section behavior over a photon energy range up to about 40 eV for

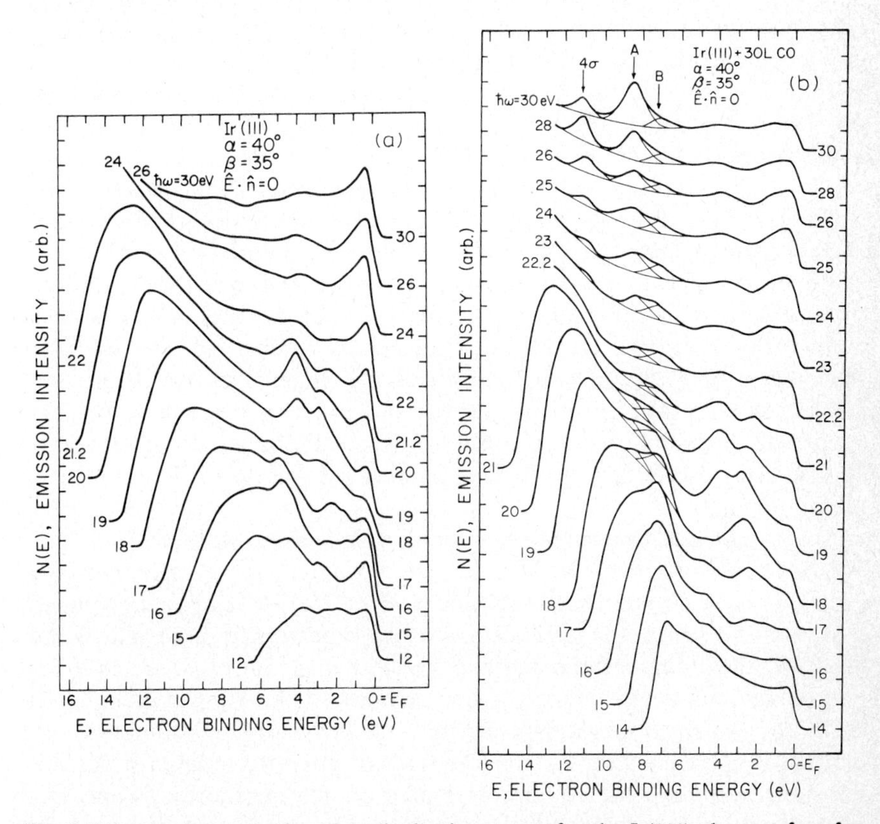

Fig. 3.30 (a) Angle-averaged energy distribution curves for the Ir(111) clean surface for photon energies between 12 and 30 eV. The photon polarization vector was in the plane of the surface, e.g. s-polarization. α and β are the angles of incidence and collector axis respectively with reference to the normal of the crystal plane (Brodén et al. 1976). (b) Angle-averaged energy distribution curves for the Ir(111) surface with a chemisorbed layer of CO at 300 K for photon energies between 12 and 30 eV. All other conditions are the same as in fig. 3.19a (Brodén et al. 1976).

Ir(100)–CO and Ir(111)–CO for different polarization modes (Rhodin et al. 1977a, Rhodin and Brucker 1977b). A selected set of angle-averaged data for s-polarization on the chemisorbed Ir(111) is shown in fig. 3.30b. Three CO-chemisorption features associated with the 1π, 5σ, and 4σ CO valence molecular orbitals are apparent. The assignment of the 1π and 5σ peaks are assumed to be in the same energy sequence as that reported for the Ni(100)–CO system (Allyn et al. 1977a). It should be noted that the precise position of the 5σ-peak is difficult to predict with certainty. A common method for determining chemisorbed orbital parentage for the gas-phase spectra is to compare the photoionization cross section dependence on photon energy for the free and chemisorbed molecule. This is done for two different polarizations for the Ir(100)–CO, the Ir(111)–CO adsorption systems and for gaesous CO in fig. 3.31. The data appear to indicate that only the 4σ-peak can be unambiquously identified in this manner. It is clear that *both* the 1π and 5σ levels are strongly perturbed upon chemisorption in this system.

Note that the intensity of the final state resonance observed at about 28 eV is always larger for the s/p- than for the s-polarization. Although this does not prove conclusively that the molecule is oriented normal to the surface, it is the expected result for that orientation. It is also noteworthy that the curves, and hence the nature of the bonding reflected in both the 1π and 5σ orbitals, are quite different for the two crystal faces as well as for the gaseous spectra. This is taken to indicate occurrence of some distinctive differences in the relative contributions of these two orbitals to the surface binding for the two different crystal orientations.

Angle-resolved measurements were also made with s- and s/p-polarized photon excitation. Polar angle variations were essentially the same for both crystal faces (Rhodin et al. 1977a). It is again concluded that both the 1π and 5σ orbitals are strongly perturbed upon adsorption and that the origin of the marked variations in the cross section for photoionization stems mainly from changes in bonding modes for a molecule oriented essentially normal to the surface within an uncertainity of 20–30 degrees. It is indicated that, depending on surface coverage, the molecule could be oriented off the normal by as much as 30 degrees. It has already been stated that complete interpretation of the cross-sectional measurements depends on the quantitative evaluation of a number of important factors including the electron structures of the molecule and of the substrate. Both need to be coupled into the geometries of the scattering process as well as the energy and polarization of the photon excitation. It is possible, however, that the observed

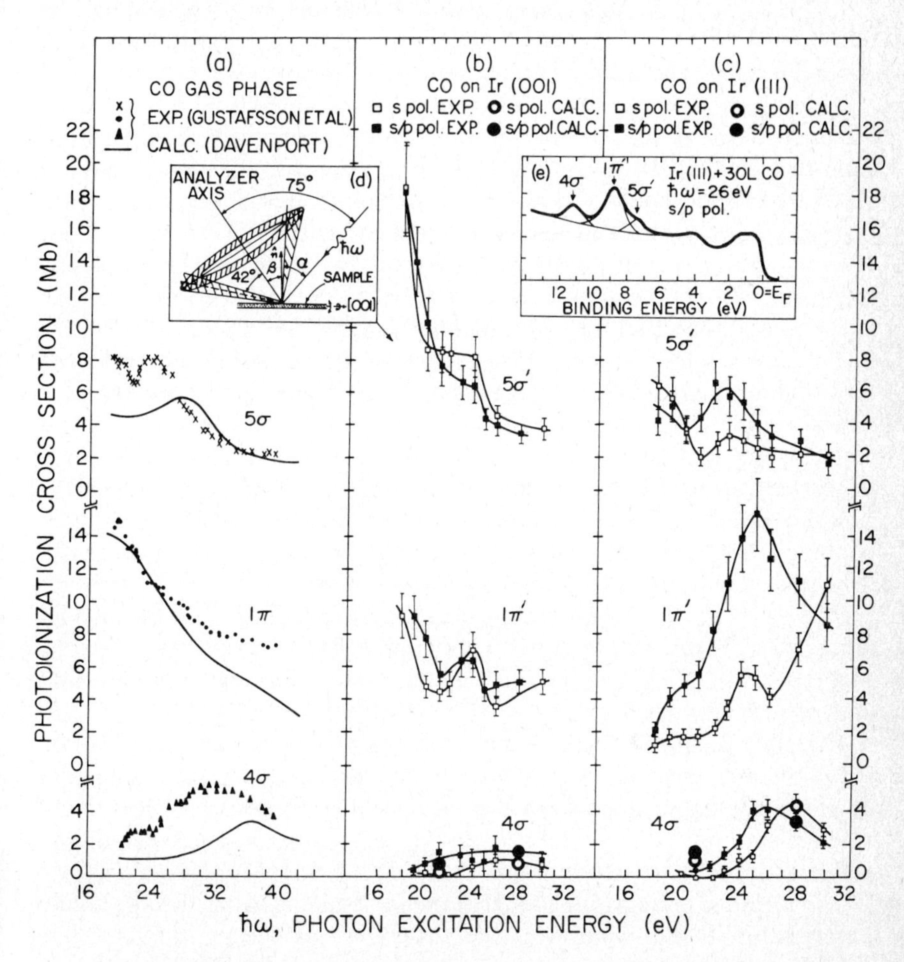

Fig. 3.31 (a) Photoionization cross sections for the 5σ, 1π, and 4σ molecular orbitals of gaseous CO. The data points are from Plummer et al. (1977). The solid curves are calculated Sw X^{α} results from Davenport (1976). Note the difference in horizontal scale compared to (b) and (c). (Plummer et al. 1977). (c) Corresponding data for Ir(111). Compare to Ir(100)–(5 × 1) (fig. 3.31b). (b) CO-induced peak areas for chemisorption on Ir(100)–(5 × 1) as a function of photon energy and polarization mode. For s-polarization, $\vec{A}$ was in the plane of the surface; for s/p-polarization, there was a substantial component of $\vec{A}$ normal to the surface (Rhodin et al. 1977a). Note vertical units relate to peak areas.

variations in photoionization with photon energy are more related to different bonding modes rather than to different orientations on the Ir(100) and Ir(111) surfaces. The latter would also depend on both crystallography and on the surface density of adsorbate.

A change in bonding with coverage is also indicated by the occurrence of a split-off peak from the 1π orbital for angle-averaged photoemission from the Ir(100)–CO adsorption system at room temperature (Rhodin et al. 1977b). The shoulder-structure is strongest at low coverages (less than 0.3 monolayers) and decreases with increasing coverage to about 0.6 monolayers until it disappears at monolayer coverage. These results correlate with the occurence of two LEED structures, a low coverage $(\sqrt{3} \times \sqrt{3})$ R30° structure and a high coverage $(2\sqrt{3} \times 2\sqrt{3})$ R30° structure (Comrie and Wienberg 1976). Thermal desorption likewise shows two binding sites at higher coverages and only one at lower coverages. It appears likely that the shoulder-structure in the UPS-spectra is associated with the higher energy bonding mode at low coverages which is swallowed up in the spectra as the coverage increases to a monolayer (Rhodin et al. 1977a). Transitions from terminal (on-top) to bridge-bonding with coverage in about exactly the same coverage range, indicated from the EELS-loss spectroscopy for the Pt(111)–CO adsorption system, discussed previously (Krebs and Luth 1977, Froitzheim et al. 1977a, b, c, Ertl et al. 1977) may apply here. In this case a strong analogy in the CO-adsorption binding characteristics appears to be valid between Pt(111) and Ir(111).

The chemisorptive bonding and reactions of nitric oxide on transition group metals, particularly the noble metals like platinum and iridium are of particular importance from the viewpoint of understanding reaction mechanisms on catalysts in automotive exhaust emissions (Kliminisch and Larson 1975). Photoemission studies for the Ir(100)–NO and Ir(111)–NO adsorption systems indicate that the molecule is partly dissociatively adsorbed at 300 K, confirming the borderline position of iridium in the periodic table with respect to NO-molecular versus dissociative adsorption (Brodén et al. 1976a, b). Three distinctively different bonding modes are indicated by the occurrence of split-off structure from the 1π orbitals in the UPS-spectra (Kanski and Rhodin 1977). One of the split-off features is concluded to indicate the presence of a skewed position of the molecule on the surface.

In this case it is likely that the adsorbed molecule is bent through the bond itself as distinct from a molecule which is simply tilted with reference to the plane of the surface. Photoelectron spectra representing cases of different degrees of rehibridization of the metal–nitrogen bond would be expected to show variations in the positions and shapes of the orbitals involved in surface bond formation such as the 5σ and 1π orbitals. The observed differences in the split-off structures characteristic of the photoemission spectra for Ir(100)–NO and Ir(111)–NO indicate the occurrence of different degrees of rehydridization (Kanski and Rhodin 1977). In addition, since a "shaped resonance" emission feature is also characteristic of photon stimulated emission from the 4σ orbital of NO, the emitted intensity should depend sensitively on the polarization of the photon beam. Recent UPS-spectral data on Ir(111)–NO at about 35 eV show substantial enhancement using s-polarized versus s/p-polarized excitation (Seabury and Rhodin 1978). This is a good indication that the adsorbed molecule is bonded in a skewed orientation to the surface on Ir(111). Finally, it is noteworthy that skewed bonding of NO due to rehybridization of the metal–nitrogen bond is not unusual in nitrosyl-stabilized metallic clusters of iridium (Cotton and Wilkinson 1972). Similar results for photoemission spectra for Ni(100)–NO corresponding to a shaped resonance reported by Loubriel et al. (1977) suggest an angle of 30 degrees between the NO molecule and the surface.

Carbon monoxide and nitric oxide on iron

The fact that iron surfaces possess unique behavior in their interactions with molecular gases is consistent with its position in the middle of the first row of the 3d-transition metals. It is proposed that borderline chemisorption behavior (Brodén et al. 1976a, b) is often an essential feature of useful catalytic processes characteristic, for example of the ammonia and Fischer–Tröpsch syntheses. The uniqueness of iron base alloys as a corrosion-resistant structural material is also connected with their unique capability to form passivated and other corrosion-resistant surface states. Reasons for this unique chemical bonding behavior of iron surfaces towards molecular gases do not respond to simple analysis. Interpretation of chemical bonding on iron is now being facilitated by the use of electron spectroscopic methods in conjunction with other surface diagnostic probes.

Carbon monoxide adsorbs molecularly at 320 K on Fe(110) (Brodén et al. 1977) at room temperature compared to the dissociative chemisorption which occurs on Fe(100) (Brucker and Rhodin 1978) at the same temperature. This distinction also occurs in a similar manner for W(110)–CO (Plummer et al. 1976) and W(100)–CO (Plummer et al. 1976) at the same temperature. A qualitative rationalization of this behavior was discussed previously based on the energy barrier for dissociation of diatomic molecules on transition metals (Brodén et al. 1976a, b). See figs. 3.21a, b, and c illustrating the spectra, the orbitals, and the tabulation respectively for application of this principle to the chemisorption of CO, NO and N_2. The borderline between molecular and dissociative chemisorption for CO at room temperature runs from elements Fe to W in the periodic table (Brodén et al. 1976a, b). Iron is clearly a borderline element since the CO-dissociation process is sensitive to both crystal face and surface defect structure.

It is not surprising, similiarly, that NO is dissociatively chemisorbed on Fe(110) (Brodén et al. 1977) at 320 K since the borderline for dissociative chemisorption of NO at room temperature runs from the elements Ni to Ir (Brodén et al. 1976a, b). Iron is far to the left in this category. The structural sensitivity for N_2-dissociative chemisorption likewise follows a similiarly consistent pattern. It is dissociatively chemisorbed on αFe(111) (Bozso et al. 1978, Brill et al. 1967) at room temperature, difficultly adsorbed on Fe(100) and not adsorbed at all on Fe(110) (Brodén et al. 1977). The borderline for dissociative chemisorption at room temperature for nitrogen runs from the elements Fe to W (or Re) (Brodén et al. 1976a, b) both of which are structure-sensitive to molecular vs. dissociative adsorption. The chemical bonding of nitrogen on iron is of particular interest because of the importance of nitrogen chemisorption in the ammonia synthesis on iron-based catalysts (Brill et al. 1976). The rate-determining step for ammonia synthesis is thought to involve the dissociative chemisorption of nitrogen (Ertl et al. 1976). The surface crystallography of the iron surface is indicated to be important in that both the ammonia production rate and chemisorption seems to occur most effectively on the Fe(111) surface. This crystal surface is just the one upon which the production of atomic nitrogen is optimized. Study of chemical bonding characteristics of NH_3 on transition metals and particularly in the Fe(111)–N_2 and the Fe(111)–NH_3 systems using electron spectroscopy (Grunze and Ertl 1977), show interesting correlations between the dependence of the preferred reaction path and chemical bonding modes on the surface crystallography for certain transition metals such as Ni(111) and Ir(111) (Purtell et al. 1978).

Carbon monoxide is partly dissociatively adsorbed on αFe(100) and more completely dissociatively adsorbed on polycrystalline iron at room temperature as indicated by detailed XPS-photoelectron emission studies on these two surfaces (Brundle 1978). It should be noted that polycrystalline iron in these studies is considered to be composed primarily of (110) facets.

Different molecular CO-binding states, observed on Fe(100) (Brundle 1978), and compared to polycrystalline iron (Yu et al. 1976b), indicate differences in bond length or orientation in the CO-chemisorbed molecule. The idea of a stretched CO bond for adsorption on polycrystalline iron can be rationalized from increased back-donation from the metal substrate into the anti-bonding $2\pi^*$ CO orbital, thus weakening it. This is consistent with the observation of a tendency towards enhanced dissociation on that surface. The LEED evidence indicates that CO adsorption does not develop into a (1×1) structure through island nucleation but stops at 0.5 monolayer coverage. Apparently the adsorbed CO becomes stabilized towards dissociation by the partly covered iron surface.

Adsorption of CO at 123 K produces another state of molecular adsorption which is essentially the same on both the αFe(100) (Brundle 1978, Rhodin and Brucker 1977a) and polycrystalline (Yu et al. 1976) surfaces. This apparently is a weakly chemisorbed state similar to that which occurs at room temperature (Brucker and Rhodin 1977) as indicated by the shifts of the binding energies of the C(1s) and O(1s) to higher values. This result is in essential agreement with UPS-photoemission studies of the Fe(100)–CO (Brucker and Rhodin 1978) adsorption system although the former is richer in the details of chemical bonding accessible to the XPS measurements. Results similar to the XPS study on Fe(110) have also been observed for UPS-photoemission on polycrystalline iron (Yu et al. 1976). It is clear that more detailed studies of the dependence modes on surface geometry for the different adsorption stages are required for a complete understanding of chemical bonding for Fe–CO. It is interesting that the Fe–CO adsorption system possesses many of the complexities of the W–CO adsorption system which also remain to be completely interpreted (Plummer et al. 1976).

Clean and partially deactivated αFe(100) surfaces show variable chemical activity towards carbon monoxide chemisorption, depending on preparation (Rhodin and Brucker 1977a). On clean αFe(100) surfaces, for example, UPS-photoelectron spectra indicate severe stretching of the CO bond at low temperatures (98–223 K) and complete dissociation at higher temperatures (300 K). Prechemisorption of C, O, P, or S will quench to varying degrees the ability of the surface to influence dissociative adsorption of CO. Chemisorption of sulfur to form Fe(100)–c(2×2)S will inhibit the surface towards dissociative CO-chemisorption at room temperature and lead to an adsorbed molecule in the unstretched state (Rhodin and Brucker 1977a). These effects can all be closely monitored by the position and intensity of the 5σ and 1π peaks which are clearly resolvable in the UPS-spectra for adsorption on Fe(100)–c(2 × 2)S but not for the spectra from clean αFe(100). There appears to be a very significant effect of the adsorbed sulfur in inhibiting the typically strong reactivity of the surface iron towards molecular interactions at room temperature.

4.5. Chemisorption and decomposition of hydrocarbons

It is particularly appropriate to discuss applications of electron spectroscopic measurements to chemical bonding on surfaces with specific reference to the interaction of simple unsaturated hydrocarbons with transition metals. Work is advancing in this area and significant interpretations are being achieved in terms of basic chemical bonding models relative to carbon–hydrogen chemistry. Consideration here will be directed towards those specific adsorption systems which appear to be reasonably understood and which illustrate principles particularly relevant to electron structure and chemical reactivity of carbon–carbon and carbon–hydrogen bonding at surfaces. The most useful general approach has been UPS-photoemission combined with EELS-loss spectroscopy and LEED. Reference will be made primarily to the adsorbates; acetylene, ethylene and benzene, and to the adsorbents; close-packed and/or (100) single crystal planes of nickel, platinum, iridium, tungsten or iron.

The main use of the photoemission spectra is to provide a chemical finger-print with which to track a given molecule or molecular fragment through a series of bonding transformations resulting from either kinetic or steady-state perturbations. Specific orbitals can often be indexed with reference to gas-phase information, and in some favorable cases qualitatively correlated with changes in peak shapes or position in terms of chemical bonding. Only in a few of the most favorable situations can the changes in peak properties be quantitatively related to changes in the nature of the bonding chemistry. Achievements in the last category are improving steadily with advances in calculational analysis relating photoemission theory to chemisorption and in the opportunity of combining complementary data on chemical bond formation from different spectroscopies.

The usefulness of electron loss spectroscopy for the characterization of intermediate molecular entities is particularly effective when coupled with the UPS-photoemission spectroscopy of hydrocarbons on reactive transition metal surfaces. This point is well illustrated by Ibach et al. (1977) for the interpretation of proposed reaction intermediates involved in the chemisorptive behavior of acetylene and ethylene on Pt(111) surfaces and the comparison to similar adsorption results for the close-packed (111) faces of other fcc metals such as nickel and iridium. Analogies from data on molecular crystallography, NMR-shifts and chemical thermodynamics derived independently for metallo–organo compounds and for ligand-stabilized metal clusters can also be useful in some cases (Muetterties et al. 1979). To achieve a more detailed understanding of surface reactions, it is required to obtain a reasonably clear concept of the chemical state of the major adsorbed species involved in the surface reaction mechanism. The energy levels shift in energy and vary in nature for hydrocarbon adsorbates to a greater degree than diatomic inorganic molecules relative to the spectra of the gaesous molecules. As previously indicated with reference to photoemission spectra in general, these shifts result from non-bonding relaxation and polarization effects as well as changes in chemical bonding. Overall relaxation shifts include final-state image charge screening as well as initial or final state changes associated with molecular polarization and charge transfer from the molecule. They have less to do directly with chemical bonding and often are difficult to assess precisely.

The greater number of bonding and non-bonding orbitals typical of adsorbed unsaturated hydrocarbon on transition metals can also facilitate the separation of these orbitals. It is generally observed that bonding occurs through the π–d orbital interactions and the π–d interaction strengths can be estimated. The surface-induced relaxation shifts are often less severe for unsaturated hydrocarbon adsorbates than for many diatomic or atomic gaseous adsorbates. Reference to the theoretical models discussed by Grimley and by Messmer (chapters 1 and 2) for a more detailed discussion is recommended.

The study of UPS-spectra associated with condensation of physical adsorption can provide a useful method in some cases to isolate the chemical bonding effects, being primarily concerned with those associated mainly with physical origins. This approach was implemented by Yu et al. (1975) in an experimental evaluation of the physical contribution to the relaxation shift in photoemission spectra of a number of hydrocarbons on an inert substrate. Another interpretative effort in photoemission spectra analysis was also made to relate the heat of chemisorption to the magnitude of the π and σ level shifts (Yu et al. 1976). Although useful correlations can be made in some cases, sufficient complications generally exist in the spectra to make this method unreliable. These complications are often due to difficulties in assessing the interactive contributions of the metal surface with the electron structure of the adsorbate and to the observed electron emission patterns.

We shall try to consider those effects which illustrate particular principles pertinent to the characterization of surface bonding of hydrocarbons. Reference should be made to the literature for more extensive reviews on various aspects of the subject (Plummer 1975a, Ibach et al. 1977, Demuth 1978a, b). It is timely to start with consideration of the molecular geometries of acetylene and ethylene on the (111) surfaces of nickel and related fcc transition metals. It was the original study by Demuth and Eastman (1974) on the adsorption of C_2H_2 and C_2H_4 on Ni(111) and by Plummer et al. (1974) on tungsten which were the forerunners of the current UPS-photoemission research on the electroemission spectra of simple hydrocarbon adsorbates.

Unsaturated hydrocarbons on fcc metals

Unsaturated hydrocarbons on the close-packed planes of the fcc metals in group VIII of the periodic table presents a useful system with which to initiate consideration. The spectra can be readily used to distinguish non-bonding π and σ orbital relaxation shifts from π orbital bonding shifts. Interesting photoemission spectral difference curves with reference to the clean metal are shown in figs. 3.32 and 3.33 for approximately monolayer coverages of benzene, and partial monolayer coverage of acetylene and ethylene on Ni(111) at 300 K using $\hbar\omega = 21.2$ eV (Demuth and Eastman 1974).

It is noted that all the orbital energies and intensities are essentially unchanged for weakly bound condensed benzene compared to the gas, except that all the ionization potentials measured relative to the vacuum level are reduced by a relaxation shift of about 1.4 eV (fig. 3.32). In addition, comparing the chemisorbed with the condensed benzene spectra, the binding energies of all the lower-lying σ orbitals are further reduced 0.3 eV, while the binding energy of the two upper-most degenerate π orbitals is increased. The latter shift is associated with π–d chemical bonding. There is, in addition, an overall attenuation in the d-band emission of the metal particularly near the Fermi level with some corresponding enhancement observed at somewhat higher binding energies (2.5 eV). The observed attenuation in the Fermi level intensity is a general feature of these spectra. It is apparently caused by the influence of the π–d bonding in the adsorbate on the d orbitals of the metal surface. Such π orbital bonding shifts are generally observed for chemisorbed unsaturated hydrocarbon adsorbates and not observed for saturated hydrocarbon adsorbates which lack π orbitals.

The adsorbate-induced difference curves for chemisorbed and condensed acetylene and ethylene, shown in fig. 3.33, are observed to

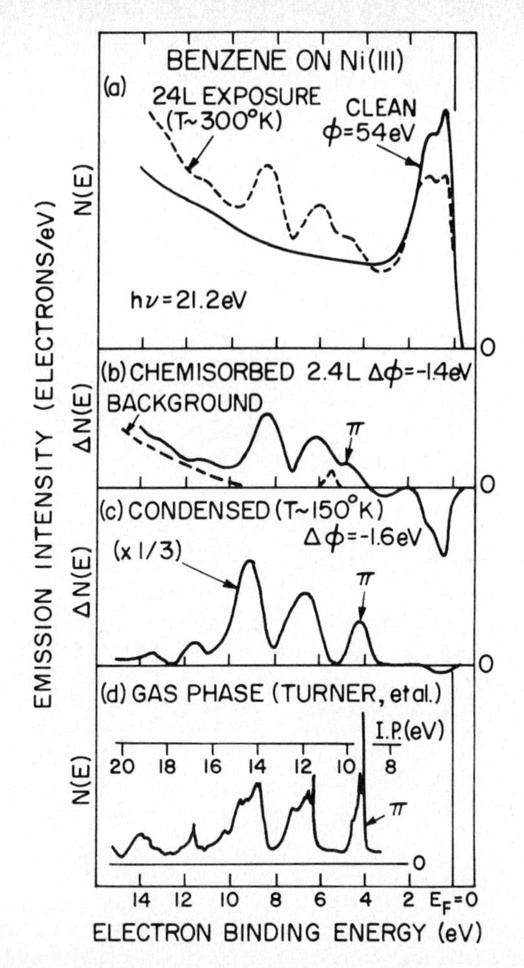

Fig. 3.32. (a) Photoemission spectra $N(E)$ for Ni(111) and with 2.4×10^{-6} Torr-seconds benzene exposure at $T \sim 300$ K, (b) adsorbate induced difference in emission, $\Delta N(E)$, for a condensed benzene layer formed at $T \sim 150$ K with a benzene pressure of 2×10^{-7} Torr, (d) gas phase photoelectron spectra for benzene. Note for adsorbates that ionization levels $\varepsilon_i = E_i$ (binding energy) $+ \phi(ni) + \Delta\phi$ (see Demuth and Eastman 1974).

have the same general behavior as the benzene adsorbate. To dramatize the downward bonding shifts of the uppermost π levels and the overall non-bonding shifts for all levels, all the energy levels are placed relative to the σ orbitals' levels of the gaseous species, shown in fig. 3.34. These data (figs. 3.32 and 3.33) have been presented in some detail to illustrate a generally used method for depicting the UPS spectral results. If the

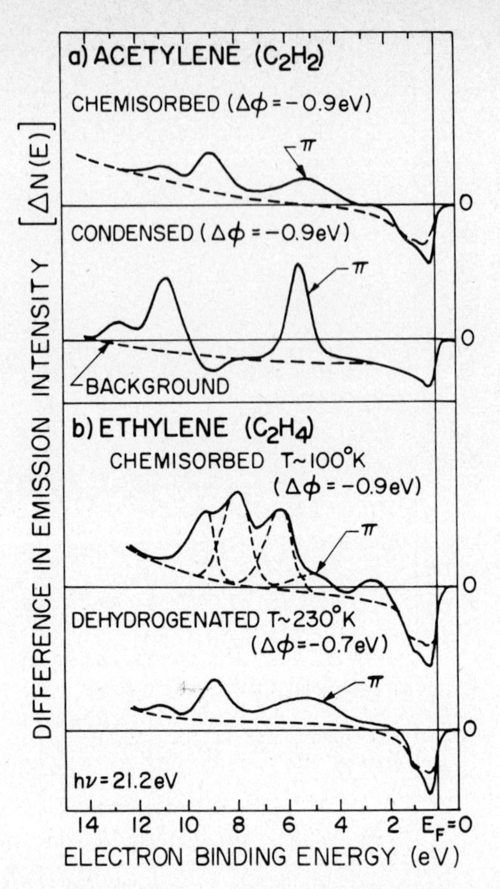

Fig. 3.33. (a) Difference in emission $\Delta N(E)$ for 1.2×10^{-6} Torr-seconds exposure to acetylene at $T\sim100$ K and for condensed acetylene formed at $T\sim100$ K with an acetylene pressure of 6×10^{-8} Torr. (b) $\Delta N(E)$ for chemisorbed ethylene (exposure of 1.2×10^{-6} Torr-seconds at $T\sim100$ K) and for dehydrogenated ethylene (obtained by warming to $T\sim230$ K or with an initial exposure at $T\sim300$ K) (see Demuth and Eastman 1974).

chemisorbed ethylene is warmed to 230 K or if the initial adsorption is done at that temperature instead of at 100 K, the effect is striking, as indicated by the lowest difference curve in fig. 3.33. A comparison of the peaks in the spectra for both ethylene and acetylene indicates that the ethylene has been partly dehydrogenated to an acetylenic complex. It is convincingly demonstrated that, except for some chemisorbed hydrogen, the difference spectra for chemisorbed acetylene is nearly identical to that for dehydrogenated ethylene in this case. A useful conclusion can be made with reference to the ionization levels of the chemisorbed

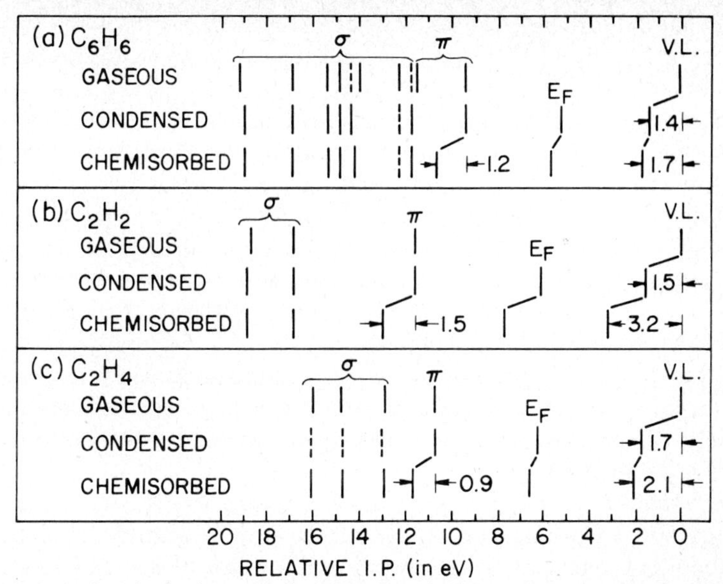

Fig. 3.34. Vertical ionization energies (IP), Fermi (E_F) and vacuum levels (V.L.) for the gaseous, condensed and chemisorbed phases of (a) benzene, (b) acetylene, and (c) ethylene, all plotted relative to the σ-orbital IP's for the gas phase. Relaxation shifts are given by the vacuum level shifts while bonding shifts are given by relevant π-orbital shifts. The dotted levels represent less certain orbital ionization energies (Demuth and Eastman 1974).

benzene as indicated in fig. 3.34. From the similarity of the levels of the former with those for gaseous and condensed benzene, it is likely that the structure and probably the aromaticity of the benzene ring is preserved upon chemisorption. There is additional information on electron structural changes during chemisorption included in the observed shifts of the ionization potentials. They are, however, rather difficult to analyze theoretically. Relaxation effects are likely to be important indirectly for all adsorbates. A useful approach, however, from the viewpoint of interpreting relaxation effects to a first order approximation, is to make a uniform correction for them in the spectra on the assumption that they have little to do directly with surface chemical bond formation.

It should be noted in this context that any modification of the geometry of the molecule upon adsorption referring to bond lengths or angles associated, for example, with state of hybridization of the carbon atoms in a hydrocarbon adsorbate, will be reflected in the UPS-spectra. The state of hybridization of unsaturated hydrocarbons chemisorbed on copper, nickel, palladium and platinum single crystal surfaces has been considered in some detail from this viewpoint by Demuth and Eastman

(1976). The molecular geometries of acetylene chemisorbed on the (111) surfaces of these metals at 80 K were determined in a systematic study from comparison of the relative photoionization energies of these species to Hartree–Fock ground state energies for distorted molecules calculated using a self-consistent-field linear-combination-of-atomic-orbitals method. The observed trends can be accounted for by the electronic structure of the substrate in terms of the component of rehybridization introduced for the degree of molecular distortion which results upon adsorption. The degree of rehybridization is negligible for adsorption on nickel but increases with the atomic number of the substrate from palladium to iridium. The molecular geometry of ethylene adsorbed on Pd, Pt (and Ir as well) is characteristic of rehybridization to a sp^3 configuration rather than the undistorted sp^2 planar arrangement. These trends are consistent with π–d bonding in all cases with a corresponding increase in its contribution as one progresses from adsorption on Cu to that on Ir.

These different observations and interpretations of unsaturated hydrocarbon adsorbates on these metals may be summed up: (i) The π–d bonding interaction commonly occurs for these adsorbates on extended surfaces of transition metals. It is similar to that which occurs in organometallic compounds. Bonding involves a mixture of occupied metallic d-states with unoccupied molecular π^*-states and occupied molecular π-states with occupied metallic d*-states to produce both dative and back-bonding components respectively. (ii) The heat of adsorption and resultant distortions of the adsorbate upon adsorption derive from this back-bonding component. The π–d bonding-interaction would be expected to increase with the spatially extended d-wave functions of the adsorbent, e.g. the atomic number for transition metals. The increase $d \rightarrow \pi$ adsorbate–adsorbent interaction is associated with a corresponding increase in the $d \rightarrow \pi^*$ component of back-bonding as one proceeds from the first to the third row of the group VIII transition metals. (iii) The degree of distortion in the chemisorbed molecule varies inversely with the carbon–carbon separation since the π–π^* level separation varies the same way.

Using this information the unperturbed π and π^* levels of acetylene and ethylene can be positioned relative to the d-states in the metal surface to indicate the separation between the initial π^*-level and the highest lying occupied d-states of the metal. The smaller the energy of the corresponding $d \rightarrow \pi^*$ interaction, the stronger the $\pi \rightarrow d$ back-bonding and the greater the distortion of the molecule upon adsorption. These qualitative observations are generally applicable for the interpretation of simple unsaturated hydrocarbon adsorbates on transition metals. A more quantitative description requires detail information of the π level ground state energies, of the $\pi \rightarrow \pi^*$ transition energies, of the d-wave functions of the surface metal atoms and of the local geometry of the molecule. More spectroscopic data are also needed, particularly on the nature of the lower-lying levels of hydrocarbon adsorbates, on structural changes in the vibrational modes upon adsorption and on more precise measurements and interpretation of energy shifts in the so-called non-bonding orbitals. Strong shifts in the low-lying C 2s-derived levels of chemisorbed hydrocarbons on nickel relative to the higher-lying σ levels have been observed (Demuth 1978a). These are concluded to indicate that the existence of an initial state potential or rehybridization effect (Demuth and Eastman 1976, Brodén and Rhodin 1976) is important in explaining the role of the high-lying valence orbital shifts on the nature of chemisorptive bonding.

The identification of reaction intermediates in reactions of chemisorbed hydrocarbons on surfaces is one of the main objectives of studies best served by both electron energy loss (Ibach et al. 1977) as well as photoemission spectroscopic methods. The cited example of ethylene dehydrogenation on nickel (Demuth and Eastman 1974) and on platinum, for example from UPS-spectra, fits into the analysis of the rate determining step for the hydrogenolysis of benzene as proposed by Clark and Young (1968). Other interesting examples of intermediate bonding configurations observed from UPS-photoemission spectra on W, Ir and Fe will be noted later. The potential usefulness, on the other hand, of observations of vibrations of surface species with electron energy loss spectroscopy, as proposed by Ibach (1977a, b) is in a more preliminary stage of development. Exploratory applications of this promising approach have already been demonstrated for the chemisorption of ethylene and cyclohexane on Ni(111) (Demuth et al. 1978c) and Pt(111) (Ibach 1977a), for the decomposition of acetylene and ethylene on Pt(111) (Ibach et al. 1977b), and for the disproportionation of CO on Ni(100) (Andersson et al. 1977). Unusual vibrational mode softening and broadening observed to occur in the chemisorption of C_2H_4 and cyclohexane on Ni(111) and Pt(111) is attributed to electronic interactions related to the dehydrogenation process (Demuth et al. 1978). The second example deals with the surface decomposition with temperature of chemisorbed C_2H_2 and C_2H_4 on Pt(111). The first step in the dehydrogenation process is concluded to indicate formation of an ethylidene group (Ibach et al. 1977).

With reference to the disproportionation of CO on Ni(100) (Andersson et al. 1977), the presence of linear and bridge-bonded surface CO indicates that the role of the surface may be to circumvent the existing symmetry restrictions with reference to the corresponding CO linear gas-phase reaction. The proposed mechanisms are supported by both the EELS-spectra and by LEED structures (Andersson et al. 1977). The combination of EELS-spectroscopy and LEED can be effectively applied to the study of many other surface catalytic processes, such as those involving the chemisorption of benzene on Ni(100) (Bertolini et al. 1978). Interpretation of the benzene bonding on this surface at room temperature has been refined by measurement of the energy shifts in the CH stretching and bending modes accompanying adsorption of C_6H_6 and C_6D_6. A tabulation of these modes, shown in table 3.4, illustrates both the complexity and subtlety in the surface bonding of the larger molecules now explorable in detail using high resolution EELS studies.

Unsaturated hydrocarbons on tungsten

A body of measurements on the hydrocarbon decomposition chemistry on the low index faces of tungsten surfaces is also being developed using

TABLE 3.4
(from Bertolini et al. 1977)

Symmetry type	Normal Vibration (a)	Description (b)	Designation (c)	C_6H_6 Vapor (c)	C_6H_6 Chemisorb (d) Ni(100) face	C_6D_6 Vapor (c)	C_6D_6 Chemisorb (d) Ni(100) face
A_{2u}	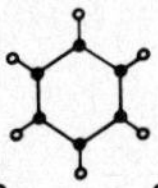	C–H⊥bending	$\nu_4^{H\perp}$	83	95	62.3	67
E_{1u}	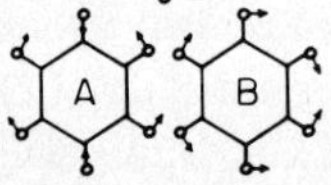 A B	C–H//bending	$\nu_{14}^{H/\!/}$	129	125	101	105
E_{1u}	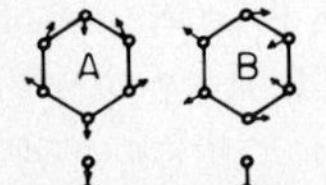 A B	C–C stretching	ν_{13}^{CC}	184	180	165	166
E_{1u}	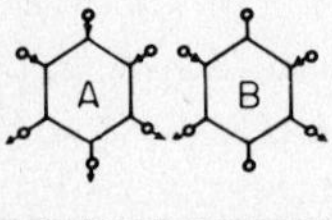 A B	C–H stretching	ν_{12}^{CH}	382	376	284	280

(a) Motions perpendicular to the plane of the molecule.
(b) Herzberg's notation is used.
(c) Infrared adsorption band.
(d) Averaged measured values.

photoemission (Plummer 1975a) and electron loss spectroscopies (Backx et al. 1977). Of major importance from the chemical bonding viewpoint is that such research provides a supplementary basis of comparison for a bcc metal. It also helps to couple the electron spectroscopy data from the great body of surface research data on tungsten. These results present the same generalized trends observed for the close-packed fcc metal surfaces but with diminished interpretative clarity associated with the greater electronic complexity of the tungsten chemistry.

The photoemission features of C_2H_4 chemisorption on W(110) at room temperature are very similar to those observed on Ni(111) in that partial dehydrogenation to C_2H_2 is observed (Plummer et al. 1976). The various peaks in the difference spectrum can be indexed with reasonable assurance in terms of the formation of C–H and C–C bonds by comparison to the gaseous spectra after applying appropriate surface-induced relaxation energy shifts (Plummer 1975a). The two hydrogen atoms produced per dehydrogenated molecule are retained on the surface depending on the coverage of C_2H_2. The thermal decomposition process for the chemisorbed ethylene followed by heating to progressively higher temperatures (300–1500°K) as monitored by UPS-spectral and LEED measurements at 300 K, indicates that a series of C–H and C–C bonds form in the thermal sequence, eventually producing a carbonized surface (Plummer 1975a).

The hydrocarbon–tungsten system may not be the most fruitful one to study using photoemission, from the viewpoint of following the evolution of hydrocarbon fragment formation, because of the complexity. It typifies, however, an interesting approach to the tracking of the bonding characteristics of carbon–hydrogen reaction intermediates at metal surfaces, especially when used in conjunction with high resolution electron energy loss vibrational measurements. Efforts made recently to trace out the carbon bonding sequences typical of acetylene chemisorption and decomposition on W(111) (Backx et al. 1977), W(110) (Backx et al. 1977) and W(100) (Froitzheim et al. 1977d) using EELS-spectra have produced rather complicated spectra indicative of multi-atom molecular surface complex formation containing C, W, and H atoms. It is concluded that no significant difference exists between the general features of the surface complexes observed for the W(111) and W(110) surfaces, and that the occurrence of the C–H fragments initiates at lower exposures for the W(111) surface, consistent with its more open lattice and higher surface reactivity. Results also confirm that complete rehybridization occurs accompanied by σ- rather than π-bonding, in contrast to that generally observed for hydrocarbon reactions on nickel surfaces. The number of observed vibrational modes indicates that the C–C bond is oriented parallel to the surface with a bond order of 0.25. The hydrocarbon–tungsten adsorption system is complex and interesting, but perhaps not likely to be among the first to be completely understood by electron spectroscopic methods.

Unsaturated hydrocarbons on iridium

The surface chemistry of hydrocarbons on platinum group metals are of special interest in both homogeneous and heterogeneous catalysis because of the practical importance of their reactions on extended surfaces as well as on ligand-stabilized metallic clusters (Muetterties et al. 1979). Although the heterogeneous reaction systems are more interesting in their complexity and intensity, the basic reaction parameters such as mobility, chemical binding and molecular structure are better defined for the metallic clusters (Muetterties 1977). The electron spectral approach to understanding the chemical bonding and molecular structure of hydrocarbons on extended metal surfaces of platinum group metals such as Pt, Ir, Ru and Rh is now being usefully reapproached in conjunction with the systematized knowledge characteristic of organometallic chemistry (Ozin 1978, Cotton and Wilkinson 1972). In this regard, the calculational analytical studies using self-consistent scattered wave formalism (Messmer chapter 2) have stimulated efforts to develop this analogy on a more quantitative theoretical basis.

It is not yet clear how far the cluster–surface analogy in chemical bond interpretation can be carried between chemisorbed surfaces and metallic clusters (Muetterties 1977). Exploratory photoemission and Auger measurements however have indicated some striking similarities in the one-electron binding energies, the shake-up spectra, the relative peak intensities and the Auger peak kinetic energies as a function of the number of metal atoms in transition metal carbonyl complexes containing these elements (Plummer et al. 1978). The strong similarity between the photoemission spectra of a multimetal carbonyl and adsorbed CO are shown in figs. 3.35a and b where the valence band spectra of $Ir_4(CO)_{12}$ and $Ru_3(CO)_{12}$ are compared to monolayer adsorption at room temperature on Ir and Ru (Plummer et al. 1978). It is evident that the structure in the d levels of the carbonyl cluster is very similar to the band structure of the chemisorbed metals. The fact that the relative intensities of the 4σ to the $(1\pi + 5\sigma)$ for CO on Ir is noticeably smaller than for CO on Ru is similar to the same behavior observed in the carbonyls (Plummer et al. 1978).

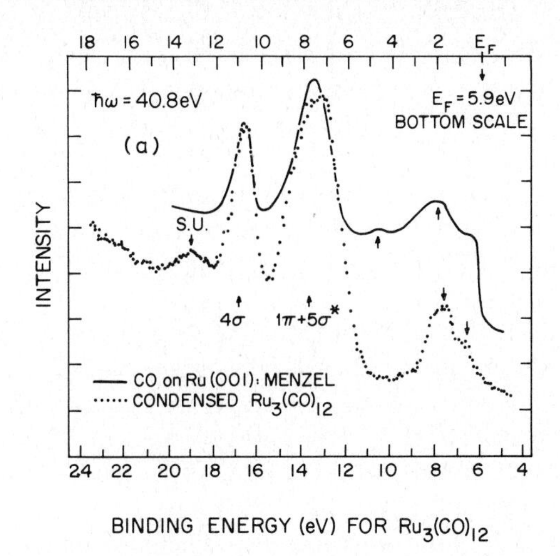

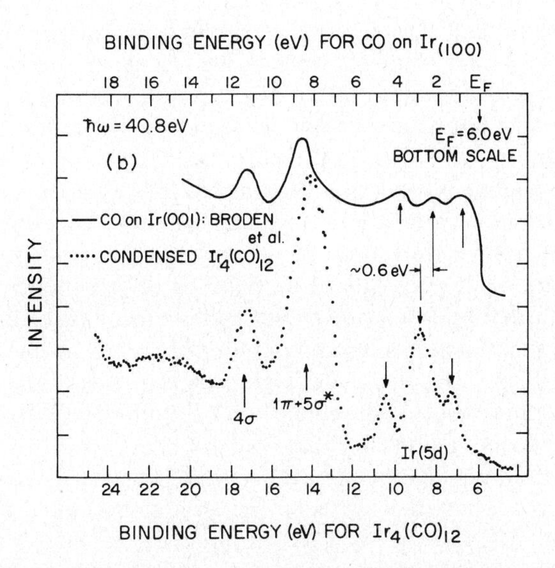

Fig. 3.35. Comparison of the 40.8 eV spectra of $Ru_3(CO)_{12}$ and $Ir_4(CO)_{12}$ with spectra of CO on Ru(001) (a) and Ir(001) (b), respectively. The differences in the energy zero between the carbonyl and the adsorption case is the work function (Plummer et al. 1978).

That the Au(100), Ir (100) and Pt(100) surfaces undergo spontaneous reconstruction under clean surface conditions from a simple orthogonal to a compressed hexagonal structure has also been rationalized, based on a localized interaction model, e.g. an atomic concept relating to their specific susceptibility to undergo modified surface valency due to the presence of the surface (Palmberg and Rhodin 1968). In addition, the availability of a clean reconstructed and normal surface provides a unique opportunity to explore differences in surface reactivity due to a small and controllable alteration in the electronic geometric surface structures (Rhodin and Brodén 1976). Bonding configurations and intermediate hydrocarbon fragments for the chemisorption and reaction of C_2H_2, C_2H_4, and C_6H_6 were studied using photoemission at $\hbar\omega = 21.2$ eV as a function of temperature and coverage on the clean normal and clean reconstructed surface of Ir(100) to evaluate the effect on chemical reactivity of small differences in the geometric and electronic structures of the substrate (Brodén et al. 1976c).

Compared with previous chemisorption measurements on Ni(111) and W(110), chemisorption spectra of acetylene on the close-packed surfaces or iridium [Ir(100) (5×1) and Ir (100) (1×1)] looks generally similar. No major differences were observed between the (111) and reconstructed Ir surfaces. One distinctively different feature is that the spectra on iridium at 300 K showed that the energy separation between peaks $3\sigma_g$ and $2\sigma_\mu$ is significantly larger than for either the gas phase or for Ni(111) or W(110). This is attributed to stretching of the C–C bond and bending of the C–H bonds upon adsorption at room temperature (Brodén and Rhodin 1976), an effect not unusual for acetylene- or ethylene- metallorganic compounds (cotton and Wilkinson 1972). It is in agreement with the rehybridization calculations and experimental spectral results previously discussed with reference to Pt(111) (Demuth 1978a, b). The larger separation is partly caused by bonding of the filled π-like orbitals of the unsaturated hydrocarbon to the metallic orbitals of the metal, and partly to back-bonding into empty antibonding π^* orbitals of the hydrocarbon. It is likely that the C–C displacement perturbs the other acetylene orbitals indirectly in agreement with the observed displacement of the spectral binding energy levels.

Chemisorption of ethylene on Ir(100) (1×1) at 300 K dehydrogenates into chemisorbed stretched actylene and surface hydrogen similar to that observed for Ni(111) and W(110) (Brodén et al. 1976c) at 300 K. In contrast to the Ir(100) (1×1) surface (see fig. 3.28) the Ir(100) (1×5) surface does not dehydrogenate to produce a stretched acetylene spectra at 300 K, but produces a spectra typical of neither ethylene nor acetylene in the normal or stretched configurations. It is likely that the C–C bond has been fragmented to produce a chemisorbed ethylidene-type radical similar to that postulated for C_2H_4 on Pt(111) (Ibach et al. 1977), although its precise nature has not been clearly determined.

Bonding appears to occur via the π-level orbitals on both the Ir(100) and Ir(111) surfaces at room temperature in agreement with other 5d as well as 3d transition metals. Systematic relaxation energy shifts are observed in the spectra but it is not possible to resolve whether these shifts are uniformly due to shielding of the photoejected electron by the electrons in the metal or to a shift in the energy levels due to the altered chemical environment of the adsorbate, e.g. to final state or to initial state relaxation effects. It is significant that the relaxation energy shift is smaller for ethylene adsorbate (2.0 eV) than for acetylene adsorbate (3.0 eV), which is in agreement with the expection that the closer the molecule is to the surface, the larger the expected relaxation shift. It appears that acetylene is positioned closer to the surface than ethylene on iridium in agreement with similar observations for hydrocarbon adsorption on nickel (Demuth 1978a, b) and iron (Brucker and Rhodin 1977, Rhodin et al. 1978).

When discussing bonding of unsaturated hydrocarbons to transition metals, useful analogies have been made to bonding patterns and molecular structures of organometallic compounds, particularly involving coordination centers of platinum, iridium, rhodium and osmium (Brucker and Rhodin 1977). Some interesting similarities in stretching and bending of bonds exist which must bear directly on the electronic and molecular character of hydrocarbon adsorbates on extended metal surfaces (Brucker and Rhodin 1977). There is also an observed temperature-dependent energetic shift of the $2\sigma_\mu$ and $3\sigma_g$ levels at about 170 K for acetylene chemisorbed on Ir(100), indicating a definite stretching and bending of the molecule at that temperature (Brodén et al. 1976c). It is generally concluded from this and other observations that the molecule does not distort discontinuously at 170 K but that there is a continuous change with temperature in the ratio of the normal to distorted configuration which becomes significant at higher temperatures.

The origin of the temperature-dependent energy shifts for ethylene on reconstructed Ir(100) is similar to that of acetylene, with the following additional modifications: (i) no geometric distortion occurs at very low temperatures (140 K), (ii) stretching and bending occurs at intermediate temperatures (140–300 K), and (iii) scission of the C–C bond to form methyl radical fragments occurs above 300 K. Close examination of the spectra provides no evidence for the alternative hypothesis that the adsorbed ethylene dehydrogenates to acetylene prior to fragmentation of the carbon bonds (Brucker and Rhodin 1977).

The reasons as to why chemisorbed acetylene is distorted on Ir(100) and Ir(111) and not on Ni(111) and W(111) surfaces are similar to those previously discussed with reference to the hybridization calculations for unsaturated hydrocarbon absorbates on Pt(111).

In summary, UPS-photoemission results for hydrocarbon adsorption on iridium single crystal low index surfaces provide interesting implications as to the nature of C–C and C–H bonding. These interpretations are similar to those for platinum, with some significant differences. Interesting qualitative and semiquantitative interpretations can be made with reference to interpretation of the orbital contributions. Unfortunately the electron interactions are more complicated for a 5d than for a 3d transition metal. The interpretations are similarly rendered more difficult. The availability of electron energy loss studies of surface vibrational modes on chemisorbed iridium would provide a useful additional clarification of the proposed models.

Unsaturated hydrocarbons on iron

The chemisorption and reaction of unsaturated hydrocarbons on clean well defined surfaces of iron are of particular interest because iron is near nickel in the periodic table. A great many surface studies have been done on iron. The surface chemical behavior of iron is characterized by great reactivity and complexity. Iron is a good catalyst for the Fischer–Tröpsch synthesis (Bond 1962, Ponec 1978) of higher molecular weight hydrocarbons. Hence a good understanding of the basic factors involved in C–C and C–H bond formation would be most pertinent. Like other bcc transition metals a rather striking difference in chemical bonding and reactivity occurs between the close-packed (110) (Brodén et al. 1977b) face and the more open (100) (Brucker and Rhodin 1977, Rhodin et al. 1978), and (111) (Grunze and Ertl 1977) faces.

Iron is known to be a borderline element with reference to dissociative chemisorption of CO at room temperature (Brodén et al. 1976a, b) so some rather striking differences in reactivity for hydrocarbons might be expected for the different crystal faces. Although much effort has been directed towards the study of hydrocarbon reactions on iron and iron alloy surfaces, most of it has not been achieved using modern methods of electron spectroscopy on well defined clean single crystal surfaces. Such samples are difficult to obtain and prepare because clean iron surfaces interact very strongly with carbon and oxygen and because of the allotropic transformation at 1183 K. The net effect is that relatively little is defined about the microscopic mechanisms involving the surface crystallography and activity of clean iron surfaces on the chemical bonding and molecular perturbation of unsaturated hydrocarbon adsorbates.

Chemisorption and reaction of acetylene and ethylene on $\alpha Fe(100)$

Saturated layers of acetylene and ethylene chemisorbed on Fe(100) at 142 K show photoemission spectra at $\hbar\omega = 21.2$ eV generally resembling those of the free molecular structure of the molecules displaced in energy positions by the expected relaxation shifts (Brucker and Rhodin 1977). The spectra as indicated in fig. 3.36 does contain small but significant perturbations of certain orbitals which are definitely related to stretching and bending of the acetylene molecule (Demuth 1978b). It is unlikely even at 142 K that the spectra solely represent molecularly chemisorbed acetylene, as evidenced by the inordinately large relaxation shift of 4.8 eV, an unusually large π-bonding shift of 2.7 eV, and a compression of the σ-levels. It is suggested that strong π–d bonding interactions leads to carbon–carbon scission during adsorption, in agreement with predictions based on parameterized molecular orbital calculations (Rhodin et al. 1978). A slow but distinct time-dependent conversion takes place at 142 K, indicated by an increase in amplitude with time of the peak at 7.5 eV. This spectral feature is attributed to the presence of chemisorbed CH fragments.

Another hydrocarbon fragment corresponding to Fe–CH is observed to build up at 4.9 eV. A second (or perhaps interspersed) layer of more weakly bonded acetylene is presumed to form over the more strongly bound layer of CH-fragments containing atomic hydrogen and carbon (Brucker and Rhodin 1977). Formation of multilayer islands growing laterally from perimeter-localized dissociation of chemisorbed acetylene molecules is postulated. The spectra characteristic of the adsorptive behavior of ethylene in fig. 3.37 bear no resemblence to either chemisorbed ethylene or to acetylene, which might result from dehydrogenation of the chemisorbed ethylene. The spectral features appear to indicate the presence of chemisorbed CH_2-fragments or methylene radicals. Evidence of the major presence of strongly bounded CH_2-fragments is provided by the essential features of the ethylene chemisorption spectra also produced by the combination of hydrogen with chemisorbed carbon (Brucker and Rhodin 1977) on a partly deactivated surface at 300 K.

It should be noted that the dissociative adsorption of ethylene can occur either from a two-hydrogen loss from a single carbon atom, as in

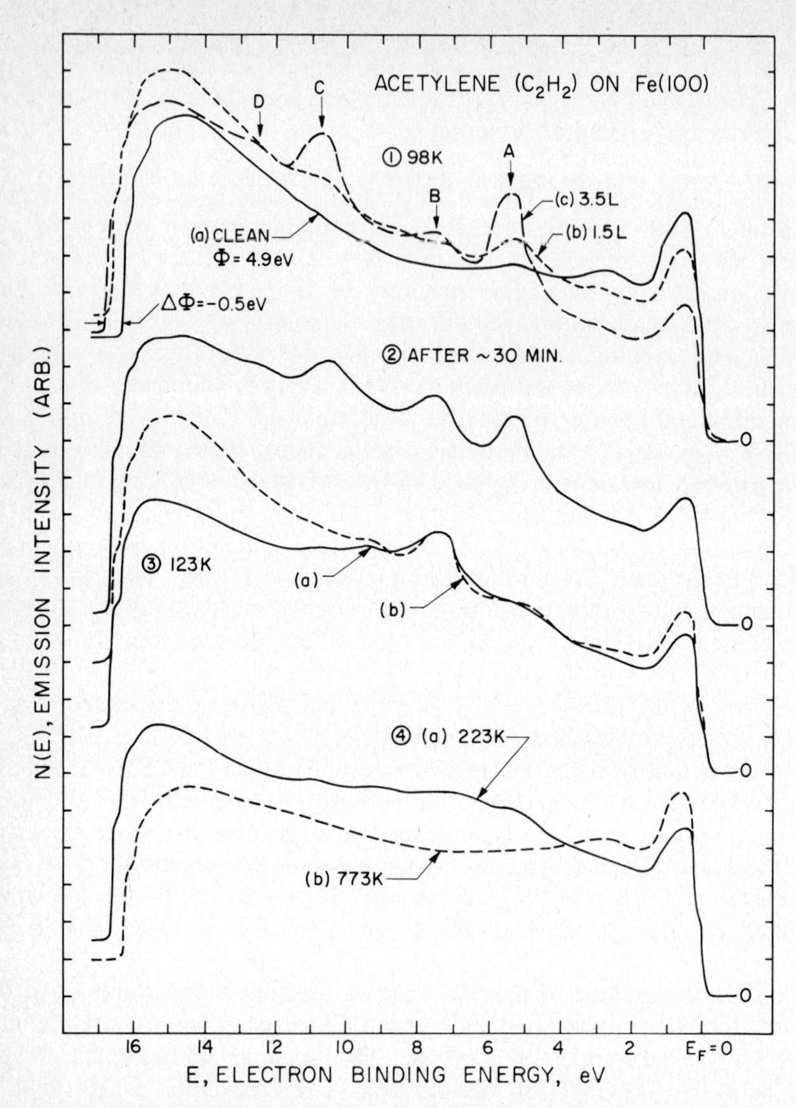

Fig. 3.36. UPS spectra ($\hbar\omega$ = 21.2 eV) for acetylene adsorbed on a clean Fe(100) surface as a function of time and temperature. (1) Sequential exposure ($1L = 10^{-6}$ Torr-sec) of acetylene onto the clean surface at 98 K. Work function change, $\Delta\phi$, indicated for saturation exposure (3.5L); (2) adsorption as in (1c) followed by 30 min. wait at 98 K with uv-light source valved off; (3a) after warming the surface in (2) to 123 K or (b) direct adsorption (6L) onto the clean surface at 123 K; (4) after warming the surface in (3a) to (a) 223 K and (b) 773 K (Brucker and Rhodin 1977).

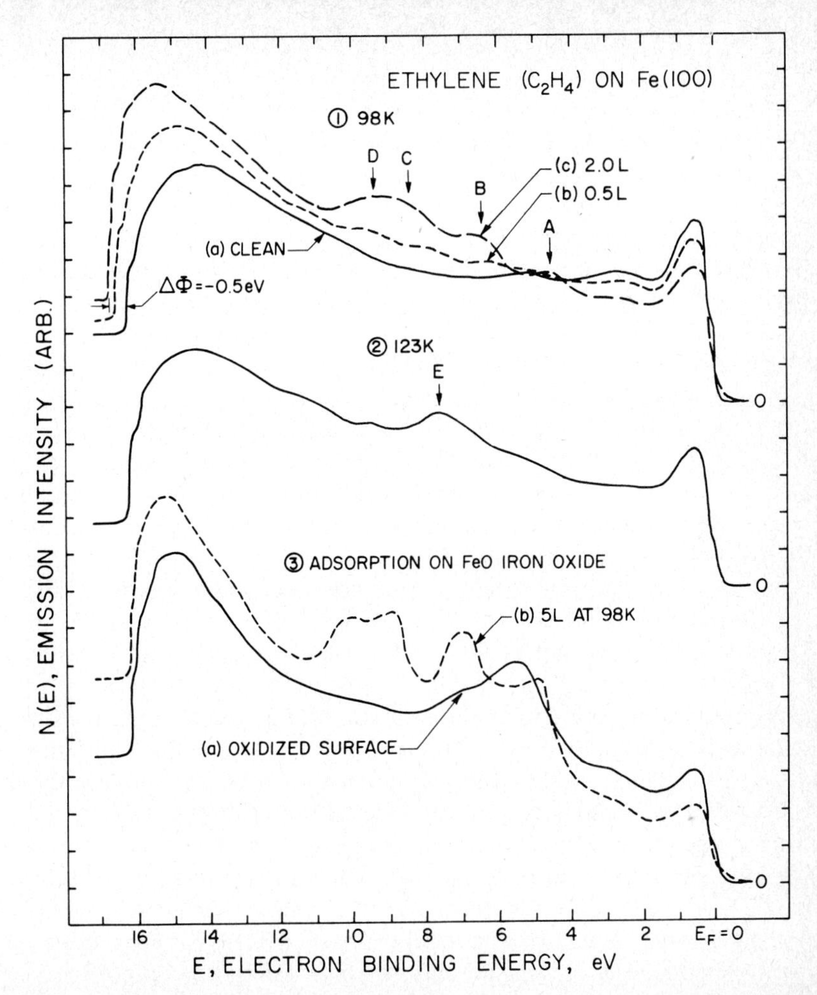

Fig. 3.37. UPS spectra ($\hbar\omega = 21.2$ eV) for ethylene adsorbed on clean and oxidized Fe(100) surfaces. (1) Sequential exposure of ethylene onto the clean surface at 98 K. Work function change indicated for saturation exposure (2L); (2) after warming the surface in (1c) to 123 K; (3) exposure of the oxidized surface (a) to 5L ethylene at 98 K (b) (Brucker and Rhodin 1977).

the cluster isomer case (Muetterties 1977), or a one hydrogen loss from each of two carbon atoms in one molecule. The presence of a surface complex such as vinylidene ($-C=CH_2$) could be identified with the peak at 11 eV. It appears that chemisorbed hydrogen molecules induce carbon bonding interactions on the carbonated (αFe – 100) surface (Rhodin et al. 1978). It should be noted in this regard that whereas

αFe(100) iron appears to possess an equally high reactivity towards both chemisorbed CO and unsaturated hydrocarbons, the geometric factor has a much more important effect on the chemisorptive behavior of the olefinic hydrocarbons. Conversion of the molecular to the fragmented state of ethylene is considered to be analogous to that for acetylene, except for a some what higher activation energy and formation of more complex hydrocarbon fragments. The valence orbitals of the various fragmented species are considered to interact with the metallic d orbitals of the metal until a deactivated layer of chemisorbed complexes forms. Ethylene molecules which are weakly chemisorbed on top of this layer at very low temperatures (below 100 K), appear to surface-diffuse and to react with uncomplexed surface sites upon warming.

The lengthening of the C–C bond and the bending of the substituent groups away from the metal surface as the π^* orbital becomes populated also occurs during the chemisorption process. These and other aspects of organometallic and metal surface binding have been discussed in detail (Brucker and Rhodin 1977).

It is apparent that the bonding interactions of these relatively simple unsaturated hydrocarbons on iron are not simple. A complete interpretation requires a careful and systematic kinetic analysis of each of the proposed bonding mechanisms. The information required for an unequivocal analysis remains to be acquired, but some of the important features by which the αFe(100) surface differs from those of the fcc metals are evident. The effect of deactivation of the surface bonds sequentially from clean iron, to carbonated iron, to oxygen-chemisorbed iron for chemisorption of both acetylene and ethylene is illustrated in the photoemission spectra of fig. 3.38. The shifts in energy positions of the various peaks with adsorption on the various surfaces are in good agreement with the experimentally observed peak separations for weakly chemisorbed acetylene and ethylene. The stretching effects are attributed mainly to the π^* back-bonding donations into the unfilled anti-bonding orbitals of the weakly chemisorbed molecules.

Trends in the observed relaxation energy shift of the orbital ionization energies run parallel to trends in the stretching, being greatest for acetylene on the clean surface, slightly less on the carbonated surface and least on the oxygen chemisorbed surface (Rhodin and Brucker 1977a). It is expected, for example, that the oxygen-covered surface is less effective in screening the photohole during ionization. It may be summed up in the following statement with respect to chemisorption of acetylene and ethylene on iron: "the greatest stretching effects occur on those surfaces showing the largest relaxation shifts". This implies that

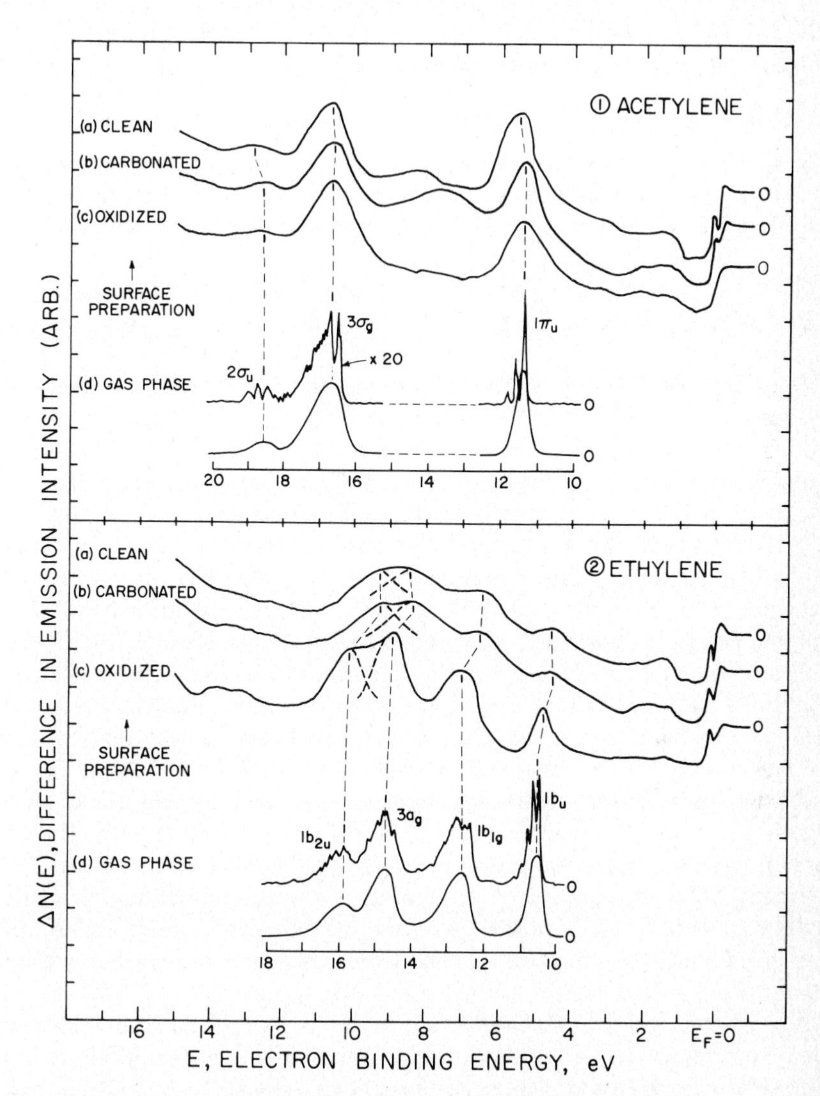

Fig. 3.38. Difference spectra for (1) acetylene and (2) ethylene adsorbed on the (a) clean, (b) carbonated, and (c) oxidized Fe(100) surfaces at 98 K. Gas phase spectra and molecular orbital designations included for comparison. Proposed deconvolutions indicated for $3a_g$ and $1b_{2u}$ ethylene levels (Brucker and Rhodin 1977).

the ability of the surface to donate electrons in π^* back-bonding interactions increases as its ability to screen the photohole. These conclusions are supported by calculations of the energy levels for various surface–molecule combinations (Rhodin et al. 1978).

Chemisorption and reaction of acetylene and ethylene on αFe(110)

It is instructive to compare the photoemission spectral interpretations for unsaturated hydrocarbon adsorbates on αFe(100) (Brucker and Rhodin 1977, Rhodin et al. 1978) with those on Fe(110) (Brodén et al. 1977b). The reactivities of these two orientations are known to differ significantly since the chemisorption of unsaturated adsorbates is expected to be particularly sensitive to surface geometry. Consideration of comparable results on clean iron films (Yu et al. 1976) will also be included on the assumption that the annealed film surfaces are likely to contain a large fraction of (110) facetted surfaces.

The UPS-photoemission spectra for acetylene adsorbed on Fe(110) at room temperature by Brodén et al. (1977b) is very similar to spectra obtained for acetylene chemisorbed under comparable conditions on Ni(111) (Demuth and Eastman 1974) and on W(110) (Plummer et al. 1974). It is concluded that the chemisorbed molecule has essentially the same geometry as in the gas-phase since the energy separation of the $2\sigma_{\mu}$ and $3\sigma_{g}$ peaks is essentially unchanged upon adsorption, e.g. 2.0 eV for the gas molecule and 2.1 eV for the adsorbate. This behavior contrasts sharply with that found for αFe(100), where fragmentation of the acetylene molecule into CH-radicals is considered to occur even at 100 K (Brucker and Rhodin 1977, Rhodin et al. 1978). It should be noted that fragmentation or distortion is known to occur also at room temperature on Ir(100) (Brodén et al. 1976c), Pt(100) (Fischer et al. 1978), and on W(100) (Waclawski and Vorburger 1976) as well as on Pt(111) (Demuth 1978a, b) and Pd(111) (Demuth 1978a, b), indicating that the surface geometry of the substrate does indeed play a more important role in chemical bonding for unsaturated hydrocarbon adsorbates than for diatomic molecules such as CO and NO (Brodén et al. 1976a, b). A rather extensive UPS-photoemission study of acetylene and ethylene chemisorption at room temperature on polycrystalline iron films by Yu et al. (1976) yields results similar to those reported on αFe(110) (Brodén et al. 1977b). In this work the UPS-photoemission spectra show many of the features of the gas-phase spectra shifted in the expected manner for relaxation effects and for π–d bonding. However, no correlation was found between the π level bonding shifts and the heats of adsorption, in agreement with previous observations that bonding contributions from other sources involving the d-electron behavior of the metal surface must also be considered (Demuth 1978a, b).

Chemisorption behavior of hydrocarbons on iron is clearly more complicated than for other group VIII metals to the right of it in the periodic table. Additional detailed studies of surface bonding on iron using angle-resolved photoemission and electron energy loss spectroscopy are greatly needed to more fully interpret bonding behavior, especially of the crystal faces with open atomic structures such as Fe(111).

5. Prospects

It is interesting to consider briefly in summary the most interesting implications of the previously discussed theoretical and experimental material.

The surface electron structure studies using photoemission from clean metal surfaces have dealt so far in depth mainly with Cu, Ag, and Au. The main emphasis has been on surface and bulk aspects of band structure, including the occurence of surface resonances and surface states. There also has been a limited amount of interesting work on clean surfaces of free electron metals like Al and Be. Sufficient electron spectroscopic studies on the surface electron structure of the 3d, 4d, and 5d transition metals have also been made to indicate that their systematic study is likely to be complicated by additional considerations relating to magnetic, relativistic, and other effects. The vibrational studies have to date been focussed primarily on overlayers. This is due mainly to instrumental features of the technique. The XPS-spectral work has been directed mainly to chemical shift and related effects in compounds. The Auger spectroscopy has been directed almost exclusively to surface compositional analysis and relatively little to the interpretation of electron structure. It is likely that angle resolved photoemission spectroscopy over a wide photon energy range will be systematically applied to the clarification of both bulk and surface band structure of clean single crystal surfaces of a variety of transition metals in the immediate future. It is fortunate that photoemission spectroscopy can be usefully applied within practical limits to the study of chemical bonding of overlayers on surfaces of transition metals without requiring a detailed understanding of the surface electron structure of the clean metals.

Atomic bonding involving overlayers has been studied sufficiently using electron spectroscopic methods to indicate that it provides an effective tool for the exploration of both surface and subsurface bonding changes associated with chemisorption and reaction. It has dealt mostly with the chalocogens on the close packed surfaces of fcc metals, primarily nickel. Studies of oxygen and hydrogen chemisorption pursued on a variety of low-index planes of the bcc metals, mainly tungsten, have produced interesting results which, in general, however, have not resulted in any broad new interpretations. It is significant from the oxidation work on aluminum and on iron that much can be learned about bonding changes during oxidation–reduction reactions from the combined use of both valence- and core-level measurements. Angle-resolved studies of these and related chemical reaction-systems are likely to provide significant insight into the nature and orientation of the chemical bonding modifications which are involved. It is to be expected that correlated measurements using both LEED and EELS will play an important supportive role in achieving validated interpretations of reaction models.

Molecular adsorption on metals, particularly transition metals, offers a fascinating area for the application of electron spectroscopic methods to the study of the role of chemical bonding in the mechanisms of surface reactions. The potential use of this approach has already been demonstrated by the bonding and orientation studies of CO and NO adsorption on the low-index planes of nickel, platinum, and iridium, using angle-resolved photoemission. Although these two molecules provide useful molecular probes for surface bonding, a very fruitful additional area for such studies is presented by the use of other inorganic molecules for probing surface reactivity, such as ammonia, cyanogen, thiophene, water, hydrogen cyanide and the hydrogen chalcogenides. In these studies combined angle-resolved UPS- and XPS-photoemission with LEED and EELS should provide unique new information on bonding and structural changes associated with formation of chemical intermediates essential to reaction mechanisms on surfaces. In this regard the use of polarized photon excitation coupled with angle-resolved photoemission measurements is likely to make significant contributions to the measurement of bond distances in molecules at surfaces. Finally, the availability of continuously variable, intense polarized photon sources in the energy range from 10 to 2000 eV utilizing new synchrotron radiation facilities will play an important role in the diagnosis of surface electron structure.

Chemisorption and decomposition of hydrocarbons on metals using electron spectroscopies offers a powerful approach to developing new insight into the electronic characteristics associated with hydrocarbon chemistry, particularly on the surfaces of transition metals. This has been well demonstrated for some of the simple unsaturated olefins on single crystal surfaces of the close packed surfaces of fcc metals like nickel, platinum, iridium and palladium, on some of the low-index faces of tungsten and of α iron as well as on the basal plane of ruthenium. Such studies have wide and obvious implications for the study of heterogeneous reaction mechanisms associated with important catalytic reactions such as the ammonia synthesis, the methane synthesis, and the Fischer–Tröpsch process.

The photoemission spectra provide a valuable method for tracking the development of specific molecular entities through a chemical process, and provides, in addition, useful information on energy level shifts and polarization effects. It may also be possible, in favorable cases, to elucidate the molecular orbitals involved in heterogeneous molecular transformations, although in some cases the possible ramifications of molecular arrangements may at times be too great for this to be achieved. The concommitant interpretation of vibrational mode effects from electron energy loss measurements (EELS) can add much clarification in the identification of hydrocarbon fragments and their role in the reaction. It should also be noted that LEED pattern analysis can be very helpful in those cases where ordered overlayers form. On the other hand, dynamical analysis of surface crystallography using LEED applied to adsorbed hydrocarbons has been limited in the past and is likely to progress slowly.

The wide variation of bonding modes and configurations provided by chemisorbed hydrocarbons makes their study using combined electron spectroscopic methods very promising. There is likely to be sustained and fruitful effort in this area in the near future.

It might be said that a quantum jump has occurred in both the experimental and theoretical understanding of chemical bonding at the vacuum–metal interface during the last ten years. It is helpful to sum up the areas which have been neglected to date and to suggest what might be likely directions for new thrusts within the framework of the electron spectroscopic approach.

It is clearly apparent that there remains much to be completed and clarified within the presently developed methods and concepts. The theory of photoemission based on a one-electron approximation from clean nearly-free-electron metal surfaces is just now emerging. Its extension beyond the simplest first-order model to transition metals remains to be achieved. An understanding of surface band structure and photoemission for these materials, including the variables of frequency and polarization of the excitation and of energy and momentum of the emission over a wide frequency range, remains to be developed. An adequate interpretation of atomic and molecular surface metal bonding must include a more complete description of the surface electron structure both before and after chemisorption. Self-consistent scattered wave treatments based on the metal cluster concept need to be more reconciled with the band structure concepts. Reflection of the electron perturbations of the metal surface during adsorption need to be correlated with bonding changes in the adsorbed molecule in terms of both the bonding and the non-bonding orbitals. Hopefully this can be extended not only to simple molecular interactions on transition metals involving CO and NO, for example, but to a reasonable extent, to more complicated adsorbed molecules such as low-molecular weight unsaturated hydrocarbons, nitrogen–hydrogen compounds, etc. In this sense most of the present work has focussed on equilibrium or steady-state systems. A natural immediate extension of this effort would be to the clarification of bonding characteristic of molecular fragments and reaction intermediates.

Some very fruitful correlations have been made with both valence and core level data on the same system in a limited number of cases. It is likely that great benefits will derive from more extensive combination of valence level information with both AES- and XPS-chemical shift data. In this sense it is expected that the presently rather undeveloped utility of AES-spectra in general will be expanded as a more useful probe of electron changes associated with surface bond formation. A great deal of information on the electronic nature of surface bonding will accrue from a better understanding also of the contributions associated with shake-up and related phenomena to the photoemission spectra. In a related vein, the study of photoemission and Auger spectra for metal-clusters stabilized with ligands of interest to chemisorption systems should provide a fruitful direction for advancing bonding interpretations, not only of the metal-cluster systems, but of the related molecular adsorbate-extended metal surfaces. The utility of angle-resolved, spectral data over a wide range of excitation energies with well-defined polarization available from synchrotron radiation sources will greatly expand both the quality and range of available spectral measurements.

There is no doubt that the effectiveness of electron spectroscopy as a probe of surface bonding is greatly implemented by the combination of two or more types of supplementary electron-structure related measurements on the same system, especially when they derive from different aspects of the surface such as, for example, valence level vs. core level excitations, surface crystallography, thermal desorption states, local surface composition and vibrational modes. In this sense, information on bond distances obtainable from angle-resolved photoemission can usefully supplement that being derived from dynamical analysis of LEED. The LEED pattern analysis in turn strongly implements interpretation of the vibrational mode behavior. The latter, in turn, contributes important credibility to the interpretation of the UPS-spectra in terms of specific changes in the nature of molecular orbitals and bonding. All are enhanced by the availability of well defined surface parameters for the initial surface, especially those relating to composition and structure.

The contribution of vibrational mode studies on surfaces has already made a significant impact. Perhaps the most striking aspect of this technique on single crystal surfaces has been the use of the electron energy loss approach. It is likely that other powerful techniques for infrared spectroscopy of molecular interactions on well defined crystal surfaces will be improved and conceived to meet the pressing need for characterization of the excitation behavior of bonding modes in molecules at metal surfaces. It is disappointing in this regard that so relatively little electron spectroscopy work has been done so far on well defined metal alloy systems, with some notable exceptions. Surface bonding studies on single-phase binary alloys using a variety of different electron spectral ranges and methods would provide fruitful enlightenment of the nature of bonding between dissimiliar metal atoms at surfaces. Elucidation of surface chemical bonding in various unitary and binary metallic systems could add much to the general basic understanding of the nature of the metallic bond itself.

The main thrust of surface chemical bonding in this discussion, directed as it is primarily to metals for the purposes of clarity and convenience, has given scant attention to the important area of surface chemical bonding on elemental and compound semiconductors. An area of challenging research lies in the application of electron spectroscopy to the analysis of surface electron structure and bonding for these interesting materials. Emphasis has tended to be placed on metal surfaces during the past five years, partly because of the great wealth of adsorption and chemical reaction phenomena associated with the chemistry of transition metals. The nature of the directed covalent bond characteristic of semiconductor solids together with the propensity of such surfaces to undergo spontaneous reconstruction make them a promising class for the study of surface chemical bonding using electron spectroscopic methods, particularly angle-resolved photoemission and vibrational loss studies. This is particularly true for the surface chemical properties of the III–V and II–VI compound semiconductors. Here the nature of bonding at metal–semiconductor, insulator–semiconductor and semiconductor–semiconductor interfaces is a particularly important area for exploration.

Acknowledgments

One author (TNR) is particularly grateful for secretarial assistance from Ms. V. Rollins and for research support from NSF grant DMR77-05078-A02 and the Cornell Materials Science Center (NSF-DMR76-81083-A01). Helpful discussions with colleagues and graduate students were also very much appreciated in the preparation of the material. Thanks are also due to the colleagues of the authors who kindly made available results and suggestions for inclusion in the manuscript. Appreciation is expressed to E.L. Muetterties, J. Wilkins, and R. Merrill for reading parts of the manuscript during preparation. We are also grateful to E. Rowe, Director, and the Staff of the Electron Storage Ring Facility, Physical Science Laboratory, Stoughton, Wisconsin, for research support in the use of synchrotron radiation for electron spectroscopy and to the following scientists for making the results of their research available to us prior to publication : R. Brundle, E.W. Plummer, T. Gustafsson, J. Demuth, G. Brodén, S. Andersson, J. Pendry, M. Traum, Z. Hurych, R. Benbow, G. Somorjai, E. Muetterties, R. Merrill, S.Y. Tong, M. Van Hove, H. Weinberg and G. Ertl.

References

Aksela, S., M. Karras, M. Pessa and E. Suoninen, 1970, Rev. Sci. Instr. **41**, 351.
Allen, C.G., P.M. Tucker and R.K. Wild, 1977, (see Dobrozemsky et al. 1977) p. 959.
Allyn, C.L., T. Gustafsson and E.W. Plummer, 1977, Chem. Phys. Lett. **47**, 127; G. Loubriel, T. Gustafsson and E.W. Plummer, private communication.
C.L. Allyn, T. Gustafsson and E.W. Plummer, 1978, Rev. Sci. Instr. to be published.
Almbladh, C.O., 1974, Nuovo Cimento **B23**, 75.
Almbladh, C.O. and P. Minnhagen, 1978, Phys. Rev. **B17**, 929.
Anderson, J. and G.J. Lapeyre, 1976, Phys. Rev. Lett. **36**, 376.
Andersson, S., 1976, Solid State Comm. **20**, 229.
Andersson, S., 1977a, (see Dobrozemsky 1977) p. 1019.
Andersson, S., 1977b, Solid State Comm. **21**, 75.
Andersson, S. 1978, to be published.
Andersson, S. and U. Jostell, 1974, Surface Sci. **46**, 625.
Andersson, S. and U. Jostell, 1975, Faraday discussions of the chemical society **60**, 255.
Andersson, S. B.I. Lundqvist and J.K. Nørskov, 1977, (see Dobrozemsky 1977) p. 815.

Apai, G., P.S. Wehner, R.S. Williams, J. Stöhr and D.A. Shirley, 1976, Phys. Rev. Lett. **37**, 1497; (see also Castex et al. 1977a) p. 210.

Bachrach, R.Z., S.A. Flodström, R.S. Bauer, S.B.M. Hagstrom and D.J. Chadi, 1977, (see Dobrozemsky et al. 1977) p. 461.

Backx, C., R.F. Willis, B. Feuerbacher and B. Fitton, 1976, (see Willis et al. 1976) p. 291.

Backx, C., R.F. Willis, B. Feuerbacher and B. Fitton, 1977a, Surf. Sci. **68**, 516.

Backx, C., R.F. Willis, B. Feuerbacher and B. Fitton, 1977b, Surf. Sci. **63**, 193.

Bagchi, A., 1977, Phys. Rev. **B15**, 3060.

Baker, J.M. and D.E. Eastman, 1973, J. Vac. Sci. Tech. **10**, 223.

Batra, I.P. and P.S. Robo, 1975, Solid State Comm. **16**, 1097.

Batra, I.P. and S. Ciraci, 1977, (see Dobrozemsky et al. 1977) p. 1141; Phys. Rev. Lett. **39**, 774.

Batterman, B.W. and N.W. Ashcroft, 1977, Cornell high energy synchrotron source CHESS, private communication.

Bertolini, J.C., G. Dalmai-Imelik and J. Rousseau, 1976, Proc. of the Int. symposium on photoemission (ESA scientific publication, Noordwijk) p. 285.

Bertolini, J.C., G. Dalmai-Imelik and J. Rousseau, 1978, Surf. Sci. **67**, 478.

Bethe, H.A., L. Maximon and F. Low, 1953, Phys. Rev. **91**, 417.

Bianconi, A., R.Z. Bachrach and S.A. Flodström, 1977, (see Castex et al. 1977a) p. 216

Billington, R. and T.N. Rhodin, 1978, to be published.

Blakely, J.M., 1975, Surface physics of materials (Academic Press, New York).

Bond, G.C., 1962, Catalysis by metals (Academic Press, London) p. 353, 371.; (see also V. Ponec, 1979, J. Cat. to be published).

Bonzel, H.P. and G. Pirug, 1976, (see Willis et al. 1976) p. 263.

Bozso, F., G. Ertl, M. Grunze and M. Weiss, 1978, J. Cat. submitted.

Brill, R., E.L. Richter and E. Ruch, 1967, Agnew. Chem. Int. Ed. **6**, 882; (see also Dumesic, J.A., H. Topsoe, S. Khammouma and M. Boudart, i975, J. Cat. **37**, 503).

Brodén, G. and T.N. Rhodin, 1976, Chem. Phys. Lett. **40**, 247.

Brodén, G., T.N. Rhodin, C.F. Brucker, R. Benbow and Z. Hurych, 1976a, Surf. Sci. **59**, 593.

Brodén, G., T.N. Rhodin, C.F. Brucker, R. Benbow and Z. Hurych, 1976b, (see Willis et al. 1976) p. 269.

Brodén, G., T.N. Rhodin and W. Capehart, 1976c, Surf. Sci. **61**, 143.

Brodén, G., G. Pirug and H.P. Bonzel, 1977a, Chem. Phys. Lett. **51**, 250.

Brodén, G., G. Gafner and H.P. Bonzel, 1977b, Appl. Phys. **13**, 333.

Brucker, C.F. and T.N. Rhodin, 1976, Surf. Sci. **57**, 523.

Brucker, C.F. and T.N. Rhodin, 1977, J. Cat. **47**, 214.

Brucker, C.F. and T.N. Rhodin, 1979, to be published.

Brundle, C.R., 1975a, (see Day 1975).

Brundle, C.R., 1975, Surf. Sci. **48**, 99.

Brundle, C.R., 1977, Surf. Sci. **66**, 581.

Brundle, C.R., Surf. Sci., 1979, to be published.

Brundle, C.R., T.J. Chuang and K. Wandelt, 1977, Surf. Sci. **68**, 459.

Brundle, C.R. and A.D. Baker, 1977, Electron spectroscopy: theory, techniques, and applications (Academic Press, London).

Carlson, T.A., 1975, Photoelectron and Auger spectroscopy (Plenum Press, New York).

Caroli, C., D. Lederer-Rozenblatt, B. Roulet and D. Saint-James, 1973, Phys. Rev. **B8**, 4552.

Castex, M.C., M.Pouey and N. Pouey, 1977a, Vacuum ultraviolet radiation physics, 5th conf. vol. 2, Montpelier, Meudon.

Castex, M.C., M. Pouey and N. Pouey, 1977b, Some of the most recent angle-resolved studies on 5d-transition metal surfaces using either fixed or variable energy sources are abstracted in Castex et al. 1977a, as follows: Au, p. 414 and 144; Pt, p. 210 and 102; and Ir, p. 259.
Chabel, Y.J. and A.J. Sievers, 1978, Appl. Phys. Lett. **32**, 90; see also J. Vac. Sci. Tech. **15**, 638.
Chang, J.J. and D.C. Langreth, 1972, Phys. Rev. **B5**, 3512.
Chang, J.J. and D.C. Langreth, 1973, Phys. Rev. **B8**, 4638.
Chattarji, D., 1976, The theory of auger transitions (Academic Press, London).
Chester, M. and J.C. Rivière, 1977, (see Dobrozemsky et al. 1977) p. 873.
Chung, M.S., and T.E. Everhart, 1977, Phys. Rev. **B15**, 4699.
Cini, M., 1976, Solid State Comm. **20**, 605.
Cini, M., 1977, Solid State Comm. **24**, 681.
Cini, M., 1978, Phys. Rev. **B17**, 2788.
Citrin, P.H., P. Eisenberger and D.R. Hamann, 1974, Phys. Rev. Lett. **33**, 965.
Citrin, P.H., P.M. Eisenberger, W.C. Mara, T. Ålberg, J. Utrianinen and E. Källne, 1974, Phys. Rev. **B10**, 1762.
Citrin, P.H. and D.R. Hamann, 1977, Phys. Rev. **B15**, 2923.
Citrin, P.H., G.K. Wertheim and Y. Baer, 1977, Phys. Rev. **B16**, 4256.
Clark, H.D. and R.D. Young, 1968, Surf. Sci. **12**, 385.
Collings, D.M. and W.E. Spicer, 1977a, Surf. Sci. **69**, 85.
Collings, D.M. and W.E. Spicer, 1977b, Surf. Sci. **69**, 114.
Comrie, C.M. and W.H. Weinberg, 1976, J. Chem. Phys. **64**, 250.
Condon, E.U. and G.H. Shortley, 1967, The theory of atomic spectra (Cambridge Univ. Press, London).
Conrad, H., G. Ertl, H. Knözinger, J. Küppers and E.E. Latta, 1976, Chem. Phys. Lett. **42**, 115.
Cooper, J. and R.N. Zare, 1968, J. Chem. Phys. **48**, 942.
Cotton, F.A. and G. Wilkinson, 1972, Advanced inorganic chemistry (Interscience, New York).
Czanderna, A.W., 1975, ed., Methods of surface analysis Vol. 1 (Elsevier, New York).
Datta, A. and D.M. Newns, 1976, Phys. Lett. **59A**, 326.
Davenport, J.W., 1976a, Phys. Rev. Lett. **36**, 945.
Davenport, J.W., 1976B, PhD thesis, Dept. of Phys. Univ. of Pennsylvania.
Davenport, J.W., 1978, J. Vac. Sci. Tech. **15**, 433.
Davenport, J.W., W. Ho and J.R. Schrieffer, 1978, Phys. Rev. **B17**, 3115.
Day, P., 1975, Electronic states of inorganic compounds: New Experimental techniques, NATO advanced study institute (Reidel, Dordrecht).
Delanaye, F., A. Lucas and G.D. Mahan, 1978, Surf. Sci. **70**, 629.
Demuth, J.E., 1977, Surf. Sci. **69**, 365.
Demuth, J.E., 1978a, Phys. Rev. Lett. **40**, 409.
Demuth, J.E., 1978b, Surf. Sci. to be published.
Demuth, J.E. and D.E. Eastman, 1974, Phys. Rev. Lett. **32**, 1123.
Demuth, J.E. and D.E. Eastman, 1976, Phys. Rev. **B13**, 1523.
Demuth, J.E., H. Ibach and S. Lehwald, 1978, Phys. Rev. Lett. **40**, 1044.
Derouane, E.G. and A.A. Lucas, 1976, Electronic structure and reactivity of metal surfaces, NATO ASI (Plenum, New York).
Dill, D., 1976, J. Chem. Phys. **65**, 1130.
Dill, D. and J.L. Dehmer, 1974, J. Chem. Phys. **61**, 1692.

Dill, D. and J.L. Dehmer, 1975, Phys. Rev. Lett. **35**, 213.
Dill, D. and S. Manson, 1977, (see Brundle and Baker 1977).
Dill, D., J. Siegel and J.L. Dehmer, 1976, J. Chem. Phys. **65**, 3158.
Dionne, N.J. and T.N. Rhodin, 1976, Phys. Rev. **B14**, 322; See also, 1974, Phys. Rev. Lett. **32**, 1131.
Dobrozemsky, R., R. Rudenauer, F.P. Viehbock and A. Breth, 1977, Proc. of the 7th Int. vacuum congress and the 3rd Int. conference on solid surfaces, Vienna.
Doniach, S. and E.H. Sondheimer, 1974, Green's functions for solid state physicists (Benjamin, New York).
Doniach, S. and M. Šunjić, 1970, J. Phys. **C3**, 285.
Dow, J.D., D.L. Smith, D.R. Franceschetti, J.E. Robinson and T.R. Carver, 1977, Phys. Rev. **B16**, 4704.
Doyen, G. and T.B. Grimley, 1976, (see Willis et al. 1976) p. 220.
Drauglis, E. and R.I. Jaffee, 1975, The physical basis for heterogeneous catalysis (Plenum, New York).
Eastman, D.E., 1971, Phys. Rev. Lett. **27**, 487; See also Eastman, 1972b.
Eastman, D.E., 1972a, Electron spectroscopy (D. Shirley, ed.) (North-Holland, Amsterdam).
Eastman, D.E., 1972b, Tech. for study of metals, vol. 6, (E. Passaglia, ed.) (Interscience, New York) p. 411.
Eastman, D.E., 1972c, (see Eastman 1972a) p. 511.
Eastman, D.E., 1972d (see Eastman 1972b) p. 447.
Eastman, D.E., 1972e (see Eastman 1972b) p. 417.
Eastman, D.E. and J.K. Cashion, 1971, Phys. Rev. Lett **27**, 1520.
Eastman, D.E., J.E. Demuth and J.M. Baker, 1974, J. Vac. Sci. Tech. **11**, 273.
Einstein, T.E., 1974, Surf. Sci. **45**, 713.
Einstein, T.E., 1975, Phys. Rev. **B12**, 1262.
Ertl, G. and J. Küppers, 1974, Low energy electrons and surface chemistry chapter 9. (Verlag Chemie, Weinheim).
Ertl, G. and K. Wandelt, 1976, Surf. Sci. **55**, 403.
Ertl, G., M. Grunze and M. Weiss, 1976, J. Vac. Sci. Tech. **13**, 314; (see also Bozso, F., G. Ertl, M. Grunze and M. Weiss, 1978, J. Cat. submitted).
Ertl, G., M. Neumann and K.M. Streit, 1977, Surf. Sci. **64**, 393.
Estrum, P.J. and J. Anderson, 1966, J. Chem. Phys. **45**, 2254.
Evans, E. and D.L. Mills, 1973, Phys. Rev. **B7**, 853.
Evans, E. and D.L. Mills, 1972, Phys. Rev. **B5**, 4126.
Fadley, C.S., 1972, Theoretical aspects of X-ray photoelectron spectroscopy, NATO ASI of Electron emission spectroscopy, Ghent.
Fadley, C.S. and S.A.L. Bergstrom 1971, Phys. Lett. **35A**, 375.
Fadley, C.S., 1978, Electron spectroscopy, theory, techniques and applications, vol. 2. (Brundle and Baker eds.) (Academic Press, London).
Fadley, C.S., R.J. Baird, W. Siekhaus, T. Novakov and S.A.L. Bergstrom, 1974, J. Electron. Spect. **4**, 93.
Fadley, C.S., R.J. Baird, L.F. Wagner and Z. Hussain, 1976, (see Willis 1976) p. 201.
Feibelman, P.J., 1975a, Phys. Rev. **B12**, 1319.
Feibelman, P.J., 1975b, Phys. Rev. Lett. **34**, 1092.
Feibelman, P.J., 1976, Phys. Rev. **B14**, 762.
Feibelman, P.J. and D.E. Eastman, 1974, Phys. Rev. **B10**, 4932.
Feibelman, P.J., E.J. McGuire and K.C. Pandey, 1977, Phys. Rev. **B15**, 2202.

Feibelman, P.J., E.J. McGuire and K.C. Pandey, 1978, Phys. Rev. **B17**, 1799.
Feuerbacher, B. and B. Fitton, 1977a, (see Ibach 1977b) p. 188.
Feuerbacher, B., and B. Fitton, 1977b, Topics in current chemistry, vol 4 (ed. Ibach) (Springer, New York) p. 151.
Feuerbacher, B., B. Fitton and R.F. Willis, 1978, eds. Photoemission and the electronic properties of surfaces (John Wiley, New York) see particularly chapters 6 (Doyen and Grimley), 7 (Liebsch), 12 (Gustafsson and Plummer), and 13 (Menzel).
Fischer, T.E., S. Keleman and H.P. Bonzel, 1978, private communication.
Flodström, S.A. and J.G. Endriz, 1973, Phys. Rev. Lett. **31**, 893.
Flodström, S.A. and J.G. Endriz, 1975, Phys. Rev. **B12**, 1252.
Flodström, S.A., G.V. Hansson, S.B.M. Hagström and J.E. Endriz, 1975, Surf. Sci. **53**, 156.
Flodström, S.A., R.Z. Bachrach, R.S. Bauer, 1976, Phys. Rev. Lett. **37**, 1282.
Flodström, S.A., R.Z. Bachrach, R.S. Bauer and S.B.M. Hagström, 1977a, (see Dobrozemsky et al. 1977) p. 869.
Flodström, S.A., R.Z. Bachrach, R.S. Bauer and S.B.M. Hagström, 1977b, (see Castex et al. 1977a) p. 225.
Fong, F.K., 1976, Radiationless processes in molecules and condensed phases (Springer-Verlag, Berlin).
Forstmann, F. and H. Stenschke, 1977, Phys. Rev. Lett. **38**, 1365.
Froitzheim, H., 1977, Topics in current chemistry, vol. 4 (Ibach, ed.) (Springer, New York) 1977, p. 205.
Froitzheim, H. and H. Ibach, 1974, Z. Phyzik, **269**, 17; see also H.S. Sar-El, 1970, Rev. Sci. Instr. **41**, 561.
Froitzheim, H., H. Ibach and S. Lehwald, 1976a, Rev. Sci. Inst. **46**, 1325 (1975); Phys. Rev. Lett. **36**, 1549.
Froitzheim, H., H. Ibach and S. Lehwald, 1976b, Phys. Rev. **B14**, 1362.
Froitzheim, H., H. Ibach and S. Lehwald, 1977a, Surf. Sci. **63**, 56.
Froitzheim, H., H. Ibach and S. Lehwald, 1977b, (see Willis et al. 1977) p. 277.
Froitzheim, H., H. Hobster, H. Ibach and S. Lehwald, 1977c, Appl. Phys. **13**, 147.
Froitzheim, H., H. Ibach and S. Lehwald, 1977d, Surf. Sci. **63**, 56.
Fuggle, J.C., E. Umbach and D. Menzel, 1976a, Solid State Comm. **20**, 89.
Fuggle, J.C., E. Umbach and D. Menzel, 1976b, (see Willis et al. 1976), p. 192.
Fuggle, J.C., and D. Menzel, 1977, (see Dobrozemsky et al. 1977) p. 1003.
Fuggle, J.C., E. Umbach, D. Menzel, K. Wandelt and C.R. Brundle, 1979, to be published.
Gadzuk, J.W., 1974a, Phys. Rev. **B9**, 1978.
Gadzuk, J.W., 1974b, Phys. Rev. **B10**, 5030.
Gadzuk, J.W., 1974c, Solid State Comm. **15**, 1011.
Gadzuk, J.W., 1975a, (see Blakely 1975).
Gadzuk, J.W., 1975b, Surf. Sci. **53**, 132.
Gadzuk, J.W., 1975c, Phys. Rev. **B12**, 5608.
Gadzuk, J.W., 1976a, Electronic structure and reactivity of metal surfaces, NATO Advanced Study Institute Namur (Derouane and Lucas, eds.) (Plenum, New York) p. 341.
Gadzuk, J.W., 1976b, Phys. Rev. **B14**, 2267.
Gadzuk, J.W., 1976c, Phys. Rev. **B14**, 5458.
Gadzuk, J.W., 1977a, Surf. Sci. **67**, 77.
Gadzuk, J.W., 1977b, J. Elec. Spect. **11**, 355.
Gadzuk, J.W., 1978, (see Feuerbacher et al. 1978).
Gadzuk, J.W. and S. Doniach, 1978, Surf. Sci. **77**, in press.
Gadzuk, J.W. and M. Šunjić, 1975, Phys. Rev. **B12**, 524.

Gallon, T.E., 1978, Electron and ion spectroscopy of solids, NATO ASI, (Fiermans, Vennik and Dekeyser, eds.) (Plenum, New York).
Gartland, P.O., 1976, (see Willis et al. 1976) p. 63.
Gartland, P.O. and B.J. Slagsvold, 1977, private communication.
Gay, J.G., J.R. Smith and F.J. Arlinghaus, 1977a, Phys. Rev. Lett. **38**, 561.
Gay, J.G., J.R. Smith and F.J. Arlinghaus, 1977b, Solid State Comm. **24**, 279.
Geiger, J., 1968, Elektronen und Festkorper, vol. 128, Sammlung (Viewig, Braunschweig).
Goldberger, M.L. and K.M. Watson, 1964, Collision theory (Wiley, New York).
Gomer, R., 1974, Advan. Chem. **27**, 211.
Gomer, R., 1975a, Topics in applied physics, vol. 4 (Springer, New York).
Gomer, R., 1975b, Solid State Phys. **30**, 94.
Greenler, R.G., 1966, J. Chem. Phys. **44**, 310.
Gross, E.P., 1967, Mathematical methods in solid state and superfluid theory (R.C. Clark and G.H. Derrick, eds.) (Plenum, New York).
Grunze, M. and G. Ertl, 1977, (see Dobrozemsky et al. 1977) p. 1137.
Guillot, C., Y. Ballu, G. Chauvin, J. LeCante, J. Paigne, P. Thiry, D. Dageneax, Y. Petroff, R. Pinchaux and R. Cinti, 1976, (see Willis 1976) p. 51.
Guillot, C., Y. Ballu, J. Paigne, J. LeCante, P. Thiry, R. Pinchaix and Y. Petroff, 1977, (see Castex et al. 1977a) p. 22.
Gumhalter, B., 1977a, Le J. de Physique **38**, 1117.
Gumhalter, B., 1977b, (see Dobrozemsky et al. 1977) p. 783.
Gumhalter, B., 1977c, J. Phys. **C10**, L 219.
Gumhalter, B. and D.M. Newns, 1975, Phys. Lett. **53A**, 137.
Gumhalter, B. and D.M. Newns, 1976, Phys. Lett. **57A**, 423.
Gustafsson, T., E.W. Plummer, D.E. Eastman and J.L. Freeouf, 1975, Solid State Comm. **17**, 391; see also J.C. Fuggle, T.E. Madey and M. Steinkilberg and D. Menzel, 1975, Phys. Lett. **51A**, 163.
Hafner, H., J.A. Simpson and C.E. Kuyatt, 1968, Rev. Sci. Instr. **39**, 33.
Hair, M.L., 1967, Infrared spectroscopy in surface chemistry, (Dekker, New York) p. 14.
Hannay, N.B., 1976, Treatise in solid state chemistry, vol. 6A (Plenum, New York).
Harris, J. and R.O. Jones, 1973, J. Phys. **C6**, 3585.
Harris, J., and R.O. Jones, 1974, J. Phys. **C7**, 3751.
Harris, J., 1975, Solid State Comm. **16**, 671.
Hawkins, D.T., 1977, Auger electron spectroscopy: a bibliography, 1925–1975 (Plenum, New York).
Heddle, D.W.O., 1971, J. Phys. **E4**, 589.
Hedin, L., 1973, Electrons in crystalline solids, International atomic energy agency, Vienna.
Heimann, P. and H. Neddermeyer, 1976, (see Willis et al. 1976) p. 59.
Heimann, P. and H. Neddermeyer, 1977, (see Castex et al. 1977a) p. 147.
Heinrichs, J., 1973, Phys. Rev. **B8**, 1346.
Herbst, J.F., 1977, Phys. Rev. **B15**, 3720.
Hercules, S.H. and D.M. Hercules, 1974 (see Kane and Larabee 1974) p. 307.
Herring, C. and M.H. Nichols, 1949, Rev. Mod. Phys. **21**, 185.
Hewson, A.C. and D.M. Newns, 1974, Japan J. Appl. Phys. Suppl. **2**, pt. 2, 121.
Hobson, J.B., 1974, Proc. 6th Int. Vacuum Congress.
Hodgson, K., H. Winick, and G. Chu, 1976 eds. *Synchrotron radiation research*, SSRP report 76/100, Stanford University, p. 7, 17, and 127, Stanford.
Horoguichi, T., and S. Nakanishi, 1974 Japan J. Appl. Phys. Suppl. 2, pt. 2, 89.

Houston, J.E., 1975, J. Vac. Sci. Tech. **12**, 255.
Hussain, S.S. and D.M. Newns, 1978, Solid State Comm. **25**, 1049.
Ibach, H., 1970, Phys. Rev. Lett. **24**, 1416.
Ibach, H. 1971, Phys. Rev. Lett. **27**, 253.
Ibach H. 1972, J. Vac. Sci. Tech. **9** 713.
Ibach, H., 1977a (see Dobrozemsky et al. 1977), p. 743.
Ibach, H., 1977b, ed. "Topics in current physics", vol. 4: Electron spectroscopy for surface analysis (Springer Verlag, New York) p. 28.
Ibach, H., H. Hopster and B. Sexton, 1977, Appl. Surf. Sci. **1**, 1.
Ilver, L. and P.O. Nilsson, 1976, Solid State Comm. **18**, 677.
Jackson, A.J., C. Tate, T.E. Gallon, P.S. Bassett and J.A.D. Matthew, 1975, J. Phys. F **5**, 363.
Jacobi, K., M. Scheffler, K. Kambe and F. Forstmann, 1977, Solid State Comm. 22, 17.
Jenkins, L.H. and D.M. Zehner, 1973, Solid State Comm. **12**, 1149.
Jones, L.H., 1963, Spectrochim. Acta. **19**, 329.
Joyner, R.W., 1977, Surf. Sci. **63**, 291.
Kane, P., and G. Larabee, 1974, eds. Characterization of solid surfaces (Plenum, New York).
Kanski, J. and T.N. Rhodin, 1977, Surf. Sci. **65**, 63.
Kar, N. and P. Soven, 1976, Solid State Comm. **19**, 1041.
Kincaid, B.M., P. Eisenberger and D.E. Sayers, 1977 private communication.
Kirschner, J., 1977, Topics in current physics, vol. 4 (Ibach ed.) (Springer, New York) p. 94.
Kliewer, K.L., 1976, Phys. Rev. **B14**, 1412.
Kliminsch, R.L. and J.G. Larson, 1975, The catalytic chemistry of nitrogen oxides (Plenum, New York) p. 63.
Koch, E., R. Haensel and C. Kunz, 1974, Vacuum ultraviolet radiation physics, 4th Conf. Hamburg (Pergamon New York).
Kotani, A. and Y. Toyozawa, 1974, J. Phys. Soc. Japan **37**, 912.
Krebs, H.J. and H. Lüth, 1977, Appl. Phys. **14**, 337.
Kumagai, H. and T. Toya, 1974, Proc. of the 2nd Int. Conf. on solid surfaces, Kyoto, March 1974, Japanese J. Appl. Phys. Suppl. **2**, pt. 2.
Kunz, C., 1974a, (see Koch, Haensel and Kunz 1974) p. 753; see also T.N. Rhodin and C.F. Brucker, Proc. 3rd Int. Conf. solid surfaces, Vienna p. 731.
Kuyatt, C.E. and E.W. Plummer, 1972, Rev. Sci. Instr. **43**, 198.
Lambe, J. and R.C. Jaklevic, 1966, Phys. Rev. Lett. **17**, 1139.
Lambe, J. and R.C. Jaklevic, 1968, Phys. Rev. **165**, 821.
Lang, N.D. 1973, Solid State Phys. **28**, 225.
Lang, N.D. and A.R. Williams, 1977, Phys. Rev. **B16**, 2408.
Langreth, D.C., 1970, Phys. Rev. **B1**, 471.
Langreth, D.C., 1971, Phys. Rev. Lett. **26**, 1229.
Langreth, D.C., 1974, Collective properties of physical systems (B.I. Lundqvist and S. Lundqvist eds.) (Academic Press, New York).
Lapeyre, G.J., J. Anderson and R.J. Smith, 1976, (see Willis et al. 1976), p. 249.
Laramore, G.E. and W.J. Camp, 1974, Phys. Rev. **B9**, 3270.
Larsen, P.K., N.V. Smith, M. Schluter, H.H. Farrell, K.M. Ho and M.L. Cohen, 1978, Phys. Rev. **B17**, 2612.
Lee, L., 1976, ed. Characterization of metal and polymer surfaces, vol. 1 (Academic Press, New York).
Leygraf, C. and S. Eklund, 1973, Surf. Sci. **40**, 609.

Li, C.H. and S.Y. Tong, 1978, Phys. Rev. Lett. **40**, 46.
Liebsch, A., 1974, Phys. Rev. Lett. **32**, 1203.
Liebsch, A., 1976, Phys. Rev. **B13**, 544.
Liebsch, A., 1977, Phys. Rev. Lett. **38**, 248.
Liebsch, A., 1978a, Angle resolved photoemission: theoretical interpretation of results, from Electron and ion spectroscopy of solids, NATO Inst. Ghent (1977) (Plenum, London).
Liebsch, A., 1978b, (see Liebsch, 1978a), p. 24.
Liebsch, A., 1978c, (see Liebsch, 1978a, p.27; (see also Feuerbacher et al. 1978) p. 221.
Lindgren, S.A. and L. Walldén, 1978, Solid State Comm. **25**, 13.
Lloyd, D.M., C.M. Quinn and N.V. Richardson, 1976, J. Phys. **C8**, 1371.
Loubriel, T. Gustafsson and E.W. Plummer, 1977, private communication.
Lucas, A.A. and M. Šunjić, 1969, Prog. Surf. Sci. **2**, 40.
Lucas, A.A. and M. Šunjić, 1972a, Progress in surface science, vol. 2 (Pergamon press, London) p. 2.
Lucas, A.A. and M. Šunjić, 1972b, Surf. Sci. **32**, 439.
Lundqvist, B.I., 1967, Phys. Kondens. Mater **6**, 193, 206.
Lundqvist, B.I., 1969, Phys. Kondens. Mater **9**, 236.
MacDonald, N.C. and C.T. Hovland, 1977, Proc. 8th Int. Conf. on X-ray and Microanalysis, Boston, Mass. p. 64a.
Madden, H.H. and J.E. Houston, 1977, Solid State Comm. **21**, 1081.
Mahan, G.D., 1970, Phys. Rev. **B2**, 4334.
Mahan, G.D., 1974, Solid State Phys. **29**, 75.
Mahan, G.D., 1977, Phys. Rev. **B15**, 4587.
Maradudin, A.A., E.W. Montrol, G.H. Weiss and I.P. Ipatova, 1971, Theory of lattice dynamics in the harmonic approximations, 2nd edn. Solid State Phys. Suppl. 3.
Martin, R.L. and D.A. Shirley, 1977, (see Brundle and Baker 1977).
Martinson, C.W.B., L.G. Petersson, S.A. Flodström and S.B.M. Hagström, 1976, (see Willis, 1976) p. 177.
McGowan, J.W. and E. Row, 1976, eds. Proc. of workshop on synchrotron radiation, Univ. Laval Quebec, sects. 9 and 10.
McHugh, J.A., 1975, "Secondary ion mass spectroscopy", in Methods of surface analysis (A.W. Czanderna ed.) (Elsevier Scientific, New York) p. 223.
Meldner, H.W. and J.D. Perez, 1971, Phys. Rev. **A4**, 1388.
Melmed, A.J. and J.H. Carroll, 1972, J. Vac. Sci. Tech. **10**, 164.
Menzel, D., 1975, (see Gomer 1975) p. 101, p. 12.
Meyer, R.J., 1977, Phys. Rev. **B15**, 2306.
Mills, D.L., 1975, Surf. Sci. **48**, 59.
Mills, D.L., 1978, "Interaction of low energy electrons with surface lattice vibrations", to be published in Surfaces and interfaces, vol 1, (Dobrozemski ed.) (Marcel Dekker, New York).
Minnhagen, P., 1976a, J. Phys. **F6**, 1789.
Minnhagen, P., 1976b, Phys. Lett. **56A**, 327.
Minnhagen, P., 1977, J. Phys. **F7**, 2441.
Morrison, S.R., 1977, The chemical physics of surfaces (Plenum, New York).
Muetterties, E.L., 1975, Bull. Sci. Chim. Belg. **84**, 959.
Muetterties, E.L., 1977, Science **196**, 839.
Muetterties, E.L., 1978, Angewandte Chemie **17**, 545.
Muetterties, E.L., J.C. Hemminger and G. Somorjai, 1971, "A coordination chemists view of surface science", Inorganic Chem. **16**, 3381.

Muetterties, E.L., T.N. Rhodin, E. Band, C. Brucker and H. Pretzer, 1979, Chemical Reviews, to be published.

Müller-Hartman, E., T.V. Ramakrishnan and G. Toulouse, 1971, Phys. Rev. **B3**, 1102.

Murday, J.S., 1975, "Review of surface physics", Naval Research Laboratory Memorandum Report #3062.

Muscat, J.P. and D.M. Newns, 1978, Progress in Surface Science, to be published.

National Research Council, 1976, "An assessment of the national need for facilities dedicated to the production of synchrotron radiation", National Academy, Washington.

Newns, D.M., 1969, Phys. Rev. **178**, 1123.

Newns, D.M., 1977, Phys. Lett. **60A**, 461.

Nilsson, P.O. and L. Ilver, 1976, (see Willis et al. 1976) p. 73.

Noziérès, P. and C.T. deDominicis, 1969, Phys. Rev. **178**, 1097.

Ozin, G.A., 1977, Catalysis Rev. **16**, 191.

Ozin, G.A., 1978, "Very small metallic and bimetallic clusters; the metal cluster–metal surface analogy in catalysis and Chemisorption processes", Catalysis Reviews.

Palmberg, P.W. and T.N. Rhodin, 1968, J. Chem. Phys. **49**, 134.

Pardee, W.J., G.D. Mahan, D.E. Eastman, R.A. Pollak, L. Ley, F.R. McFeely, S.P. Kowalczyk and D.A. Shirley, 1975, Phys. Rev. **B11**, 3614.

Pendry, J.B., 1976, (see Willis et al. 1976).

Pendry, J.B., and D.J. Titterington, 1977, Comm. on Phys. **2**, 31.

Penn, D.R., 1976, Phys. Rev. **B13**, 5248.

Penn, D.R., 1977, Phys. Rev. Lett. **38**, 1429.

Persson, B.N.J., 1977, Solid State Comm. **24**, 573.

Petererssson, L.G. and R. Erlandson, 1977, (see Castex et al. 1977a) p. 159.

Pignocco, A.J. and G.E. Pellisier, 1965, J. Electrochem. Soc. **112**, 1188.

Plummer, E.W., Topics in applied physics, vol. 4, 1975a, (Springer, New York) p. 143.

Plummer, E.W., 1975b, (see Gomer 1975) p. 193.

Plummer, E.W., 1975c, (see Gomer 1975a) p. 200.

Plummer, E.W., 1977, (see Dobrozemsky 1977) p. 647.

Plummer, E.W. and A.E. Bell, 1972, Vac. Sci. Tech. **9**, 583.

Plummer, E.W. and T. Gustafsson, 1977, Science **198**, 165.

Plummer, E.W., B.J. Waclawski and T.V. Vorburger, 1974, Chem. Phys. Lett. **28**, 510.

Plummer, E.W., B.J. Waclawski, T.V. Vorburger and C.E. Kuyatt, 1976, Prog. Surf. Sci. 7, 149.

Plummer, E.W., T. Gustafsson, W. Gudat and D.E. Eastman, 1977, Phys. Rev. **A15**, 2339; see also T. Gustafsson, E.W. Plummer, D.E. Eastman and W. Gudat, 1978, Phys. Rev. to be published.

Plummer, E.W., W.R. Salaneck and J.S. Miller, 1978, Phys. Rev. **B18**, 1673; see also H. Conrad, G. Ertl, H. Knozinger, J. Küppers and E.E. Lata, 1976, Chem. Phys. Lett. **42**, 115 and Surf. Sci. **65**, 235.

Ponec, V., 1978, J. Catalysis, to be published.

Powell, C.J., 1973, Phys. Rev. Lett. **30**, 1179.

Powell, C.J., 1974, Surf. Sci. **44**, 29.

Powell, C.J., 1977, "The national measurement system for surface properties", U.S. Department of Commerce, NBSIR75–945, March (1977).

Pritchard, J. and M.L. Sims, 1970, Trans. Faraday Sci. **66**, 427.

Propst, F.M. and T.C. Piper, 1967, J. Vac. Sci. Tech. **4**, 53.

Pryce, M.H.L., 1966, Phonons in perfect lattices and in lattices with point imperfections (R.W.H. Stevenson ed.) (Plenum, New York).

Purtell, R., C.Seabury, R. Merrill and T.N. Rhodin, unpublished work.

Rabalais, J.W., Principles of Ultraviolet photoelectron spectroscopy, 1977, (Wiley, New York).
Raether, H., 1975, Springer Tracts in Modern Physics, vol. 38 (Springer, Berlin, Heidelberg, New York) p. 85.
Read, F.H., J. Comer, R.E. Imhof, J.N.H. Brunt and E. Harting, 1974, J. Elect. Spectros. **4**, 293.
Rhodin, T.N., and D.L. Adams, 1975a, "Adsorption of gases on solids" in Treatise on solid state chemistry, vol. **6A** (Hannay ed.) (Plenum, New York) ch. 5, p. 105.
Rhodin, T.N. and D. Adams, 1975b, (see Rhodin and Adams, 1975a) p. 90.
Rhodin, T.N. and G. Brodén, 1976, Surf. Sci. **60**, 466.
Rhodin, T.N. and C. Brucker, 1976a, Characterization of metal and Polymer surfaces, vol. 1 (L. Lee ed.) (Academic Press, New York).
Rhodin, T.N. and C.F. Brucker, 1976b, for reviews of recent photoemission work on Ir(111) and Ir(100) using angle-integrated and angle-resolved photoemission with the photon continuum up to 40 eV at the electron storage ring, see also T.N. Rhodin and C.F. Brucker 1976a.
Rhodin, T.N. and C.F. Brucker, 1977a, Solid State Comm. **23**, 275.
Rhodin, T.N. and C.F. Brucker, 1977b, Proc. of the 3rd Int. Conf. on solid surfaces, Vienna (Dobrozemsky et al. eds.) p. 731.
Rhodin, T.N., C. Seabury, M. Traum and Z. Hurych, 1977a, (see Castex et al. 1977a) p. 259.
Rhodin, T.N., J. Kanski and C. Brucker, 1977b, Solid State State Comm. **23**, 723.
Rhodin, T.N., C.F. Brucker and A.B. Anderson, 1978, J. Phys. Chem. **82**, 894.
Rhodin, T.N., J. Kanski, C. Brucker, C. Seabury and G. Brodén, 1979, to be published.
Rojo, J.M. and A.M. Baro, 1976, J. Phys. **C9**, L543.
Rosencwaig, A., 1975, Physics Today, September, p. 23.
Roy, D. and J.D. Carette, 1971, Can. J. Phys. **49**, 2138.
Roy, D. and J.D. Carette, 1977, Topics in current physics, vol. 4 (Ibach ed.) (Springer, New York) p. 13.
Rudd, M.E., 1972, in Low energy electron spectroscopy (K.E. Siever ed.) (Wiley, New York) p. 17.
Saas, J.K., H. Laicht and S. Stucki, 1976, (see Willis et al. 1976) p. 83.
Sampson, J.A.R., 1967, Techniques of vacuum ultraviolet spectroscopy (Wiley, New York).
Sawatzky, G.A., 1977, Phys. Rev. Lett. **39**, 504.
Schaich, W.L. and N.W. Ashcroft, 1971, Phys. Rev. **B3**, 2452.
Scheffler, M., K. Kambe and F. Forstmann, 1976, (see Willis 1976) p. 227.
Scheffler, M., K. Kambe and F. Forstmann, 1978, Solid State Comm. **25**, 93.
Schmidt, L.D., 1975, (see Gomer 1975a) p. 63.
Schmitz, W. and W. Mehlhorn, 1972, J. Phys. **E5**, 64.
Schönhammer, K. and O. Gunnarsson, 1977, Solid State Comm. **23**, 691.
Schröder, W., and J. Hölz, 1977, Solid State Comm. **24**, 777.
Seabury, C., and T.N. Rhodin, 1978, to be published.
Sewall, P.B., D.J. Mitchell and M. Cohen, 1972, Surf. Sci. **33**, 535.
Shigeishi, R.A., and D.A. King, 1976, Surf. Sci. **58**, 484.
Shirley, D.A., 1972, ed. Electron spectroscopy (North-Holland, Amsterdam).
Sickafus, E.N., 1977, Phys. Rev. **B16**, 1436.
Siegbahn, K., 1976, Electron spectroscopy for solids, surfaces, liquids, and free molecules, University of Uppsala – Institute of Physics, Publication 940, Uppsala.
Siegbahn, K., 1977, (see West 1977).

Silcox, J., 1977, Inelastic scattering as an analytical tool, in Scanning electron spectroscopy, vol. 1, Proc. of Workshop, Chicago, I11, p. 393.
Simmons, G.W. and D.J. Dwyer, 1975, Surf. Sci. **48**, 373.
Simpson, J.A., 1964, Rev. Sci. Instr. **35**, 1698; see also C.E. Kuyatt, J.A. Simpson, 1967, Rev. Sci. Instr. **38**, 103.
Slater, J.C. and K.H. Johnson, 1972, Phys. Rev. **B5**, 844.
Smith, R.J., J. Anderson and G.J. Lapeyre, 1976, Phys. Rev. Lett. **37**, 1081.
Smith, R.J. J. Anderson and G.J. Lapeyre, 1977, J. Vac. Sci. Tech. **14**, 384.
Šokčević, D., Z. Lenac, R. Brako and M. Šunjić, 1977, Z. Physik **B28**, 273.
Spicer, W.E., 1970, J. Res. Nat. Bur. Stand. **A74**, 397.
Stern, E.A., D.E. Sayers, J.G. Dash, H. Schechter and B. Bunker, 1977, Phys. Rev. Lett. **38**, 767.
Šunjić, M. and A. Lucas, 1976, Chem. Phys. Lett. **42**, 462.
Šunjić, M. and D. Šokčević, 1974, Solid State Comm. **15**, 165.
Šunjić, M. and D. Šokčevič, 1976, Solid State Comm. **18**, 373.
Šunjić, M., G. Toulouse and A.A. Lucas, 1972, Solid State Comm. **11**, 1629.
Šunjić, M., Z. Crljen and D. Šokčević, 1977, Surf. Sci. **68**, 479.
Thomas, S., 1973, Solid State Comm. **13**, 1593.
Tompkins, H.G., 1975, Infrared reflection-absorption spectroscopy, in Methods of surface analysis (A. Czanderna ed.) (Elsevier Scientific, New York).
Tong, S.Y. and M.A. Van Hove, 1976, Solid State Comm. **19**, 543.
Tong, S.Y., C.H. Li and A.R. Lubinsky, 1977, Phys. Rev. Lett. **39**, 498.
Tung, C.J. and R.H. Ritchie, 1977, Phys. Rev. **B16**, 4302.
Van Hove, M.A., 1978, (see chapter 4 of this volume) "Surface crystallography and bonding".
Waclawski, B.J. and T.V. Vorburger, 1976, Bull. Am. Phys. Soc. **21**, 940.
Wallace, S., D. Dill and J.L. Dehmer, 1978, Phys. Rev. **B17**, 2004.
Wallis, R.F., 1968, ed. Localized excitation in solids (Plenum, New York).
Wang, K.L., 1972, J. Phys. **E5**, 1193.
Watts, C.M.K., 1972, J. Phys. **F2**, 574.
Weeks, S.P. and E.W. Plummer, 1977, Chem. Phys. Lett. **48**, 601.
Weeks, S.P. and E.W. Plummer, 1977, Solid State Comm. **21**, 695.
Weeks, S.P., E.W. Plummer and T. Gustafsson, 1976, (see Willis et al. 1976) p. 267.
Wehner, P.S., G.Api, R.S. Williams, J. Stöhr, and D.A. Shirley, 1977, (see Castex et al. 1977a) p. 165.
West, A.R., 1977, Molecular spectroscopy (Heydon, London).
Williams, G.P., and C. Norris, 1976, (see Willis et al. 1976) p. 55.
Williams, P.M., P. Butcher and J. Wood, 1976, Phys. Rev. **B14**, 3215.
Williams, A.R. and N.D. Lang, 1978, Phys. Rev. Lett. **40**, 954.
Willis, R.F. and B. Feuerbacher, 1978, Photoemission from surfaces, p. 281. (B. Feuerbacher, B. Fitton and R. Willis eds.) (Wiley, New York).
Willis, R.F., B. Feuerbacher, B. Fitton and C. Backx, 1976, Proc. of the Int. Symp. on photoemission, Noordwijk, Sept. 1976.
Wuilleumeier, F.J., 1976, Photoionization and other probes of many-electron interactions, NATO ASI (Plenum, New York).
Yu, K.Y., J.C. McMenamin and W.E. Spicer, 1975, J. Vac. Sci. Tech. **12**, 286.
Yu, K.Y., W.E. Spicer, I. Lindau, P. Pianetta, and S.F. Lin, 1976, Surf. Sci. **57**, 157.
Yue, J.T. and S. Doniach, 1973, Phys. Rev. **B8**, 4578.

Appendix A: Summary of measurement techniques that have been used to characterize surfaces.

TABLE A.1
Outline of surface-characterization techniques
Detected particle

	Photon	Electron	Neutral	Ion	Phonon	*E*/*H* Field
Photon	ATR	AEAPS	LMP	LMP		
		AEM	PD	PD		
	ELL	AES				
	ESR	PEM				
	EXAFS	PES				
	IRS	SEE				
	LS	UPS				
	MOSS	XEM				
	NMR	XES				
	SRS	XPS				
	XRD					
Electron	APS	AEPS	ESDN	ESDI		
	BIS	AEM	SDMM			
	CL	AES				
	CIS	DAPS				
	EM	EELS				
	SXAPS	HEED				
	SXES	IS				
		LEED				
		SEE				
		RHEED				
		SEM				
		SLEEP				
		STEM				
		TEM				
Neutral	NIRS	SEE	MBRS		ΔH_{ADS}	
			MBSS			
			$\sigma(\pi)$			
Ion	GDOS	IMXA	ISD	GDMS		
	IIRS	INS	SDMM	IMMA		
	IIXS	SEE		ISD		
				ISS		
				RBS		
				SIIMS		
				SIMS		
Phonon	ES	TE	FD	SI	ASW	
	TL					
	EL	FEES	FDM	FIM		CPD
		FEM	FDS	FIM–APS		MS
		ITS		FIS		SC

Table A.1. shows a summary of the many techniques that have been used to characterize surfaces; this summary has been adapted from one prepared recently by Murday (1975). The key to the various acronyms is given in table A.2.

TABLE A.2
Key to acronyms shown in Table A.1

Acronym	Meaning	Acronym	Meaning
AEPS	– Auger-electron appearance-potential spectroscopy	FDM	– Field-desorption microscopy
AEM	– Auger-electron microscopy	FDS	– Field-desorptionspectroscopy
AES	– Auger-electron spectroscopy	FEM	– Field-emission microscopy
APS	– Appearance-potential spectroscopy	FEES	– Field-electron energy spectroscopy
ASW	– Acoustic surface-wave measurements	FIM	– Field-ion microscopy
ATR	– Attenuated total reflectance	FIM–APS	Field-ion microscope-atom probe spectroscopy
BIS	– Bremsstrahlung isochromat spectroscopy	FIS	– Field-ion spectroscopy
CIS	– Characteristic isochromat spectroscopy	GDMS	– Glow-discharge mass spectroscopy
CL	– Cathodoluminescence	GDOS	– Glow-discharge optical spectroscopy
COL	– Colorimetry: ir, visible, uv, X-ray, and γ-ray absorption spectroscopy	HEED	– High-energy electron diffraction
CPD	– Contact potential difference (work-function measurements)	IIRS	– Ion-impact radiation spectroscopy
DAPS	– Disappearance-potential spectroscopy	IIXS	– Ion-induced X-ray spectroscopy
ΔH_{ADS}	– Heat of adsorption measurements	IMMA	– Ion microprobe mass analysis
EL	– Electroluminescence	IMXA	– Ion microprobe X-ray analysis
ELL	– Ellipsometry	INS	– Ion-neutralization spectroscopy
EELS	– Electron energy-loss spectroscopy	IRS	– Internal reflectance spectroscopy
EM	– Electron microprobe	IS	– Ionization spectroscopy
ES	– Emission spectroscopy	ISD	– Ion-stimulated desorption
ESDI	– Electron-stimulated desorption of ions	ISS	– Ion-scattering spectroscopy
ESDN	– Electron-stimulated desorption of neutrals	ITS	– Inelastic tunneling spectroscopy
ESR	– Electron-spin resonance	LEED	– Low-energy electron-diffraction
EXAFS	– Extended X-ray absorption fine structure	LMP	– Laser microprobe
FD	– Flash desorption	LS	– Light scattering
		MBRS	– Molecular-beam reactive scattering

TABLE A.2 (cont.)

MBSS	– Molecular-beam surface scattering
MOSS	– Mössbauer spectroscopy
MS	– Magnetic saturation
NIRS	– Neutral impact radiation spectroscopy
NMR	– Nuclear magnetic resonance
PD	– Photodesorption
PEM	– Photoelectron microscopy
PES	– Photoelectron spectroscopy
RBS	– Rutherford backscattering spectroscopy
RHEED	– Reflection high-energy electron diffraction
SC	– Surface capacitance
SDMM	– Scanning desorption molecule microscopy
SEE	– Secondary-electron emission
SEM	– Scanning electron microscopy
SI	– Surface ionization
$\sigma(p)$	– Adsorption isotherm measurements
SIIMS	– Secondary-ion imaging mass spectroscopy
SIMS	– Secondary-ion mass spectroscopy
SLEEP	– Scanning low energy electron probe
SRS	– Surface reflectance spectroscopy
STEM	– Scanning transmission electron microscopy
SXAPS	– Soft X-ray appearance – potential spectroscopy
SXES	– Soft X-ray emission spectroscopy
TE	– Thermionic emission
TEM	– Transmission electron microscopy
TL	– Thermoluminescesce
UPS	– Ultraviolet photoemission spectroscopy
XEM	– Exoelectron microscopy
XES	– Exoelectron spectroscopy
XPS	– X-ray photoemission spectroscopy
XRD	– X-ray diffraction (glancing incidence)

Appendix B: Definitions relevant to the electron spectroscopy of solid surfaces.

Autoionization: Non-radiative transition akin to Auger process but arising from excitation of outer atomic orbitals rather than from vacancy in core shell.

CDA: Cylindrical deflector analyzer.

Characteristic energy losses: Inelastic losses suffered by electrons passing through solid or gas at high pressure; such losses are characteristic of the material they pass through; in photoelectron spectra they are found as satellite structure on the low-energy side of the main peak.

CLS: Core level spectroscopy: photoelectron spectroscopy of the inner shells; this field involves the study of the binding energies of core electrons in atoms and molecules by X-ray photoelectron spectroscopy, which may be perturbed by the chemical environment.

CMA: Cylindrical mirror analyzer.

CNDO: Complete neglect of differential overlap: a semi-emperical method for calculating molecular orbitals; other modifications are INDO, MINDO, SPINDO.

Configuration interaction: Mixing of excited configurations of an atom or molecule with ground state wave function; an important mechanism for creating excited states.

Coster-Kronig transition: An Auger process involving electrons from the same principal shell as the initial vacancy: e.g. $L_{II} - L_{III}M$.

Crystal potential, molecular potential: Potential felt by core electrons in atom i for all other atoms and ions in molecule or solid; $V = \Sigma_j(q_i/r_{ij})$.

Electron shake-up and electron shake-off: Excitation of a bound electron into an excited state (shake-up) or continuum (shake-off) as the result of a sudden change in central potential (such as photoejection of a shielding electron).

ESCA: Electron spectroscopy for chemical analysis: (usually restricted to X-ray photoelectron spectroscopy).

FWHM: Full width at half maximum of a spectral peak (usual definition of resolution).

Hartree-Fock: Ab initio wave function calculations based on self-consistent procedure of having each electron move in potential of nuclear charge plus other electrons.

Helmholtz coils: A set of parallel coils through which electric current is passed in order to create a magnetic-field-free volume.

Hückel MO: A semi-empirical molecular orbital calculation.

IEE: Induced electron emission: (covers the general field of electron spectroscopy, including Auger and photoelectron spectroscopy).

Ion neutralization spectroscopy: The study of autoionization by ion impinging on a solid surface.

Jahn–Teller splitting: When a molecule possesses a high degree of symmetry, ejection of a photoelectron will destroy that symmetry, leading to a breakdown in degeneracy and more than one final electronic state.

Koopmans' theorem: Eigenvalues can be used as binding energies on the assumption that all other orbitals do not relax or are frozen upon the removal of an electron from a given orbital.

Lone-pair orbitals: A lone-pair orbital is a pair of molecular orbitals made up of two electrons that are not strongly involved in molecular bonding but are highly localized on a given atom.

MO: Molecular orbital.

Multiplet or exchange splitting: Splitting of photoelectron peaks due to coupling of unpaired spins created by photoelectron ejection with incompletely filled valence orbital.

Plasmon loss: One of the principal sources of inelastic scattering in solids by electrons in the range 0.1–10 keV.

Relaxation energy: The difference in energy between the eigenvalue and adiabatic binding energy.

Rydberg series: A series of excited atomic or molecular levels that are hydrogeneic in nature; they converge to an ionization limit.

SDA: Spherical deflector analyzer.

SIMS: Secondary ion mass spectrometry.

Spin–orbit splitting: Splitting in photoelectron spectra due to coupling of the spin and orbital angular momentum; splitting becomes most pronounced when orbitals of heavier elements are involved.

SMA: Spherical mirror analyzer.

SR: Synchrotron radiation: continuous radiation source created by the acceleration of high-energy electrons in a vacuum such as in a synchrotron; a most valuable source of quantized excitation for photoelectron spectroscopy.

UHV: Ultra-high vacuum.

VLS: Valence level spectroscopy: photoelectron spectroscopy of the outer shells; this field of photoelectron spectroscopy is primarily involved in directly determing the binding energies of electrons in the valence shell; it deals with both solids and gases, and use is made of both uv and X-ray sources.

X_α scattering: Method used for ab inito molecular orbital calculation, usually using muffin-tin potential (see chapter 2).

CHAPTER 4

SURFACE CRYSTALLOGRAPHY AND BONDING

M.A. VAN HOVE*,**

Department of Chemical Engineering, California Institute of Technology, Pasadena, California 91125, *USA*

* Supported in part by the Donors of the Petroleum Research Fund administered by the American Chemical Society under Grant No. 6809–AC5,7 and by the Army Research Office (Durham) under Grant No. DAHCO4–75–0170.

** Present address: Institut für Kristallographie, Universität München.

Contents

The nature of the surface chemical bond
Edited by Th.N. Rhodin and G. Ertl

1. Introduction

The geometrical location of atoms at a crystal surface is a fundamental quantity in the description of many surface properties and processes. In particular, the geometry is clearly basic to the understanding of bonding. It can also be very useful knowledge in the interpretation of results produced by various experimental techniques that investigate bonding, including for example many electron spectroscopic techniques.

The use of low-energy electron diffraction (LEED) has produced a growing body of known crystal surface structures, from which it is now possible to start to extract various trends and tentative rules concerning the atomic (and to a lesser extent molecular) bonding at surfaces. It is the purpose of this chapter to relate the present situation in this respect. No attempt will be made at interpreting the results in terms of any particular bonding theory.

The surface geometrical information used in this chapter is mainly due to the comparison of experimental and theoretical LEED intensity vs. voltage curves, whether by the "dynamical" (Pendry 1974) or the "data averaging" (Lagally et al. 1971) or the "Fourier transform" methods (Adams and Landman 1977, Cunningham et al. 1977). Such studies are based on preliminary information (such as chemical identity, cleanliness, adsorption coverage, etc.) obtained from a host of other surface analysis techniques (Auger electron spectroscopy, work function measurements, desorption kinetics, etc.). Ion scattering, ion neutralization spectroscopy and ion channelling are starting to produce surface crystallographic information, which at the time of writing in part agrees (for Ni(001) + c(2 × 2)O, cf. Brongersma and Theeten 1976; for Ni(110) + p(2 × 1)O, cf. Heiland and Taglauer 1972; for Ni(001) + c(2 × 2)S and p(2 × 2)S, cf. Hagstrum and Becker 1977) and in part disagrees (for Ni(001) + c(2 × 2)O, cf. Hagstrum and Becker 1971; for Pt(111), cf. Davies et al. 1975) with LEED results. Too few results from these techniques are available to judge their reliability. Furthermore some small-cluster calculations are predicting surface structures that partly agree with LEED results (Walch and Goddard 1977; Anderson 1977). It is unfortunate that at this time little overlap exists between the results discussed here for metal surfaces and those available from small clusters of metal atoms: their comparison would have been quite fruitful and will hopefully be possible in the near future. We shall, however, be able to compare bond lengths at surfaces with those in bulk materials and in molecules.

Concerning the quality and accuracy of LEED structural results, one can state that the results are for the most part consistent with each other and with known independent facts (bond lengths in particular). They are reproducible and very few are challenged by other workers. Nevertheless it must be pointed out that some results are of lower reliability than others and that some are non-unique in that there may occasionally remain some ambiguity between two or three different geometries. Such doubts will be mentioned where appropriate. For the sake of completeness it must be said that some failures to find a structure have occurred.

As a general rule, lengths perpendicular to the surface are reliable to within ± 0.1 Å; those parallel to the surface are in principle reliable to within ± 0.2 Å, but positions of high two-dimensional symmetry are often assumed to be chosen by atoms, thus removing this uncertainty of ± 0.2 Å. Incidentally, thermal vibration amplitudes are often of the order of 0.1 Å at room temperature; the atomic positions we refer to are the time-averaged positions.

As far as depth information is concerned, the LEED technique is sensitive to only a slice of surface of about 5 Å thickness, comprising typically three or four atomic layers. This slice will define the "surface" in our discussion. The kinds of crystal surfaces that we shall deal with are mostly bare (i.e. "clean") metals and these metals covered by regular arrays of atomic and molecular adsorbates (from sub-monolayers to multilayers), whose structural periodicity is simply related to that of the substrate metal. We shall also briefly review the present state of other surface structures.

A number of individual aspects of the geometry of such surfaces are relevant to bonding considerations. They include the registry (location parallel to the surface), the bond lengths, the bond directions and the implied charge transfers; furthermore, it should be instructive to examine the systematics of these quantities as a function of crystal face, of substrate material, of adsorbate properties, of atomic environment and of coverage. We shall discuss these questions for the bare metal surface in sect. 2 and for simple atomic adsorption on metals in sect. 3; sect. 4 will be concerned with the effect on the substrate of adsorption, including adatom penetration. Co-adsorption and molecular adsorption will be dealt with in sect. 5, while we shall briefly discuss other materials in sect. 6, concluding in sect. 7.

In the following text, individual references to work published on each surface structure determination will not normally be given. Instead the reader will find the references grouped by individual surface in the appendix.

2. Bare (clean) metal surfaces

The bare (i.e. clean) metal surfaces subdivide into "unreconstructed" and "reconstructed" surfaces. The structure of reconstructed surfaces (usually characterized by a surface superlattice) has not been determined by LEED calculations, but in a number of cases it is believed to be due to a top atomic layer having chosen hexagonal close-packing, even though the underlying lattice does not have hexagonal symmetry (Ignatiev et al. 1971). This would indicate a tendency to maximize the number of nearest neighbors and/or to smoothen the surface on an atomic scale.

The unreconstructed surfaces whose structures have been determined exhibit the atomic arrangement of the bulk up to and including the topmost atomic layer, within certain cases a top-layer relaxation perpendicular to the surface. We shall now examine the systematics of these results as a function of bulk lattice type and of exposed face.

2.1. Surface vs. bulk registries

We may ask whether the outermost few atomic layers of any given bare metal surface are positioned according to the space lattice that is prevalent in the bulk of the same metal. This question is particularly relevant for the registry (i.e. lateral position along the surface) of surface layers on the close-packed faces of hexagonal close-packed (hcp) and of face-centered cubic (fcc) metals. These materials have an atomic arrangement in the bulk that is customarily described by the sequences ...ABABAB... and ...ABCABC..., respectively, in terms of the registries of the individual close-packed atomic layers, cf. fig. 4.1. The close-packed faces of these materials, labelled hcp (0001) and fcc (111), respectively, would exhibit the following termination if no rearrangement occurs: ABABAB... and ABCABC..., respectively (the surface being at the left end of these sequences). The topmost layer in both lattices can however shift to a different registry without affecting the nearest-neighbor relationships (but changing the distances to the second-nearest and more-distant neighbors), thereby yielding the following sequences: CBABAB... and CBCABC..., respectively.

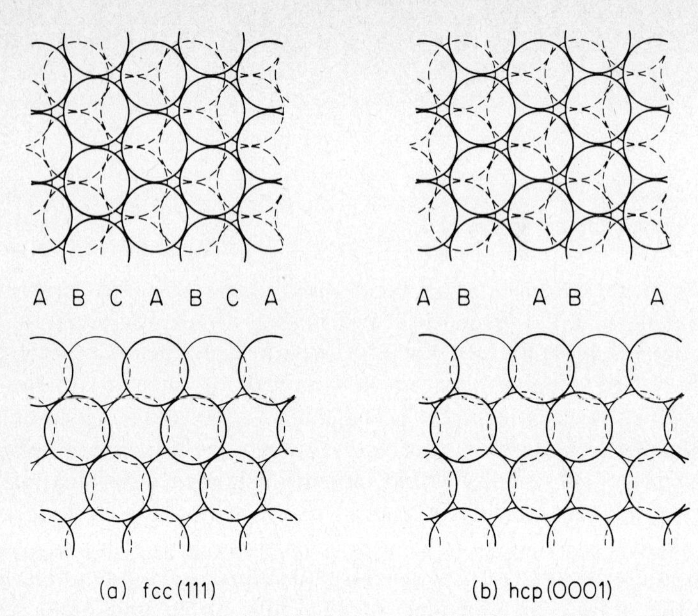

Fig. 4.1. (a) fcc (111) and (b) hcp (0001) faces, in top view (top diagrams) and side view (bottom diagrams). Letters A, B, C refer to different layer registries. Thin lines represent atoms beyond cut of view (dashed for hidden segments).

These shifts correspond to a surface-localized transition between hcp and fcc arrangements. It is well known that the difference in structural energy between the hcp and fcc lattices is relatively small, so that the presence of a surface could conceivable induce a change from one lattice type to the other, at least locally. No such occurrence has been observed, either for hcp (0001) surfaces (Be, Ti, Co in the low-temperature phase, Zn and Cd have been examined) or for fcc (111) surfaces (Al, Co in a high-temperature phase, Ni, Cu, Ag, Ir, Pt and Au have been examined).

A different situation where a registry change of the topmost layer is easily conceivable is that of the body-centered cubic (bcc) face (110). As is illustrated in fig. 4.2, a top-layer atom of this surface has 2 nearest-neighbors in the next layer (and 4 in the top layer), according to the positions implied by the bulk lattice. However top-layer atoms might roll over to a position that adds another nearest neighbor in the second layer, thereby increasing the total number of nearest neighbors from 6 to 7. This registry shift appears not to take place in the materials that have been investigated (Na, Fe and W). It seems therefore that the energy gained by increasing the number of nearest neighbors by one is insufficient to

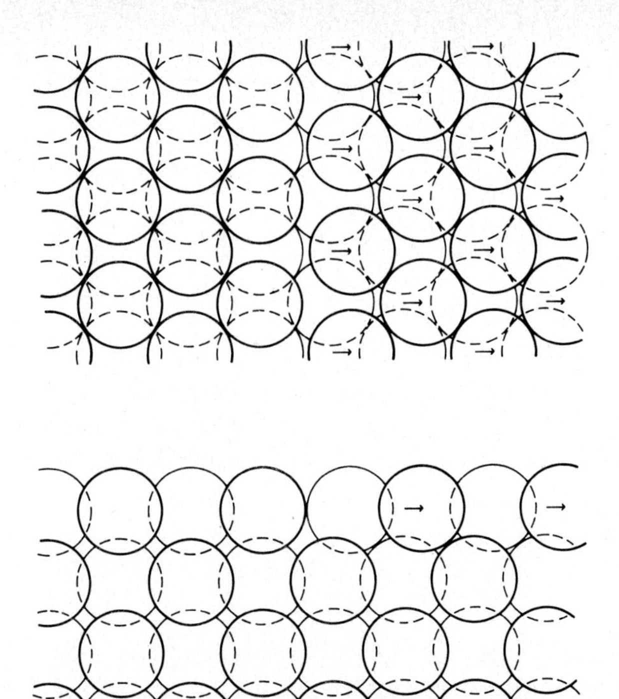

Fig. 4.2. bcc (110) face: top view in top diagram, side view in bottom diagram. Topmost layer sits in bulk-like position at left, in higher-coordination position at right (arrows indicate required relaxation).

offset the energy loss suffered in the bending of the bonds. This is consistent with the fact that the bulk structure of bcc materials already shows relatively strong bond-directionality effects that may be responsible for reducing the number of nearest neighbors from the close-packed 12 to 8.

Surface registry changes seem less likely on other metal surfaces, in view of the lack of reasonable alternative sites for surface atoms to move to. This intuitive feeling is confirmed by the structural results obtained on fcc (001) faces (Al, Co, Ni, Cu, Rh and Ag), on fcc (110) faces (Al, Ni and Ag*) and on bcc (001) faces (Fe, Mo and W), cf. fig. 4.3. Summarizing, the various bare metal faces we have mentioned show

* These surfaces exhibit some inconsistencies in the comparison of LEED theory and experiment: both Al(110) and Ni(110) show a few irreducible discrepancies while for Ag(110) one study (Zanazzi et al. 1977) has major problems (some beams give a very poor fit), whereas an independent study (Moritz 1976) actually shows very good agreement between theory and experiment.

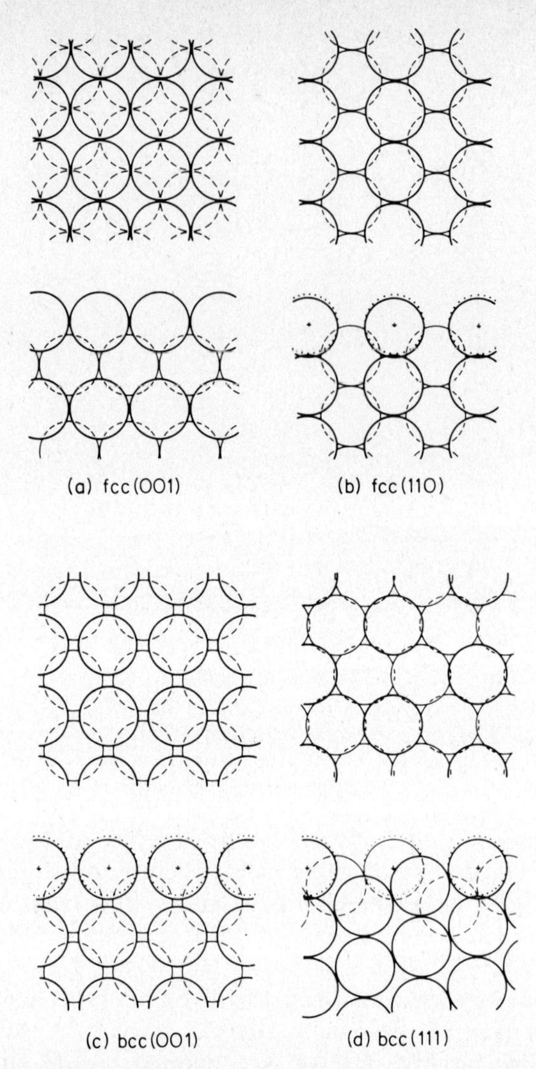

Fig. 4.3. (a) fcc (001), (b) fcc (110), (c) bcc (001) and (d) bcc (111) faces, each in top view (top diagrams) and side view (bottom diagrams). Arrows indicate atomic relaxations; dotted circles outline unrelaxed atomic positions.

that a surface atom generally maximizes the number of nearest neighbors in choosing its bonding site; however, for a bcc (110) face this is not the case: there bond directionality may be the more important factor. Reconstruction, as on fcc (001) (Ir, Pt, Au; cf. Kesmodel and

Somorjai 1976, Ignatiev et al. 1971) may possibly increase the number of nearest neighbors even more, if the hypothesis of a hexagonal close-packed top layer is correct.

2.2. Surface vs. bulk bond lengths

Interatomic distances are known to depend on a number of factors, such as number of nearest neighbors, bond order, ionicity, etc. (Pauling 1960). Therefore one can expect differences between bulk and surface bond lengths. Before we analyze the results on surfaces, the accuracy of the LEED method must be recalled. As mentioned earlier, the accuracy for distances perpendicular to the surface is of the order of 0.1 Å (some workers claim a somewhat higher accuracy in some instances); this corresponds, in the cases that we shall discuss, to a relative uncertainty of about 5% in the topmost interlayer spacing, and in view of the angle that bonds make with respect to the surface plane (typically about 35°) this translates to a much smaller uncertainty of about 2% in the bond lengths at the surface (given the assumption that the registry is not affected by any uncertainty apart from thermal vibrations).

The results of LEED structural analyses indicate, despite a few exceptions, a clear correlation between an observed bond length contraction and what we shall term "surface roughness" on an atomic scale. In this context, the hcp (0001) and fcc (111) faces shall be considered to be smooth, while the bcc (110) and fcc (001) faces are less smooth and the fcc (110), bcc (001) and bcc (111) faces are rough, cf. fig. 4.3: this roughness means that surface atoms stick out of the surface to some extent, producing a lumpy or ridged appearance.

On the rough faces, one observes a small but definite contraction of the bond length between atoms of the topmost and the next layers: for fcc (110) one observes a contraction by 3 to 4% for Al, 1.5% for Ni and 2% for Ag (as mentioned earlier, one of two studies on Ag(110) has some difficulties in matching theory and experiment), while the corresponding values for bcc (001) are 1.5% for Fe, 4% for Mo and 2 to 4% for W; Fe(111) may have a bond length contraction by 1.5%. In these structures the shortened bonds are directed at about 35° to the surface plane: the small changes in bond length therefore correspond to rather larger (and mostly easily detectable) contractions of the topmost interlayer spacing, ranging from 4 to 15% (0.06 to 0.42 Å). It must be stated that the contraction examined here may possibly depend on the surface cleanliness, cf. sect. 4.

The other metal faces show a tendency towards a very small bond length contraction (the uncertainty of about 2% however being larger than the contraction, which is mostly less than 1%). This applies, in order of increasing smoothness, to fcc (001) faces (Al, Co*, Ni, Cu, Rh** and Ag), to bcc (110) faces (Na, Fe and W), to fcc (111) faces (Al†, Co in its high-temperature phase, Ni, Cu, Ag, Ir, Pt‡ and Au) and to hcp (0001) faces (Be, Ti, Co*, Zn and Cd).

The correlation between bond length contraction and surface roughness has various possible explanations. One of these is the theoretical evidence that the uneven surface of the metallic electron cloud tends to smooth itself out as if there were a surface tension, thereby electrostatically drawing the emerging atoms into the surface (Finnis and Heine 1974). Another explanation rests on the enhanced asymmetry of the environment of surface atoms on rough faces, which makes the resultant attractive force to the nearest neighbors larger, by virtue of the absence of many atoms on the outside. Furthermore the reduced number of nearest neighbors in rough faces increases the number of bonding electrons in the fewer available bonds, thereby presumably reducing their length. These explanations are all consistent with the observation that adsorbed atoms tend to cancel the bond length contraction observed on rough surfaces, cf. sect. 4.

Note that, in contrast to the evidence from registries, there is no clear evidence in these bond length contractions to indicate that the bcc lattice is more rigid in its bond directions than the fcc lattice, since both bcc (001) and fcc (110) show similar contractions, implying similar changes in the bond angles.

It should be stated that there is a contradiction in these bond length contractions with predictions of bond length elongations from lattice-dynamics calculations (see e.g. Cheng et al. 1974). Electrostatic effects appear to be able to resolve this contradiction (Ma et al. 1977).

If one thinks of atomic steps on a surface (such as occur on high Miller-index faces, as opposed to the low Miller-index faces that we consider here), it is interesting to observe that especially the fcc (110) (and to a lesser extent also the bcc (111) and bcc (001) surfaces) can be seen as a limiting case of a stepped surface: here the steps have a mono-atomic depth as well as a mono-atomic height. The observed bond length contraction may therefore conceivably extend to the structurally

* For Co(001) and Co(0001) a bond length contraction by about 1.5% is reported.

** Rh(001) actually shows an increase in bond length by about 1%.

† For Al(111) are reported by different workers: a contraction by 1%, an expansion by 1.5% and twice no change in the bond length.

‡Ion-channelling indicates expansion of the bond length, but refinements are expected to reduce that expansion.

similar step edges of high-index faces, cf. similar conclusions in Besocke and Wagner 1975.

3. Atomic adsorption on metal surfaces

We now consider adsorption of single adsorbate species on the metal surfaces already mentioned in sect. 2. The influence of such adsorption on the structure of the metal substrate and the case of adatom penetration into the substrate will be specially dealt with in sect. 4. Co-adsorption and molecular adsorption will be discussed in sect. 5.

As in the previous section, we shall investigate the location of top-layer atoms, in this case foreign adatoms, at a metal surface, considering in particular registries, bond lengths, bond angles, and now also the associated charge transfers.

3.1. Adsorbate registries (adsorption sites)

Let us first inquire whether an adatom chooses its adsorption site on the basis of the number of nearest metal neighbors that it can have, in particular whether the number of nearest metal neighbors is maximized. The answer appears to be affirmative from the weight of the evidence: very few exceptions are known. In all the following cases the number of nearest metal neighbors is maximized*:

fcc Al(001) + c(2 × 2)Na,
fcc Ag(111) + ($\sqrt{3} \times \sqrt{3}$)R30°I,
fcc Cu(001) + c(2 × 2)N**,
bcc Fe(001) + c(2 × 2)S,
bcc Mo(001) + p(1 × 1)Si,
fcc Ni(001) + c(2 × 2)O†,
fcc Ni(111) + p(2 × 2)O,
fcc Ni(111) + p(2 × 2)S,
fcc Ni(001) + p(2 × 2)Se,
fcc Ni(001) + p(2 × 2)Te,
bcc W(110) + p(2 × 1)O.
fcc Ag(001) + c(2 × 2)Cl,
fcc Ag(001) + c(2 × 2)Se,
fcc Cu(001) + p(2 × 2)Te,
bcc Mo(001) + c(2 × 2)N,
fcc Ni(001) + c(2 × 2)Na,
fcc Ni(001) + p(2 × 2)O†,
fcc Ni(001) + c(2 × 2)S,
fcc Ni(001) + c(2 × 2)Se,
fcc Ni(001) + c(2 × 2)Te,
hcp Ti(0001) + p(1 × 1)Cd,

The periodicities and adsorption geometries of these surfaces are shown

*For the notation cf. Wood 1964.

**Poor agreement between theory and experiment.

†For these structures, a Generalized-Valence-Bond cluster calculation confirms the LEED result for the adsorption site and the bond length (Walch and Goddard 1977).

in figs. 4.4 and 4.5, respectively. A particularly noteworthy case is that of bcc W(110) + p(2 × 1)O. Although the level of agreement between theory and experiment is not in this case sufficient to definitely exclude the possibility that the oxygen adatoms choose an adsorption site with 2 nearest metal neighbors (the site prescribed by extending the bcc lattice outside of the substrate), it is nevertheless most likely that the oxygen atoms prefer the 3-fold coordinated sites that maximize the number of nearest metal neighbors.

To date one clear exception to the maximization of the number of nearest metal neighbors is known: fcc Ni(110) + p(2 × 1)O, also illustrated in fig. 4.5. Unlike sulfur, on this Ni(110) surface (cf. below), the oxygen does not choose the deep "center" site, but rather the "short bridge" site, providing only 2-fold coordination.

Two boundary cases exist. For bcc Fe(001) + p(1 × 1)O, the adatom does choose the hollow adsorption site, but the bond length to the nearest second-layer Fe atom appears to be slightly smaller than that to the 4 top-layer Fe atoms (2.07 ± 0.06° Å vs. 2.09 ± 0.02 Å): the coordination number is strictly 1, but the uncertainties in the bond lengths may

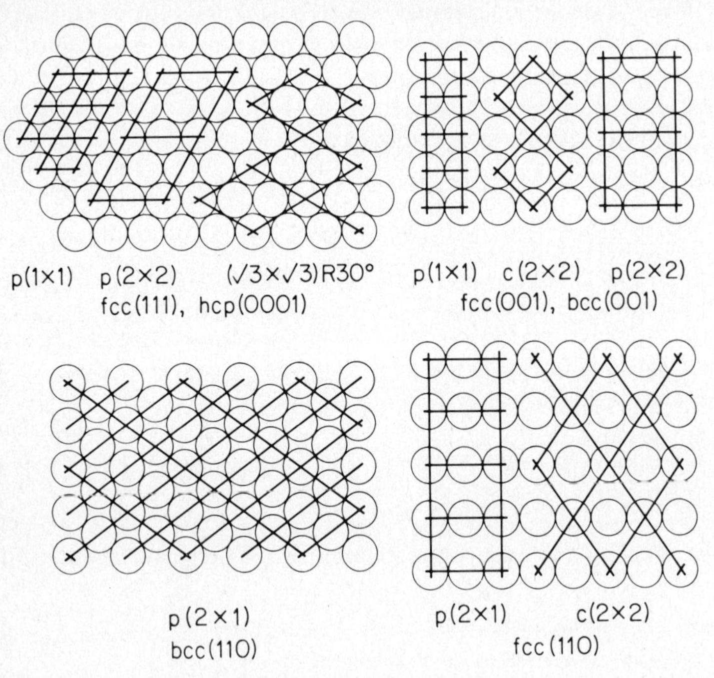

Fig. 4.4. Surface lattices: circles represent top substrate atoms, lines represent some superlattices due to ordered overlayers.

make it 5 or 4. A more diverging case is presented by fcc Ni(110) + c(2 × 2)S, where the adatom again chooses the hollow site, but now is definitely closer to a second-layer Ni atom than to the 4 nearest top-layer Ni atoms (2.17 ± 0.1 Å vs. 2.35 ± 0.04 Å).

It is interesting to consider this maximization principle of the number of nearest metal neighbors in the light of face, substrate and adsorbate dependence, since this shows how insensitive the maximization is with respect to large changes in the chemical and geometrical environment.

Face dependence: O satisfies the maximization principle on Ni(111) and Ni(001), but not on Ni(110), whereas S satisfies it on each of these faces.

Substrate dependence: The maximization principle is satisfied for: N on Cu(001) and Mo(001); O on Fe(001), Ni(111), Ni(001) but not on Ni(110)) and W(110); Na on Al(001) and Ni(001); S on Fe(001) and Ni(001); Se on Ni(001) and Ag(001); Te on Ni(001) and Cu(001).

Adsorbate dependence: The maximization principle is satisfied for: O and S on Fe(001); O, Na, S, Se and Te on Ni(001); N and Si on Mo(001); Cl and Se on Ag(001).

It will be observed that adsorption sites with many nearest neighbors are usually also sites of high symmetry. Therefore we may say that adsorbate atoms appear to favor sites of high symmetry. There is only one exception to this preference: in W(110) + p(2 × 1)O the oxygen chooses a site that has one mirror plane instead of a site that has 2 orthogonal mirror planes; this may be related to the fact that the overlayer as a whole already has low symmetry (only a 2-fold axis of rotation). Even Ni(110) + p(2 × 1)O satisfies the choice of highest symmetry: no other site has higher symmetry (though several have the same symmetry, two orthogonal mirror planes).

If we now also take the second and deeper substrate layers into consideration, we may in particular wonder whether the adsorbate atoms choose an adsorption site consistent with a continuation of the substrate lattice. It appears from the available results that the bulk lattice is in fact often continued into the overlayer, as if a substrate atom rather than a foreign atom had adsorbed, despite differing bond lengths. This bulk lattice continuation is satisfied by nearly all the examples listed at the beginning of this section. Ni(110) + p(2 × 1)O again is an exception. Ti(0001) + p(1 × 1)Cd is also an exception, belonging to an interesting class of surfaces. These are the hcp (0001) and fcc (111) surfaces which we once more shall describe by the registry sequences ABABA... and ABCABC..., respectively (surface at left). Using lower case letters for overlayers, the continuation of the bulk lattice into the overlayer would imply sequences bABABA... and cABCABC..., respectively. These sequences are indeed found for fcc Ni(111) + p(2 × 2)S and fcc Ag(111) + ($\sqrt{3} \times \sqrt{3}$)R30°I. For Ni(111) + p(2 × 2)O it could not be determined whether the sequence is cABCABC... or bABCABC... On the other hand, hcp Ti(0001) + p(1 × 1)Cd was found to have the deviating sequence cABABA..., meaning that the cadmium atom is repelled by second-nearest Ti neighbors. For this system, a second overlayer of

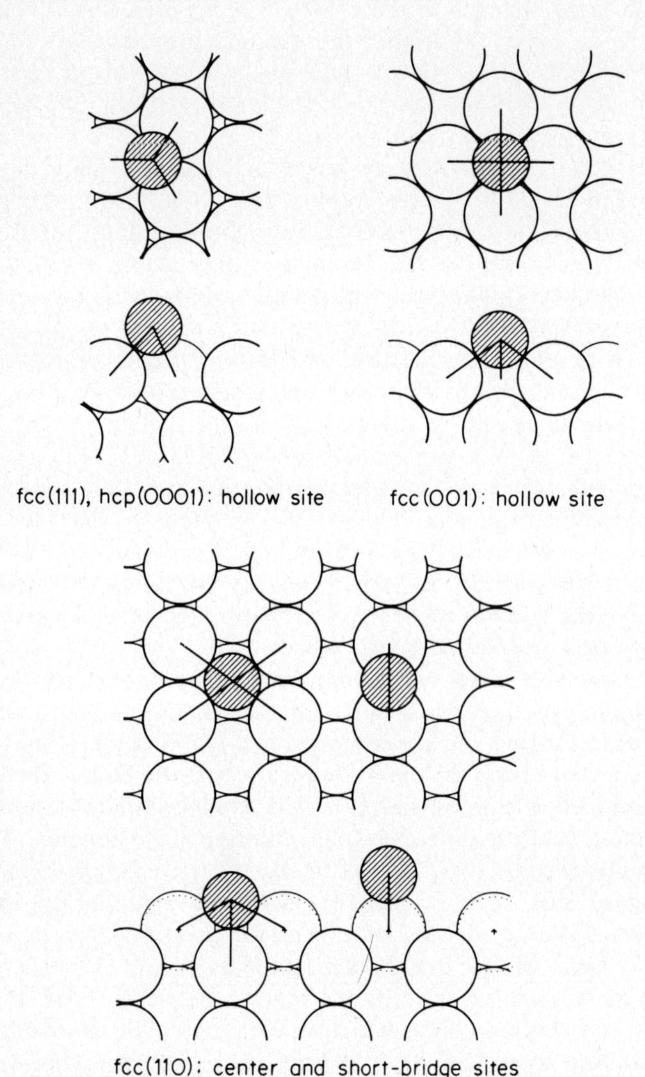

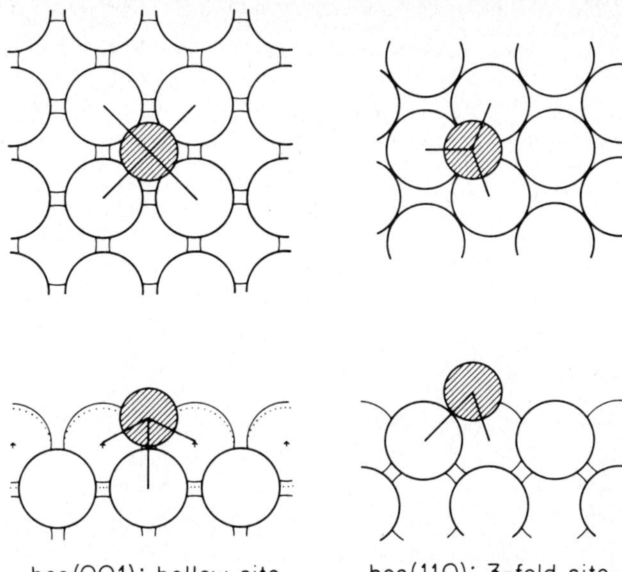

Fig. 4.5. Adsorption geometry for several surface types. Adatoms are hatched. Dotted outlines represent bare substrate atomic positions. Top diagrams give top view, bottom diagrams give side view.

cadmium was subsequently deposited on top of the first monolayer: the registry sequence was then found to be acABABA. . . , which on further cadmium adsorption is expected to produce the sequence cacABABA. . . , because adsorption of yet more layers was found to reproduce the bulk hcp lattice of cadmium (the Ti and Cd atomic radii are nearly equal). Incidentally, this metal–metal interface is seen to be structurally asymmetrical between the two metals: the only fcc region (. . . cAB. . .) involves two Ti layers and only one Cd layer.

A further question concerning the adsorption registry is whether it depends on adsorption coverage, i.e. on density of adatoms: this is relevant to the effects of adatom–adatom interactions. The situation is illustrated by a limited set of results, namely those for quarter-monolayer and half-monolayer adsorption of O, S, Se and Te on Ni(001) in p(2 × 2) and c(2 × 2) periodicities: the adsorption site is found not to depend on coverage in these cases (the nearest adatom–adatom distances are 4.90 and 3.46 Å for the two coverages, respectively, compared with a largest adatom diameter of about 2.7 Å for Te). A different aspect is presented by the multi-overlayer of cadmium on Ti(0001), where adsorption on top of an adsorbate also does not affect the initial adsorption site. (Co-adsorption of chemically different adsorbates will be discussed in sect. 5).

The adsorption site seems to be only slightly influenced by the adatom valence. Almost no site variation is observed whether one considers such chemically different adatoms as N, O, Na, Si, S, Cl, Se, Te and I. Somewhat surprisingly, even divalent oxygen on W(110) chooses a 3-fold rather than a 2-fold coordinated adsorption site. An exception is again Ni(110) + p(2 × 1)O, as described above. Similarly the metal adatom Cd behaves "abnormally", cf. above. This relative insensitivity of the bonding site to valence is particularly intriguing. For example, one wonders how a divalent atom such as S bonds to its 4 nearest Ni neighbors in the Ni(001) + c(2 × 2)S system. One attractive suggestion (applicable to many adsorbate systems) uses the concept of resonance (Pauling 1960): the adatom bonds primarily to 2 neighbors (in the case of divalent adatoms), but this double bond resonates between the several available neighbors. This kind of bonding is probably facilitated by the fact that the outward-oriented t_{2g} or e_g orbitals of the substrate atoms usually point roughly in the direction of the adatoms. The resonance idea seems to be borne out by recent cluster calculations for oxygen adsorbed on Ni(001) (Walch and Goddard 1977).

3.2. Adsorption bond lengths and bond angles

The uncertainty in the bond lengths between adsorbate atoms and the nearest metal neighbors, as obtained from LEED, is again determined by the inherent uncertainty of about 0.1 Å for distances perpendicular to the surface. We may assume that the registry of the adsorbate atom does not have any uncertainty associated with it. Then the 0.1 Å uncertainty translates, after correction for the bond orientations, into uncertainties in the bond lengths of 0.04 to 0.09 Å, corresponding to relative uncertainties of 2% to 4%. This level of accuracy may be just sufficient to examine bond-order effects and the like, which often induce bond length variations much in excess of 0.1 Å (Pauling 1960).

In table 4.1, we compare some known bond lengths for each pair of atoms encountered in LEED analyses, by arranging the bond lengths in order of increasing value.

Our first observation about the observed adsorption bond lengths is that most of them fall within the range of known bond lengths for the same pair of atoms in molecules and bulk solids. Exceptions are Ni(111) + p(2 × 2)S, which shows a slightly small bond length, and Mo(001) + c(2 × 2)N with an abnormally large bond length (this surface does not give good agreement between theory and experiment).

TABLE 4.1

Bond lengths for surface adsorption, compared with known values in molecules and bulk materials. Asterisks denote surfaces where the adsorption affects the substrate geometry enough to be detected by LEED.

Atom pair	Environment	Bond length (Å)	Refs.[a]
Cd–Ti	sum of radii	3.01	
	Ti(0001) + p(1 × 1)Cd	3.08 ± 0.03	
	Ti(0001) + p(1 × 1)Cd + p(1 × 1)Cd	3.13 ± 0.06	
Cl–Ag	$Cs_2AgAuCl_6$	2.36	TID
	sum of radii	2.43	
	Ag(001) + c(2 × 2)Cl	2.67 ± 0.06	
	AgCl (NaCl structure)	2.77	WY
I–Ag	AgI molecule	2.54	TID
	AgI bulk	2.80	
	Ag(111) + c($\sqrt{3} \times \sqrt{3}$)R30°I	2.80	
	$[Ag_4I_6]$	2.85	TID
N–Cu[b]	... $Cl_3CuNNCH_3$ ... (large molecule)	1.993	TID
	Cu(001) + c(2 × 2)N	2.02 ± 0.05	
	$[Cu(NH_3)_2]^{2+}$	2.03	TID
	$K_2Pb[Cu(NO_2)_6]$	2.11	SR
N–Mo	sum of radii	2.11	
	various molecules	2.29–2.33	TID
	*Mo(001) + c(2 × 2)N	2.45 ± 0.05	
Na–Al	sum of radii	2.82–3.00	
	Al(001) + c(2 × 2)Na	2.86–2.90 ± 0.08	
Na–Ni	Ni(001) + c(2 × 2)Na	2.84 ± 0.08	
	sum of radii	2.80–3.10	
O–Fe	*Fe(001) + p(1 × 1)O nearest neighbor	2.07 ± 0.06	
	*Fe(001) + p(1 × 1)O next nearest neighbor	2.09 ± 0.02	
	$FeCl_2H_8O_4$	2.09	TID
	FeO (NaCl structure)	2.15	WY
O–Ni	Ni(111) + p(2 × 2)O	1.88 ± 0.06	
	$Ni(C_5H_7O_2)_2$	1.90	TID
	*Ni(110) + p(2 × 1)O	1.92 ± 0.04	
	Ni-chelate complexes	1.84–2.06	
	Ni(001) + c(2 × 2)O	1.98 ± 0.05	
	Ni(001) + p(2 × 2)O	1.98 ± 0.05	
	NiO (NaCl structure)	2.08	WY
	$NiC_{10}H_{14}N_2O_8$, H_2O	2.03–2.18	TID

TABLE 4.1 (cont.)

Atom pair	Environment	Bond length (Å)	Refs.[a]
O–W	$[WO_4]^{2-}$	1.75–1.81	TID
	$BaWO_4$	1.95	SR
	W(110) + p(2 × 1)O	2.08 ± 0.07	
	WOF_4	2.11	SR
	sum of radii	2.12	
S–Fe	$Fe_{0.35}ZrS_2$ bulk	1.99	SR
	sum of radii	2.28	
	*Fe(001) + c(2 × 2)S	2.30 ± 0.06	
	$Fe_{0.35}ZrS_2$ bulk	2.30	SR
	Fe_7S_8 bulk	2.44	SR
S–Ni	Ni(111) + p(2 × 2)S	2.02 ± 0.06	
	Ni–chelate complexes	2.10–2.23	
	*Ni(110) + c(2 × 2)S to 2nd substrate layer	2.17 ± 0.10	
	γNiS bulk	2.18	
	Ni(001) + c(2 × 2)S	2.19 ± 0.06	
	Ni(001) + p(2 × 2)S	2.19 ± 0.06	
	Ni_3S_2 bulk	2.28	
	Ni(110) + c(2 × 2)S to 1st substrate layer	2.35 ± 0.04	
	αNiS bulk	2.38	
	NiS_2 bulk	2.34–2.42	
Se–Ag	$AgCrSe_2$ bulk	2.46–2.80	SR
	sum of radii	2.60–2.69	
	Ag_2Se bulk	2.62–2.86	SR
	Ag(001) + c(2 × 2)Se	2.80 ± 0.07	
Se–Ni	Ni(001) + c(2 × 2)Se	2.28 ± 0.06	
	Ni–chelate complex	2.32	
	Ni(001) + p(2 × 2)Se	2.34 ± 0.07	
	Ni_3Se_2 bulk	2.36	
	$Ni_{0.55}Se_{0.08}Te_{0.37}$ bulk	2.31–2.44	SR
	Ni_3Se_4 bulk	2.47	
	NiSe bulk	2.50	
	$NiRh_2Se_4$ bulk	2.50–2.52	SR
	$NiSe_2$ bulk	2.49–2.53	
Si–Mo	*Mo(001) + p(1 × 1)Si	2.51 ± 0.05	
	sum of radii	2.53	
Te–Cu	Cu(001) + p(2 × 2)Te	2.48 ± 0.10	
	$Cu_8Cu_{4-x}Te_6$ bulk	2.51–2.76	SR
	Cu_4Te_3 bulk	2.65	
	Cu_2Te bulk	2.67	
	$Cu_{4-x}Te_2$ bulk	2.65–2.76	SR

TABLE 4.1 (cont.)

Atom pair	Environment	Bond length (Å)	Refs.[a]
Te–Ni	Ni(001) + $p(2\times2)$Te	2.52–2.58 ± 0.08	
	$NiTe_2$ bulk	2.58–2.59	
	Ni(001) + $c(2\times2)$Te	2.59 ± 0.07	
	NiTe bulk	2.64	
	$Ni_{0.55}Se_{0.08}Te_{0.37}$bulk	2.54–2.85	SR

[a] Where no references are listed, those of the appendix apply; TID = Tables of Interatomic Distances; SR = Structure Reports; WY = Wyckoff.
[b] For Cu(001) + c(2 × 2)N LEED data averaging predicts a large bond length of 2.32 Å, while more reliable dynamical calculations yield the more reasonable value listed here (despite residual discrepancies); such multiple results are known to occur when less accurate calculations are used.

Restricting ourselves to surfaces, we may present the results of table 4.1 in a different form by defining an effective radius of the adatom as being equal to the bond length less the substrate hard-ball radius. Such a hard-ball picture may be useful in exhibiting trends. Table 4.2 groups these effective radii by adatom species in ascending order for any substrate and includes for comparison the Pauling radius; the coordination number to the metal atoms is also given.

In these series of bond lengths and radii one can observe that the adsorption bond lengths fit into the familiar general trend according to which bond lengths increase with the coordination number. There are however exceptions to this trend that may in part be ascribed to the limited accuracy of the LEED determinations but probably also to more complex factors involving the bonding mechanism itself. A detailed study of these bond lengths and radii should shed some light on the mechanism of chemical bonding at surfaces. A partial study has already been carried out (Madhukar 1975), suggesting the applicability of long-established concepts (bond order, valency saturation, resonating bond, etc.).

A few further general observations can be made. It is evident that the observed adsorption bond lengths do not in general exhibit strong charging (ionic) effects (cf. also subsect. 3.3). One should also notice the small observed coverage dependence of bond lengths, as evidenced by the p(2 × 2) (quarter monolayer) and c(2 × 2) (half monolayer) coverages of O, S, Se and Te on Ni(001): this merely confirms that adatom–adatom interactions are weak in relation to the metal–adatom bond.

When one examines the various bond angles that the LEED results

TABLE 4.2
Effective radii of adsorbates, cf. text.

Adatom species	Situation	Coordination number	Effective radius (Å)
Cd	Pauling radius		1.48
	Ti(0001) + p(1 × 1)Cd	3	1.63 ± 0.03
	Ti(0001) + p(1 × 1)Cd + p(1 × 1)Cd	3	1.68 ± 0.07
Cl	Pauling radius		0.99
	Ag(001) + c(2 × 2)Cl	4	1.23 ± 0.06
I	Pauling radius		1.33
	Ag(111) + (√3 × √3)R30°I	3	1.36
N	Pauling radius		0.70
	Cu(001) + c(2 × 2)N	4	1.05 ± 0.05
	Mo(001) + c(2 × 2)N	4	1.09 ± 0.05
Na	Al(001) + c(2 × 2)Na	4	1.45 ± 0.08
	Ni(001) + c(2 × 2)Na	4	1.60 ± 0.08
	Pauling radius		1.85
O	Ni(111) + p(2 × 2)O	3	0.64 ± 0.06
	Pauling radius		0.66
	Ni(110) + p(2 × 1)O	2	0.68 ± 0.04
	W(110) + p(2 × 1)O	3	0.72 ± 0.07
	Ni(001) + c(2 × 2)O	4	0.74 ± 0.05
	Ni(001) + p(2 × 2)O	4	0.74 ± 0.05
	Fe(001) + p(1 × 1)O	1 + 4[a]	0.84 ± 0.06
S	Ni(111) + p(2 × 2)S	3	0.78 ± 0.06
	Ni(110) + c(2 × 2)S	1 + 4[b]	0.93 ± 0.10
	Ni(001) + c(2 × 2)S	4	0.95 ± 0.06
	Ni(001) + p(2 × 2)S	4	0.95 ± 0.06
	Pauling radius		1.04
	Fe(001) + c(2 × 2)S	4	1.06 ± 0.06
Se	Ni(001) + c(2 × 2)Se	4	1.04 ± 0.06
	Ni(001) + p(2 × 2)Se	4	1.10 ± 0.07
	Pauling radius		1.13
	Ag(001) + c(2 × 2)Se	4	1.36 ± 0.07
Si	Mo(001) + p(1 × 1)Si	4	1.15 ± 0.05
	Pauling radius		1.17
Te	Cu(001) + p(2 × 2)Te	4	1.21 ± 0.10
	Ni(001) + p(2 × 2)Te	4	1.28 ± 0.08
	Ni(001) + c(2 × 2)Te	4	1.35 ± 0.07
	Pauling radius		1.39

[a] 1 nearest neighbor + 4 next-nearest neighbors 0.02 Å farther away.
[b] 1 nearest neighbor + 4 next-nearest neighbors 0.18 Å farther away.

imply, it is rapidly apparent that they are primarily determined by the bond lengths: the hard-ball picture seems to be a useful concept. It appears that, as in molecules and solids, bond bending is energetically rather more favorable than bond stretching. In particular, the bond directions implied by substrate orbitals of t_{2g} and e_g symmetry are not well respected: for example, on fcc (001) surfaces, one might predict adsorption bonds oriented at 45° to the surface plane, whereas they in fact have directions of between 27° and 47° with respect to the surface plane, depending directly on the size of the atoms. Nevertheless, if one considers the metal–adatom–metal bond angle, one does find a fairly constant value for O and S adsorption on Ni (the only two cases where enough data are available to look for such a trend): the Ni–O–Ni bond angle is 77.5° on Ni(001) (between two touching Ni atoms), 81.0° on Ni(110) (short-bridge site) and 82.5° on Ni(111), while the Ni–S–Ni bond angle is 63.7° on Ni(110) (at least for one of two such angles available at the center site), 69.0° on Ni(001) (between two touching Ni atoms) and 75.7° on Ni(111). On the other hand, a Generalized-Valence-Bond cluster calculation for O on Ni(001) (Walch and Goddard 1977) predicts bonding to two non-touching Ni atoms (and essentially the same bond length as LEED predicts), which corresponds to a much larger Ni–O–Ni bond angle of 126° instead of 77.5; such an angle is not present on the other faces of nickel. (However, the reliability of such conclusions is mitigated by the fact that the GVB calculation predicts a different bonding site for S on Ni(110) than LEED does: the long-bridge site rather than the center site, cf. Walch and Goddard 1977.)

3.3. Work function changes and charge transfers

The change of the work function of a surface on adsorption of an overlayer can be ascribed to a dipole layer resulting from electric charge transfers between substrate and adsorbate atoms. If one knows the work function change and the atomic positions, one can then obtain an estimate of the amount of charge transfer induced by the adsorption through the relation $\Delta\phi = 4\pi e\sigma d$, where $\Delta\phi$ is the work function change, e the electronic charge, σ the surface charge density and d the dipole length perpendicular to the surface. For d we may, for the sake of argument, take the component of the bond length perpendicular to the surface, while σ is related to the known adsorption coverage. We can now ask, within this simple model, what fraction $\Delta e/e$ of an electron transferred at each adsorbate site through a distance equal to the bond

TABLE 4.3
Implied charge transfers $\Delta e/e$ as described in text. Coverage is normalized to 1 for an overlayer that has one adatom per substrate unit cell; d is the overlayer spacing, i.e. the component of the bond length perpendicular to the surface.

Adsorbate	Substrate	Coverage	d(Å)	$\Delta\phi$(eV)	$\Delta e/e$(%)
I	Ag(111)	0.33	2.25	−0.5	−2.6
N	Mo(001)	0.5	1.02	1.05	11[a]
Na	Al(001)	0.5	2.07	−1.45	−6.3
Na	Ni(001)	0.5	2.23	−2.5	−7.7
O	Ni(111)	0.25	1.20	~0.7	~7.0
O	Ni(001)	0.25	0.90	0.22	3.3
O	Ni(001)	0.5	0.90	0.36	2.7
O	Ni(110)	0.5	1.46	0.46	3.1
O	W(110)	0.5	1.25	0.7	4.4
S	Ni(111)	0.25	1.40	~1.0	~8.4
S	Ni(001)	0.25	1.30	0.24	2.5
S	Ni(001)	0.5	1.30	0.38	2.0
Se	Ni(001)	0.25	1.55	0.08	0.7
Se	Ni(001)	0.5	1.45	−0.07	−0.3
Te	Ni(001)	0.25	1.80	−0.29	−2.2
Te	Ni(001)	0.5	1.90	−0.43	−1.5

[a] This surface presents difficulties in the comparison of LEED theory and experiment.

length produces the observed work function change. One finds in this way values for $\Delta e/e$ that do not exceed about 11%, even for alkali adsorbates, cf. table 4.3. Such small values can be qualitatively understood in terms of mutually destructive dipoles, and this is confirmed by the observed coverage dependence of the implied charge transfers: the charge transfer per adatom is reduced when the adsorption coverage is increased.

But in one case (Se on Ni) the work function changes and charge transfers switch their signs on a variation of the coverage between quarter-monolayer and half-monolayer. Note also that Te on Ni produces charge transfers opposite in sign to O and S on Ni. Furthermore, O and S on Ni(111) seem to have a much larger charge transfer than on other faces of Ni, for no obvious reason. The chalcogen series O, S, Se and Te appears to have a complicated behavior in terms of charge transfer. One hopes that such peculiarities will, at least, be explained by detailed bonding theories.

4. Adsorption affecting the substrate

In this section we shall review those surfaces for which LEED has detected a clear modification of the substrate structure due to adsorption.

Our first observation is that for those metal surfaces that in the bare state have essentially the bulk structure (i.e. very little bond length change at the surface), adsorption does not affect the metal–metal bond lengths (within the uncertainties of the method). On the other hand, for surfaces that, bare, exhibit a top-layer contraction, adsorption does alter metal–metal bond lengths, usually increasing their value to about their bulk metal value. This situation applies to bcc Fe(001) + c(2 × 2)S, fcc Ni(110) + p(2 × 1)O, fcc Ni(110) + c(2 × 2)S*, bcc Mo(001) + c(2 × 2)N** and bcc Mo(001) + p(1 × 1)Si, whose structures can be found further described and illustrated in sect. 3. These results may indicate that an excess of electronic charge located, for the bare surface, in the shortened metal–metal bonds, is now used instead to bond the adatom, thereby approximately restoring the bulk metal–metal bond length. Also adsorption submits the top metal atoms to a less asymmetrical set of bonding forces than on the bare surface, so that an environment more like that of the metal bulk is simulated. Some such mechanism seems to be operating even though most of these overlayers have only half-monolayer coverage and one of them (O on Ni) does not even involve the adsorption site prescribed by the substrate lattice.

Recent LEED data suggest that H adsorption on bcc W(001) (Lee et al. 1977) has the same effect as described above; in fact the contracted metal–metal bond appears to gradually recover its bulk length as the H coverage increases monotonically (however, some averaging between bare and covered adsorption sites may be at work here, giving the impression of a gradual bond length change with coverage, rather than a sudden change at each adsorption site).

From these results it should be clear that surface contamination can play a significant perturbing role in the determination of atomic positions at bare metal surfaces (and any other surfaces).

A stronger effect takes place with bcc Fe(001) + p(1 × 1)O. Here the topmost Fe atoms are drawn outwards by 0.11 Å from their bulk positions (i.e. by 0.17 Å from their contracted bare-surface positions,

*For this surface each S atom sits closer to a second-layer Ni atom than to the 4 nearest first-layer Ni atoms.

** This surface does not give very good agreement between theory and experiment and has a somewhat anomalously large Mo–N bond length.

implying a bond length expansion of almost 3%), cf. fig. 4.6a. The O adatom seems to sit slightly closer to a second-layer Fe atom than to the 4 nearest first-layer Fe atoms (bond lengths 2.07 ± 0.06 Å and 2.09 ± 0.02 Å, respectively), but this probably is not the cause of the large observed effect, since the above-mentioned fcc Ni(110) + c(2 × 2)S has a similar situation with an even greater difference between the corresponding bond lengths (2.17 ± 0.1 Å and 2.35 ± 0.04 Å, respectively), but no expansion beyond the bulk dimensions. Whatever the cause, O on Fe(001) produces a surface structure that tends towards that of bulk FeO.

A similar expansion of the top metal layer appears to occur for fcc Ni(001) + p(2 × 2)C, although the structure of that surface remains otherwise largely unknown. With Ni(110) + p(2 × 1)H, the adsorbed H atoms may cause rows of surface Ni atoms to pairwise move together slightly, reducing their mutual distance by about 0.2 Å and contracting the bond length to the second-layer Ni atoms by about 4% (the position of the H atoms remains unknown).

Compound formation by adsorption has been observed by LEED in the case of hcp Ti(0001) + p(1 × 1)N, whose structure is shown in fig. 4.6b. The small adatoms penetrate just under the first layer of Ti, which as a result moves outward by only 0.11 ± 0.05 Å. The adatoms settle in octahedral holes already available between the first and second layers of

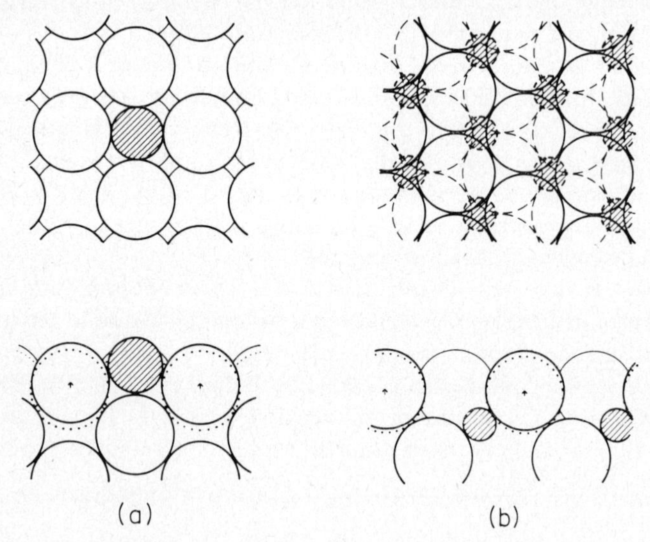

Fig. 4.6. (a) bcc Fe(001) + p(1 × 1)O structure; (b) hcp Ti(0001) + p(1 × 1)N structure. Dotted outlines represent bare-surface atomic positions.

the Ti(0001) surface. No lateral motion is involved and the result is a three-layer thick slice of the bulk TiN compound (of NaCl structure), the dimensions agreeing within the uncertainty of the method. For comparison, bulk samples of MgO and NiO (also of the NaCl structure, but exposing the (001) face instead of the (111) face) were also found to have surfaces with dimensions equal to those of the bulk.

A structure that needs confirmation is fcc Cu(001) + c(2 × 2)O, that from application of the LEED averaging technique is concluded to be reconstructed, each O adatom taking the place of a topmost Cu atom. This would be an example of oxidation. A similar structure may occur for Ni(001) + c(2 × 2)O in addition to the overlayer structure described in sect.3 (Hagstrum and Becker 1971).

5. Co-adsorption and molecular adsorption

One small family of surfaces created by co-adsorption of two different atomic species has been structurally investigated. The substrate is fcc Ni(001) which is not structurally affected by the adsorption. The adatoms are S and Na, deposited sequentially in that order, each in either half-coverage c(2 × 2) or quarter-coverage p(2 × 2) ordered overlayers. With a half monolayer of each species, cf. fig. 4.7a, the position of the S atoms in hollow sites is not affected by the addition of the Na atoms; the Na atoms choose the unoccupied hollow sites on the substrate, where they have 4 nearest S neighbors with a Na–S bond length of 2.76 ± 0.1 Å (compared with 2.735–3.38 Å in a number of bulk compounds). The Na atoms are 0.2 Å farther away from the substrate than in the absence of S, an increase by 0.15 Å of the Ni–Na bond length. Halving the Na coverage, leaving that of S unchanged, does not affect these results, cf. fig. 4.7b, indicating little charge effect in the bonding. This last impression is confirmed by work function measurements. For half-monolayer coverage of both species the work function change relative to the bare substrate is −2.65 eV (compared with −2.55 eV in the absence of S), while halving the Na coverage yields −2.85 eV; this halving therefore induces a charge transfer of the order of 1% of an electron between two atoms.

With a quarter monolayer of both S and Na, again the position of the S atoms is insensitive to the addition of Na atoms, cf. fig. 4.7c, and again the Na atoms choose unoccupied hollow sites on the substrate, but only those sites that provide the closest contact with S atoms, rather than the sites that allow closer contact with the substrate. So an attractive force

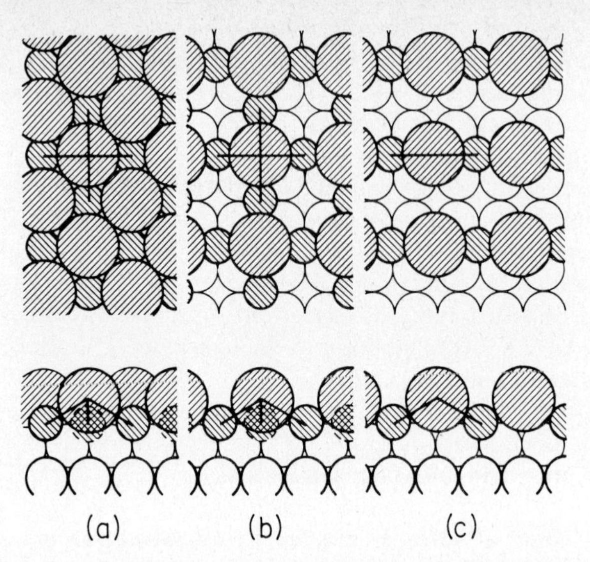

Fig. 4.7. Top diagram gives top view, bottom diagram gives side view of (a) fcc Ni(001) + c(2 × 2)S + c(2 × 2)Na, (b) fcc Ni(001) + c(2 × 2)S + p(2 × 2)Na and (c) fcc Ni(001) + p(2 × 2)S + p(2 × 2)Na. Smallest circles represent S atoms, largest circles Na atoms.

acts between the co-adsorbed species. Again the Na–S bond length is 2.76 ± 0.1 Å, even though the number of nearest S atoms if now reduced from 4 to 2. The work function change (relative to the bare substrate) is now −3.10 eV, so that again little charge effect is seen in the bonding, despite the fact that Na and S could be expected to have strong ionic character. The mutual destruction of parallel dipoles seems to play a significant role here, as in all cases of high-coverage adsorption discussed previously.

Molecular adsorption has been structurally investigated on two systems, fcc Pt(111) + p(2 × 2)C_2H_2 and fcc Ni(001) + c(2 × 2)CO. The adsorbed acetylene is reported to exist in two states, the initial "metastable" state and a "stable" state obtained after heating. In the stable state, the acetylene appears to be in molecular form, oriented parallel to the surface and centered over a 3-fold hollow site of the substrate (in a cABCABC... stacking sequence, in the symbolism of sect. 2), as illustrated in fig. 4.8. The relevant Pt–C distances are 2.25 ± 0.1 Å and 2.59 ± 0.1 Å; the geometrically most analogous structure (as far as the Pt–C bond neighborhood is concerned) is the complex $Os_3(CO)_{10}(C_2Ph_2)$ with a Os–C bond length of 2.22 Å (Os has virtually the same radius as Pt). These Pt–C distances assume a C–C bond length of 1.20 Å, although

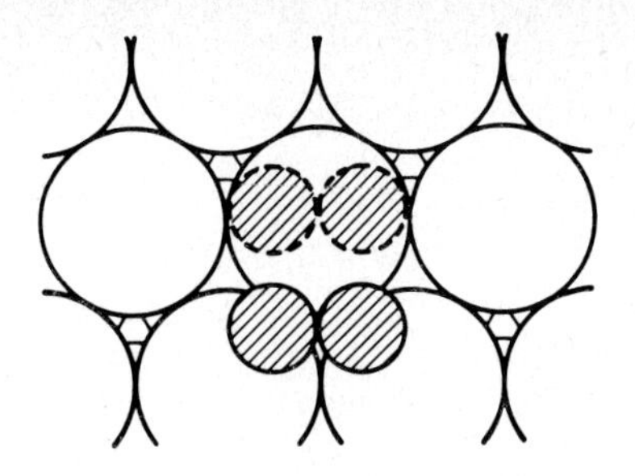

Fig. 4.8. Structure of fcc Pt(111) + p(2 × 2)C_2H_2 in the stable state (solid lines) and the metastable state (dashed lines). The positions of the H atoms are left undecided.

the calculation could not exclude an elongation of the C–C bond to the double-bond distance 1.34 Å. The hydrogen atoms were essentially ignored at first in the calculation (because of their weak scattering strength and for computational economy); their location may be hard to determine without some independent information (from chemistry, photoemission, infrared spectroscopy, etc.).

The metastable acetylene adsorption state appears to have a similar average Pt–C bond length, but a different registry (the molecular axis passing 0.2 Å away from a top Pt atom, in a top view, cf. fig. 4.8.

The second molecular adsorption system, fcc Ni(001) + c(2 × 2)CO, yields undissociated carbon monoxide molecules, in which the C atoms are located at a distance of 1.8 Å directly above top-layer Ni atoms, while the O atoms appear to be 0.95 Å above the plane of C atoms. The LEED calculations are not very sensitive to the registry of the O atoms and the usual C–O bond length of 1.15 Å is therefore consistent with the LEED evidence, if one assumes a Ni–C–O bond angle of about 150°, i.e. a registry difference between C and O atoms of about 0.65 Å (in which direction the CO would be tilted is unknown). For comparison the molecule Ni$(CO)_4$ has a Ni–C bond length of 1.83 Å and a C–O bond length of 1.15 Å.

It may be premature to extract lessons about molecular bonding from so few examples.

6. Brief overview of other surface structures

Surface structures known from LEED but not dealt with in previous sections, can be classified into semiconductors, layer compounds and ionic compounds.

The group IV semiconductor silicon (diamond structure) presents intriguing reconstructions of the surface, as evidenced by superlattice formation. The Si(001) p(2 × 1) surface structure has, despite several attempts, not been found to correspond to any of the popular models that have been presented over the years. Indications are, however, that the surface may adopt a structure based on horizontal zigzag rows of atoms (Jona et al. 1977). For Si(111) p(2 × 1) the proposed model consisting of alternatively raised and lowered top-layer atoms may prove to be the backbone of the actual structure (Forstmann 1976). The structure of an unreconstructed Si(111) surface has however been successfully determined (Shih et al. 1976): the outermost Si atoms move into the surface by about 0.12 Å, thereby reducing the three equal nearest-neighbor distances by about 2%, which one might ascribe to increased bond charging after the breaking of the fourth tetrahedral bond (now a dangling bond pointing straight out of the crystal). A Generalized-Valence-Bond cluster calculation is in agreement with this result (Redondo et al. 1976).

A similar topmost bond length contraction is believed to occur on the compound semiconductor face ZnO(0001), which exposes only zinc atoms (the bulk has the wurtzite structure). In this case ionic forces can be invoked to account for a contraction. On the non-polar face ZnO($10\bar{1}0$) it seems likely that the small top-layer Zn atoms are pulled into the surface by about 0.3 Å, while the large top-layer oxygen atoms might penetrate by about 0.1 Å, thereby reducing the bond lengths by corresponding amounts. Ionic forces and ion size effects could explain such contractions.

GaAs(110), with a bulk zincblende structure, exhibits on its surface zigzag rows of alternating Ga and As atoms: these infinitely long rows appear to be tilted by roughly 30° with respect to their bulk orientation parallel to the surface, reducing the bond lengths somewhat and pushing the top As atoms outward, while pulling the top Ga atoms inward. Bond charge redistribution and rehybridization may account for this structure.

For GaAs(001), exposing a termination consisting of only As atoms, no deviation from the bulk structure appears to occur. ZnSe(110), also a zincblende material, appears to behave similarly. These results on ZnO(10$\bar{1}$0) and ZnSe(110) are preliminary and are currently under closer investigation.

In part also semiconducting, a number of layer compounds have been studied by LEED. The cleavage planes defined by the wide Van der Waals gaps between layers provide surfaces that were found not to deviate from the bulk structure, except for a slight contraction by about 2% of the topmost Mo–S bond in the MoS_2; otherwise, MoS_2, $NbSe_2$, Na_2O, TiS_2 and $TiSe_2$ show no deviation, whether in the bond lengths, in the layer separations or in the stacking sequence of the individual hexagonal planes and of the triple-plane layers.

With the insulating ionic compounds MgO and NiO, LEED analyses show that the mixed-atom (001) faces have no detectable change from the bulk NaCl structure: there is no apparent contraction towards the bulk of either the positive or the negative ions. LiF(001) may exhibit some contraction, but requires a closer study.

7. Conclusion

Notable general features that emerge from surface crystallographic studies are:

- no deviation is found from the bulk layer stacking (registries) at a bare metal surface (not counting reconstructed surfaces);
- bond length contractions occur at bare metal surfaces that are "rough": bcc (111), bcc (001) and fcc (110);
- atomic adsorption on metals takes place preferentially (not always) in sites of high symmetry and high coordination to metal atoms, often continuing the substrate lattice out into the overlayer, with little dependence on adatom valence, radius, coverage, substrate identity or face;
- adatom–metal bond lengths tend to agree best with bulk compound values, though with substantial scatter, part of which may be real and related to valence, bond order, coverage, etc.;
- adsorption bond angles seem to be usually determined by a hard-ball packing geometry, but some constancies in metal–adatom–metal bond angles are apparent;
- charge transfers (as implied by work function changes) are small ($\lesssim 10\%$ of an electron) and show a complicated behavior as function of face and coverage;

– adsorption tends to elongate underlying metal–metal bonds in cases where these bonds are already shortened on the bare surface.

Mechanisms that might explain some of these features include:
– asymmetric binding forces on surface atoms;
– bond-order and coordination dependence of bond lengths;
– bond resonance;
– bond rehybridization;
– partial mutual cancellation of dipoles.

Note added in proof

The difficulties mentioned in one analysis of Ag(110) have been removed (Maglietta et al. 1977). New ion-scattering results for Pt(111) indicate no expansion or contraction (Van der Veen et al. 1977, Bøgh and Stensgaard 1977).

The chemisorption position for Ni(111) + (2×2)2H has been found to be both types of threefold hollow sites (in a graphite-like arrangement) with Ni–H bond lengths of 1.84 ± 0.06 Å (Van Hove et al., to be published).

A good fit between theory and experiment for Si(001)p(2×1) is now found for a surface model based on the Schlier–Farnsworth pairing idea, allowing relaxations down to the fifth layer from the surface (Appelbaum and Hamann, to be published, Mitchell and Van Hove, to be published, Tong and Maldonado, to be published).

References

Adams, D. and U. Landman, 1977, Phys. Rev. B, to appear.
Anderson, A.B., 1977, J. Chem. Phys. **66**, 2173.
Appelbaum, J.A. and D.R. Hamann, to be published.
Besocke, K. and H. Wagner, 1975, Surf. Sci. **52**, 653.
Bøgh, E. and I. Stensgaard, 1977, Proc. 7th IVC and 3rd ICSS, Vienna, p. A2757.
Cheng, D.J., R.F. Wallis and L. Dobrzynski, 1974, Surf. Sci. **43**, 400.
Cunningham, S.L., C.-M. Chan and W.H. Weinberg, 1977, Phys. Rev. B, to appear.
Finnis, M.W. and V. Heine, 1974, J. Phys. **F4**, L37.
Forstmann, F., 1976, private communication.
Ignatiev, A., A.V. Jones and T.N. Rhodin, 1971, Surf. Sci. **30**, 573.
Kesmodel, L.L. and G.A. Somorjai, 1976, Accounts of Chemical Research **9**, 392.
Lagally, M., T.C. Ngoc and M.B. Webb, 1971, Phys. Rev. Lett. **26**, 1557.
Lee, B.W., A. Ignatiev, S.Y. Tong and M. Van Hove, 1977, J. Vac. Sci. Technol. **14**, 291.
Ma, S.K.S., F.W. de Wette and G.P. Alldredge, 1977, Bull. Am. Phys. Soc. **22**, 355.
Madhukar, A., 1975, Sol. St. Comm. **16**, 461.

Maglietta, M., E. Zanazzi, F. Jona, D.W. Jepsen and P.M. Marcus, 1977, J. Phys. **C10**, 3287.
Mitchell, K.A.R. and M.A. Van Hove, to be published.
Pauling, L., 1960, The nature of the chemical bond, 3rd edition (Cornell University Press, New York).
Pendry, J.B., 1974, Low energy electron diffraction (Academic Press, London).
Redondo, A., W.A. Goddard III, T.C. McGill and G.T. Surratt, 1976, Sol. St. Commun. **20**, 733.
Structure reports, 1956 ff., (Ed. A.J.C. Wilson) (Oosthoek, Utrecht).
Tables of interatomic distances and configurations in molecules and ions, 1958 ff., (Ed. H.J.M. Bowen) (The Chemical Society, London).
Tong, S.Y. and A.L. Maldonado, to be published.
Van der Veen, J. F., R.G. Smeenk and F.W. Saris, 1977, Proc. 7th IVC and 3rd ICSS, Vienna, p. 2512.
Van Hove, M.A., G. Ertl, K. Christmann, R.J. Behm and W.H. Weinberg, to be published.
Walch, S.P. and W.A. Goddard III, 1977, to be published.
Wood, E.A., 1964, J. Appl. Phys. **35**, 1306.
Wyckoff, R.G.W., 1963 ff., Crystal structures (Interscience, Wiley, New York).

Appendix

In this appendix are listed some relevant references to structural determinations, grouped by individual surfaces; these are listed alphabetically, counting only the letters of the chemical elements. Short comments are added to the references where appropriate.

Ag(111), (001) and (110)

Forstmann, F., 1974, Jap. J. of Appl. Phys., Suppl. 2, Part 2, 657: (111).
Jepsen, D.W., P.M. Marcus and F. Jona, 1973, Phys. Rev. **B8**, 5523: (001).
Moritz, W., 1976, Doctoral Thesis (University of Munich): (001), (110).
Ngoc, T.C., M.G. Lagally and M.B. Webb, 1973, Surf. Sci. **35**, 117: (111), by averaging.
Zanazzi, E., F. Jona, D.W. Jepsen and P.M. Marcus, 1977, J. Phys. **C10**, 375: (110).

Ag(001) + c(2 × 2)Cl

Zanazzi, E., F. Jona, D.W. Jepsen and P.M. Marcus, 1976, Phys. Rev. **B14**, 432.

Ag(111) + (√3 × √3)R30°I

Forstmann, F., W. Berndt and P. Büttner, 1973, Phys. Rev. Lett. **30**, 17.

Ag(001) + c(2 × 2)Se

Ignatiev, A., F. Jona, D.W. Jepsen and P.M. Marcus, 1973, Surf. Sci. **40**, 439.

Al(111), (001) and (110)

Boudreaux, D.S. and V. Hoffstein, 1971, Phys. Rev. **B3**, 2447: (111), (110).

Jepsen, D.W., P.M. Marcus and F. Jona, 1972, Phys. Rev. **B6**, 3684: (111), (001), (110).

Laramore, G.E. and C.B. Duke, 1972, Phys. Rev. **B5**, 267: (111), (001), (110).

Martin, M.R. and G.A. Somorjai, 1973, Phys. Rev. **B7**, 3607: (111), (110).

Masud, N., C.G. Kinniburgh and J.B. Pendry, 1977, J. Phys. **C10**, 1: (001), (110) by MEED.

Tait, R.H., S.Y. Tong and T.N. Rhodin, 1972, Phys. Rev. Lett. **28**, 553: (001), (110).

Al(001) + c(2 × 2)Na

Hutchins, B.A., T.N. Rhodin and J.E. Demuth, 1976, Surf. Sci. **54**, 419.

Van Hove, M., S.Y. Tong and N. Stoner, 1976, Surf. Sci. **54**, 259.

Au(111)

Soria, F., J.L. Sacedón, P.M. Echenique and D.J. Titterington, 1977, Surf. Sci. **68**, 448.

Be(0001)

Strozier, J.A. and R.O. Jones, 1971, Phys. Rev. **B3**, 3228.

Cd(0001)

Shih, H.D., F. Jona, D.W. Jepsen and P.M. Marcus, 1976, Commun. on Phys. **1**, 25.

Co(0001), (111) and (001)

Lee, B.W., R. Alsenz, A. Ignatiev and M.A. Van Hove, 1977, Bull. Am. Phys. Soc. **22**, 357: hcp (0001), fcc (111).

Maglietta, M., E. Zanazzi and F. Jona, 1977, Bull. Am. Phys. Soc. **22**, 355: fcc (001).

Cu(111) and (001)

Adams, D.L. and U. Landman, 1977, Phys. Rev. **B15**, 3775: (001), by Fourier-transform deconvolution.

Capart, G., 1971, Surf. Sci. **26**, 429: (001).

Kleiman, G.G. and J.M. Burkstrand, 1975, Surf. Sci. **50**, 493: (001), by averaging.

Laramore, G.E., 1974, Phys. Rev. **B9**, 1204: (111), (001).

Marcus, P.M., D.W. Jepsen and F. Jona, 1972, Surf. Sci. **31**, 180: (001).

Pendry, J.B., 1971, J. Phys. **C4**, 2514: (001).

Pendry, J.B., 1974, Low energy electron diffraction (Academic Press, London): (001).

Cu(001) + c(2 × 2)N

Burkstrand, J.M., G.G. Kleiman, G.G. Tibbetts and J.C. Tracy, 1976, J. Vac. Sci. Technol. **13**, 291: by averaging.

Burkstrand, J.M., S.Y. Tong and M.A. Van Hove, 1978, to be published.

Cu(001) + c(2 × 2)O

McDonnell, L., D.P. Woodruff and K.A.R. Mitchell, 1974, Surf. Sci. **45**, 1: by averaging.

Cu(001) + p(2 × 2)Te

Salwén, A. and J. Rundgren, 1975, Surf. Sci. **53**, 523.

Fe(110), (001) and (111)

Feder, R., 1973, Phys. Stat. Sol. **58**, K137: (001), relativistic.

Feder, R. and G. Gafner, 1976, Surf. Sci. **57**, 45: (110), relativistic.

Legg, K.O., F. Jona, D.W. Jepsen and P.M. Marcus, 1977, J. Phys. **C10**, 937: (001).

Shih, H.D., F. Jona, D.W. Jepsen and P.M. Marcus, 1977, Bull. Am. Phys. Soc. **22**, 357: (111).

Fe(001) + p(1 × 1)O

Legg, K.O., F. Jona, D.W. Jepsen and P.M. Marcus, 1975, J. Phys. **C8**, L492.

Fe(001) + c(2 × 2)S

Legg, K.O., F. Jona, D.W. Jepsen and P.M. Marcus, 1976, Bull. Am. Phys. Soc. (II), **21**, 319.

GaAs(110), (001)

Duke, C.B., A.R. Lubinsky, B.W. Lee and P. Mark, 1976, J. Vac. Sci. Technol. **13**, 761: (110).

Lubinsky, A.R., C.B. Duke, B.W. Lee and P. Mark, 1976, Phys. Rev. Lett. **36**, 1058: (110).

Mark, P., G. Cisneros, M. Bonn, A. Kahn, C.B. Duke, A. Paton and A.R. Lubinsky, 1977, to be published: (110).

Mrstik, B.J., M.A. Van Hove and S.Y. Tong, 1977, Bull. Am. Phys. Soc. **22**, 356: (110), (001).

Ir(111)

Chan, C.-M., S.L. Cunningham, M.A. Van Hove and W.H. Weinberg, 1977, Bull. Am. Phys. Soc. **22**, 357.

Chan, C.-M., S.L. Cunningham, M.A. Van Hove and W.H. Weinberg, 1977, Surf. Sci. **66**, 394.

LiF(001)

Laramore, G.E. and A.C. Switendick, 1973, Phys. Rev. **B7**, 3615.

MgO(001)
Kinniburgh, C.G., 1975, J. Phys. **C8**, 2382.
Kinniburgh, C.G., 1976, J. Phys. **C9**, 2695.

Mo(001)
Ignatiev, A., F. Jona, H.D. Shih, D.W. Jepsen and P.M. Marcus, 1975, Phys. Rev. **B11**, 4787.

Mo(001) + c(2 × 2)N
Ignatiev, A., F. Jona, D.W. Jepsen and P.M. Marcus, 1975, Surf. Sci. **49**, 189.

MoS_2(0001)
Mrstik, B.J., R.Kaplan, T.L. Reinecke, M. Van Hove and S.Y. Tong, 1977, Phys. Rev. **B15**, 897.
Van Hove, M.A., S.Y. Tong and M.H. Elconin, 1977, Surf. Sci. **65**, 539.

Mo(001) + p(1 × 1)Si
Ignatiev, A., F. Jona, D.W. Jepsen and P.M. Marcus, 1975, Phys. Rev. **B11**, 4780.

Na(110)
Echenique, P.M., 1976, J. Phys. **C9**, 3193.

Na_2O(111)
Andersson, S., J.B. Pendry and P.M. Echenique, 1977, Surf. Sci. **65**, 539.

$NbSe_2$(0001)
Mrstik, B.J., R. Kaplan, T.L. Reinecke, M. Van Hove and S.Y. Tong, 1977, Phys. Rev. **B15**, 897.
Van Hove, M.A., S.Y. Tong and M.H. Elconin, 1977, Surf. Sci. **64**, 85.

Ni(111), (001) and (110)
Adams, D.L. and U. Landman, 1977, Phys. Rev. **B15**, 3775: (001), by Fourier-transform deconvolution.
Demuth, J.E., P.M. Marcus and D.W. Jepsen, 1975, Phys. Rev. **B11**, 1460: (111), (001), (110).
Laramore, G.E., 1973, Phys. Rev. **B8**, 515: (111), (001).
Ngoc, T.C., M.G. Lagally and M.B. Webb, 1973, Surf. Sci. **35**, 117: (111), by averaging.
Pendry, J.B., 1969, J. Phys. **C2**, 2283: (111).
Tait, R.H., S.Y. Tong and T.N. Rhodin, 1972, Phys. Rev. Lett. **28**, 553: (001), (110).

Ni(001) + p(2 × 2)C
Van Hove, M.A. and S.Y. Tong, 1975, Surf. Sci. **52**, 673.

Ni(001) + c(2 × 2)CO

Andersson, S. and J.B. Pendry, 1978, Surf. Sci. **71**, 75.

Ni(110) + p(2 × 1)H

Demuth, J.E., 1977, J. of Colloid and Interface Science, **58**, 184.

Ni(001) + c(2 × 2)Na

Andersson, S. and J.B. Pendry, 1973, J. Phys. **C6**, 601.
Andersson, S. and J.B. Pendry, 1975, Solid St. Commun. **16**, 563.
Demuth, J.E., D.W. Jepsen and P.M. Marcus, 1975, J. Phys. **C8**, L25.

NiO(001)

Kinniburgh, C.G. and J.A. Walker, 1977, Surf. Sci. **63**, 274.

Ni(001) + c/p(2 × 2)O

Brongersma, H.H. and J.B. Theeten, 1976, Surf. Sci. **54**, 519; c(2 × 2), by ion scattering.
Demuth, J.E., D.W. Jepsen and P.M. Marcus, 1973, Phys. Rev. Lett. **31**, 540: c(2 × 2).
Hagstrum, H.D. and G.E. Becker, 1971, J. Chem. Phys. **54**, 1015: c(2 × 2), by ion-neutralization spectroscopy.
Van Hove, M.A. and S.Y. Tong, 1975, J. Vac. Sci. Technol. **12**, 230: p(2 × 2).

Ni(110) + p(2 × 1)O

Heiland, W. and E. Taglauer, 1972, J. Vac. Sci. Technol. **9**, 620.
Marcus, P.M., J.E. Demuth and D.W. Jepsen, 1975, Surf. Sci. **53**, 501.

Ni(001) + c/p(2 × 2)S

Demuth, J.E., D.W. Jepsen and P.M. Marcus, 1973, Phys. Rev. Lett. **31**, 540: c(2 × 2).
Groupe d'Etude des Surfaces, 1975, Surf. Sci. **48**, 577: c(2 × 2).
Hagstrum, H.D. and G.E. Becker, 1977, J. Vac. Sci. Technol. **14**, 369: c and p(2 × 2), by ion-neutralization spectroscopy.
Van Hove, M. and S.Y. Tong, 1975, J. Vac. Sci. Technol., **12**, 230: p(2 × 2).

Ni(111) + p(2 × 2)S and Ni(110) + c(2 × 2)S

Demuth, J.E., D.W. Jepsen and P.M. Marcus, 1974, Phys. Rev. Lett. **32**, 1182.
Marcus, P.M., J.E. Demuth and D.W. Jepsen, 1975, Surf. Sci. **53**, 501: reviews adsorption of chalcogens on nickel.

Ni(001) + c/p(2 × 2)S + c/p(2 × 2)Na

Andersson, S. and J.B. Pendry, 1976, Phys. Rev. **C9**, 2721.

Ni(001) + c/p(2 × 2)Se/Te

Demuth, J.E., D.W. Jepsen and P.M. Marcus, 1973, Phys. Rev. Lett. **31**, 540: c(2 × 2)Se and Te.

Demuth, J.E., D.W. Jepsen and P.M. Marcus, 1973, J. Phys. **C6**, L307: c(2 × 2)Te.

Demuth, J.E., P.M. Marcus and D.W. Jepsen, 1974, Phys. Rev. Lett. **32**, 1182: c(2 × 2)Te.

Van Hove, M. and S.Y. Tong, 1975, J. Vac. Sci. Technol. **12**, 230: p(2 × 2) Se and Te.

Pt(111)

Davies, J.A., D.P. Jackson, J.B. Mitchell, P.R. Norton and R.L. Tapping, 1975, Phys. Lett. **54A**, 239: by ion channelling.

Kesmodel, L.L. and G.A. Somorjai, 1975, Phys. Rev. **B11**, 630.

Kesmodel, L.L., P.C. Stair and G.A. Somorjai, 1977, Surf. Sci. **64**, 342.

Pt(111) + p(2 × 2)C_2H_2

Kesmodel, L.L., P.C. Stair, R.C. Baetzold and G.A. Somorjai, 1976, Phys. Rev. Lett. **36**, 1316.

Kesmodel, L.L., R.C. Baetzold and G.A. Somorjai, 1977, Surf. Sci. **66**, 299.

Rh(001)

Frost, D.C., K.A.R. Mitchell, F.R. Shepherd and P.R. Watson, 1977, Proc. 7th IVC and 3rd ICSS, Vienna, p. A2725.

Si(111)

Forstmann, F., 1976, private communication: (2 × 1) structure.

Shih, H.D., F. Jona, D.W. Jepsen and P.M. Marcus, 1976, Phys. Rev. Lett. **37**, 1622: (1 × 1) structure.

Si(001) p(2 × 1)

Jona, F., H.D. Shih, D.W. Jepsen and P.M. Marcus, 1977, Bull. Am. Phys. Soc. **22**, 357.

Jona, F., H.D. Shih, A. Ignatiev, D.W. Jepsen and P.M. Marcus, 1977, J. Phys. **C10**, L67.

Tong, S.Y. and M.A. Van Hove, 1977, unpublished.

Ti(0001)

Shih, H.D., F. Jona, D.W. Jepsen and P.M. Marcus, 1976, J. Phys. **C9**, 1405.

Ti(0001) + p(1 × 1)Cd multilayers

Shih, H.D., F. Jona, D.W. Jepsen and P.M. Marcus, 1976, Commun. on Phys. **1**, 25.

Ti(0001) + p(1 × 1)N

Shih, H.D., F. Jona, D.W. Jepsen and P.M. Marcus, 1976a, Phys. Rev. Lett. **36**, 798.

Shih, H.D., F. Jona, D.W. Jepsen and P.M. Marcus, 1976b, Surf. Sci. **60**, 445.

TiS_2 and $TiSe_2$(0001)

Mrstik, B.J., S.Y. Tong and M.A. Van Hove, 1978, to be published.

W(110) and (001)

Debe, M.K., D.A. King and F.S. Marsh, 1977, Surf. Sci. **68**, 437: (001).

Feder, R., 1974, Phys. Stat. Sol. (b) **62**, 135.

Feder, R., 1976, Phys. Rev. Lett. **36**, 598: (001).

Lagally, M.G., J.C. Buchholz and G.C. Wang, 1975, J. Vac. Sci. Technol. **12**, 213: (110), by averaging.

Lee, B.W., A. Ignatiev, S.Y. Tong and M.A. Van Hove, 1977, J. Vac. Sci. Technol. **14**, 291: (001).

Van Hove, M.A. and S.Y. Tong, 1976, Surf. Sci. **54**, 91.

W(110) + p(2 × 1)O

Lagally, M.G., J.C. Buccholz and G.C. Wang, 1975, J. Vac. Sci. Technol. **12**, 213: by averaging.

Van Hove, M.A. and S.Y. Tong, 1975, Phys. Rev. Lett. **35**, 1092.

Zn(0001)

Unertl, W.N. and H.V. Thapliyal, 1975, J. Vac. Sci. Technol. **12**, 263: by averaging.

ZnO(0001) and ($10\bar{1}0$)

Duke, C.B. and A.R. Lubinsky, 1975, Surf. Sci. **50**, 605: (0001).

Duke, C.B., A.R. Lubinsky, B.W. Lee and P. Mark, 1976, J. Vac. Sci. Technol. **13**, 761: ($10\bar{1}0$).

Duke, C.B., A.R. Lubinsky, S.C. Chang, B.W. Lee and P. Mark, 1977, Phys. Rev. **B15**, 4865: ($10\bar{1}0$).

ZnSe(110)

Duke, C.B., A.R. Lubinsky, M. Bonn, G. Cisneros and P. Mark, 1977, J. Vac. Sci. Technol. **14**, 294.

Mark, P., G. Cisneros, M. Bonn, A. Kahn, C.B. Duke, A. Paton and A.R. Lubinsky, 1977, J. Vac. Sci. Technol. **14**, 910.

CHAPTER 5

ENERGETICS OF CHEMISORPTION ON METALS

G. ERTL

Institut für Physikalische Chemie, Universität München, Munich, F.R. Germany

Contents

The nature of the surface chemical bond
Edited by Th.N. Rhodin and G. Ertl

1. Introduction

Langmuir (1918) was probably the first one to express explicitly the existence of similarities between the nature of adsorption on solids and "normal" chemical forces. Later work (Benton and White 1930, Taylor 1931) led to the distinction between physical adsorption (mainly van der Waals interaction) and chemisorption, i.e. true chemical bond formation between adsorbate and surface. Whereas the understanding of the chemical bond (at least in small molecules) made rapid progress and has already achieved a rather high level of sophistication our understanding of the nature of the chemisorption bond is still relatively poor and only quite recently was significant progress made both from the experimental and theoretical point of view. The ultimate goal for a complete description of the (electronic ground state) properties of a chemisorption system on a uniform single-crystal surface can be considered as consisting of the following steps:

(i) An evaluation within the Born–Oppenheimer approximation of the (multi-dimensional) energy of the electronic ground state for a single particle interacting with the surface, $E_n(\boldsymbol{r}_j)$, as a function of all the nuclear coordinates of the atoms within the adsorbate and the surface. In principle not even the positions of the metal surface atoms may be regarded to remain fixed since in cases of strong chemisorption the substrate geometry may change, i.e. the surface reconstructs.

(ii) An analysis of multi-particle effects by determining interaction potentials between adsorbed particles as a function of their mutual configuration. A rigorous treatment of this problem would certainly at least be comparable in complexity to the first topic.

(iii) A proper combination of the solutions for problems (i) and (ii) by the methods of statistical mechanics.

A solution of this formidable task would yield a microscopic interpretation for the (equilibrium) structural and thermodynamic properties of a chemisorption system. The treatment of dynamic processes would certainly be still more complicated, since additional problems such as energy transfer, the eventual break-down of the adiabatic approximation etc. would come into play.

Whereas current theories of chemisorption are revealed to be more or less successful in determining orbital and ionization energies at fixed geometry the present task would comprise the evaluation of the total energy with an accuracy of <0.1 eV as a function of all nuclear coordinates. The usual approach for a descriptive solution of this problem consists in splitting the system into coordinates which are kept fixed and a very small number (probably only one) of so-called "reaction" coordinates which are variable during the process under consideration. The usefulness of the qualitative aspects emerging from such an approach then has to be tested with experimental results which on the other hand serve to supply the data for the parameters incorporated into these models. Such a procedure may assist to rationalize the experimental findings and probably to recognize existing trends and general principles. This situation is probably best illustrated by a schematic one-dimensional potential diagram as proposed by Lennard-Jones (1932) for the dissociative chemisorption of a diatomic molecule approaching a certain point on the surface along a line perpendicular to the surface plane (fig. 5.1): The potential curve for the interaction of the molecular species X_2 exhibits only a relatively shallow minimum at a rather large distance from the surface caused by van der Waals-type physical adsorption. If on the other hand the molecule is dissociated in the gas phase (dissociation energy E_D) and the two atoms are brought to the surface, the corresponding potential curve will exhibit a deeper minimum which is

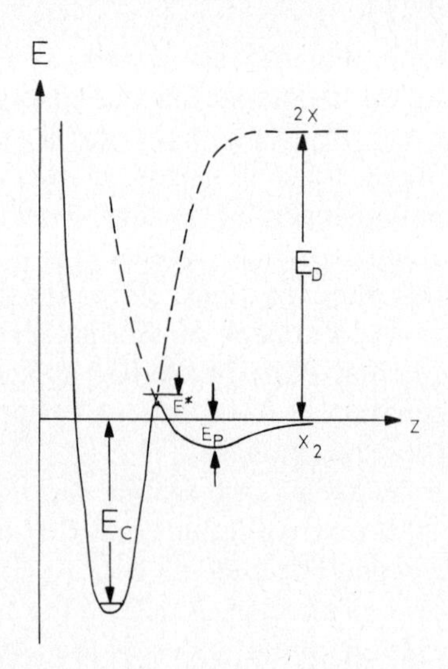

Fig. 5.1. Potential diagram for dissociative chemisorption of a diatomic molecule.

closer to the surface and which is caused by the formation of the chemisorption bond. It thus becomes evident that a diatomic molecule when adiabatically reacting with the surface will follow the full line in the potential diagram and will be dissociated at the intersection of both potential curves. The height of this barrier determines the activation energy for dissociative chemisorption. E_p is the energy of physical adsorption, E_c the energy for (dissociative) chemisorption of the molecule. The energy of the M–X bond is then given by $E_b = \frac{1}{2}(E_c + E_D)$. Experimental evidence for this picture and a discussion of the factors which are influencing these properties will form the main part of this chapter and will be dealt with in sects. 2–5, 7 and 8.

Once a particle has reached the point of minimum energy (see fig. 5.1) this by no means includes a full description of the energetics of the single-particle chemisorption bond. A single crystal surface exhibits a two-dimensional periodic structure and consequently E_c will vary periodically on the surface since the bond energy is a function of the location of the adparticle within the unit cell of the substrate lattice ("energy profile", cf. sect. 5). The situation along a particular crystallographic orientation on the surface is illustrated schematically in fig. 5.2. The variation $E_{b,max} - E_{b,min} = E^{\ddagger}_{diff}$ is usually considerably smaller than $E_{b,max}$ itself and is the activation energy for surface diffusion in this direction. As a consequence the adsorbed particle will perform many diffusion steps on the surface before it desorbs again. Two important cases may be distinguished (see e.g. Clark 1970, de Boer 1953).

(a) $E^{\ddagger}_{diff} \leqslant kT$: Nonlocalized adsorption. The adsorbed particles are highly mobile on the surface and may be considered like a two-dimensional gas rather than being fixed to sites with minimum energy on the surface.

(b) $E^{\ddagger}_{diff} > kT$: Localized adsorption. The translational degrees of freedom of the adsorbed particles are more or less transformed into

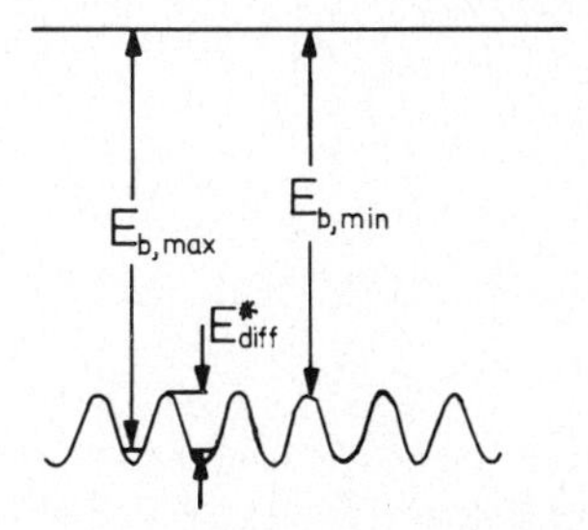

Fig. 5.2. Periodic variation of the potential of an adsorbed particle along a surface direction.

vibrational modes centered at the sites of minimum potential energy ("adsorption sites"). Surface diffusion is achieved by hopping from one potential minimum to a neighbouring one. To a good approximation the adsorbed particles may be considered to occupy the points of a two-dimensional lattice exhibiting the symmetry and periodicity of the substrate lattice. This situation appears to hold for many chemisorption systems. If $E^{*}_{\rm diff} \gg kT$ the hopping frequency becomes very small so that the adsorbed layer is practically immobile and will not be able to achieve its equilibrium configuration in reasonable periods of time. This equilibrium configuration at a given coverage will be governed by interactions between adsorbed particles: fig. 5.2 will never hold for an ensemble of many adsorbed particles but is strictly applicable only for a single particle interacting with a clean surface. The simplest way to overcome this difficulty consists in the assumption that every adsorption site can be occupied by only one adparticle. This is equivalent to a hard-wall interaction potential and forms the basis of the Langmuir adsorption isotherm. A more realistic model describes this effect in terms of pairwise interaction potentials between adsorbed particles as reproduced in fig. 5.3. As will be outlined in sect. 9 these interactions may

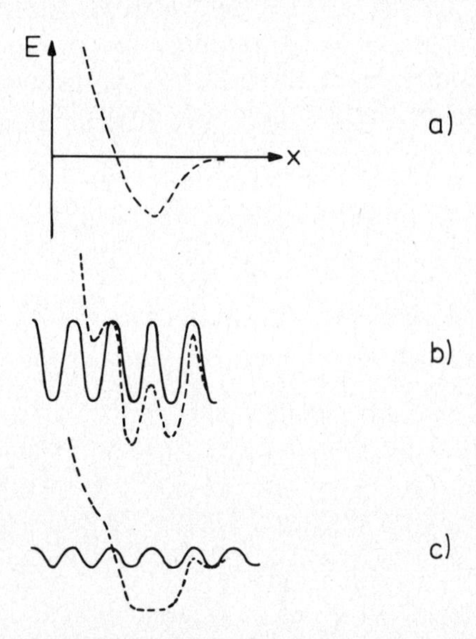

Fig. 5.3. (a) Pairwise interaction potential between adsorbed particles as a function of their distance. (b) and (c) Modulation of the periodic single-particle potential (full line) by pairwise interaction leading to effective potentials (dashed lines) characteristic for lattice-gas (b) and incoherent (c) adsorbate configurations.

be attractive as well as repulsive and depend of course again on the direction on the surface. As a consequence an adsorbed particle will modify the potential seen by a neighbouring second particle and vice versa in a manner drawn in fig. 5.3b, c. As a result the surface will no longer be energetically uniform but will exhibit an induced heterogeneity, some consequences of which will be discussed in sect. 10.

No attempt will be made to give a complete review of the existing experimental material on the different aspects of the energetics of chemisorption on metals, nor to describe the various experimental techniques presently in use. For this purpose the reader is referred to the large amount of available literature.

2. How many metal atoms are involved in the chemisorption bond?

2.1. Relations to complex chemistry

There is now increasing evidence that chemisorption may be regarded as a relatively highly localized phenomenon, i.e. the adparticle is essentially coupled to only a small number of neighbouring metal atoms. In fact, theoretical modelling of chemisorption by cluster theories (see ch. 2), receives its justification from this effect. However, although these cluster theories usually may provide a reasonably good description of the electronic properties (as probed by electron spectroscopic techniques) they still fail to yield numbers for the binding energies with acceptable accuracy (e.g. about ±10%). This task can only be solved by the application of even more rigorous theories which on the other hand are restricted to even smaller molecules (e.g. ensembles containing only 1 or 2 metal atoms) in order to keep the computational efforts tractable. Calculations by Goddard (1977), for example, on the bond dissociation energies for H, CO and other ligands attached to a single Ni atom show in fact a remarkable resemblance to experimental data for the chemisorption energies on extended single crystal surfaces, and thus offer intriguing perspectives even for the theoretical determination of activation energies and reaction paths in heterogeneous catalysis. This result is really striking insofar as a simple comparison of the ionization potential of a single Ni atom (7.6 eV) with that of a metallic Ni surface (5.0 eV = work function) demonstrates marked differences in the electronic properties of these systems.

Comparisons of binding energies in diatomic molecules with chemisorption energies were already made in the earlier literature (Takaishi

1958, Ehrlich 1959). From the limited (and not always reliable) data available at this time however, no clear relation between these quantities could be established. As pointed out by Ehrlich (1959), such a correlation could be expected only if the valence properties of a metal atom in the crystalline state maintained some fixed relation to those of the same atom in the gas phase which, for example, could be tested by comparing the cohesion energies with the dissociation energies of the corresponding diatomic molecules M_2. It was shown that for some of the transition metals (Cr, Mn, Ni, Cu, Ag, Au) this ratio was, in fact, relatively constant whereas strong variations resulted among the main group elements, from which it was concluded that a comparison of dissociation energies for diatomic M–L molecules with the chemisorption energies of the corresponding systems would not be very fruitful (Ehrlich 1959). Some more recent data for hydrogen are reproduced in table 5.1, which at a first glance show no agreement with each other. The best agreement is found for nickel which however is believed to occur by accident. On the other hand not even the trend in adsorption energies is correctly reflected by the dissociation energies of the corresponding diatomics. The ratio of the listed values of the dissociation energies of the diatomics over those for the corresponding chemisorption systems ranges between 0.95 and 1.5. This is a remarkable result insofar as both sets of data are at least of the same order of magnitude, indicating that the single atom approach to chemisorption would certainly not yield the correct numerical values, but probably already contains some of the essential ingredients for a qualitative description of the chemisorption bond.

In fact such an approach has been successfully applied for a long time with respect to chemisorption of CO. Following the work of Eischens and Pliskin (1958), many systems were studied by means of infrared spectroscopy which revealed the occurrence of bands for the C–O stretching vibration in frequency ranges around 1500 and 2050 cm^{-1} which were ascribed to "bridge" and "linearly" bound CO molecules in analogy to a carbonyl compounds exhibiting quite similar vibrational modes. While until quite recently infrared spectroscopy allowed only the detection of the C–O vibration in chemisorbed CO, the (even more interesting) M–C vibration mode also becomes observable with high resolution electron-energy loss spectroscopy (Ibach 1972). In a recent study with CO/Ni(100), Andersson (1977) reported a value of 2100 cm^{-1} for the frequency of the C–O vibration and of 470 cm^{-1} for the M–CO mode. This compares rather well with the corresponding data in $Ni(CO)_4$: $\nu_{C-O} = 2057\ cm^{-1}$, $\nu_{Ni-C} = 422$ and $459\ cm^{-1}$ (Adams 1967). It has

TABLE 5.1
Bond dissociation energies of diatomic molecules and strength of the corresponding M–H chemisorption bond (kcal/mole)

	Ni–H	Cu–H	AG–H	Au–H	Pt–H	Al–H
Molecule	60[a]	66[a]	53[a]	74[a]	83[a]	67[b]
Chemisorption	63[c]	56[a]	<52[e]	<52[e]	57[f]	44[g]

[a] Goyden, A.G., 1968, Dissociation energies and spectra of diatomic molecules (Chapman and Hall, London).
[b] de B. Darwent, B., 1970, NSRDS-NBS 31.
[c] Christmann, K., G. Ertl, O. Schober and M. Neumann, 1974, J. Chem. Phys. **60**, 4528.
[d] Balooch, M., M.J. Cardillo, D.R. Miller and R.E. Stickney, 1974, Surface Sci. **46**, 358.
[e] Hayward, D.O. and B.M.W. Trapnell, 1964, Chemisorption (Butterworths, London) p. 234.
[f] Christmann, K., G. Ertl and T. Pignet, 1976, Surface Sci. **54**, 365.
[g] Gunnarsson, O., H. Hjelmberg and B.I. Lundqvist, 1976, Phys. Rev. Lett. **37**, 292. (Theoretical value. Experiments indicate that H_2 chemisorption is endothermic, i.e. $E_{M-H} < 52$ kcal/mole.)

thus to be concluded that even the shape of the metal–CO potential curve near the equilibrium separation (which determines the M–C vibration frequency) is quite similar for CO chemisorbed on a surface and in a system containing only a single Ni atom.

A direct comparison of the CO chemisorption energy with the M–CO bond dissociation energy in monatomic carbonyl compounds is so far only possible with Ni. Adsorption energies between 27 and 30 kcal/mole on the three most densely packed Ni planes (Christmann et al. 1974a) compare well with the 35 kcal/mole dissociation energy of $Ni(CO)_4$ (Cotton et al. 1959). Average bond energies for a series of transition metal carbonyls are listed in table 5.2. These values are all in the same range as the known values for CO chemisorption energies on transition metals. (Adsorption of CO on W has been extensively studied but turned out to be very complex so that no *single* number for the CO chemisorption energy can be given at present (Gomer 1975).) Nevertheless, there are still important differences: UPS spectra from $Ni(CO)_4$ and from CO adsorbed on Ni look qualitatively similar but differ with respect to the relative peak positions indicating different electronic properties (Conrad et al. 1976a). The adsorption energies for CO on Pd or Pt are about 20% higher than on Ni, whereas – in contrast to $Ni(CO)_4$ – the congeners $Pd(CO)_4$ and $Pt(CO)_4$ are thermally very unstable and have been discovered only recently (Huber et al. 1972, Darling et al. 1972).

It has thus to be concluded also that with CO the essential features for the chemisorption bond may be found already with systems containing only a single metal atom – this also justifies a description of the CO chemisorption bond in terms of the well-known donor–acceptor mechanism for bonding in carbonyls (Blyholder 1964a, Doyen and Ertl 1974) – but that on the other hand quantitative agreement might only be found with compounds containing several metal atoms linked together and thus being more "metallic". Species like Ni_2CO and Ni_3CO were recently formed by using matrix-isolation techniques (Hulse et al. 1976). The observed CO stretching frequencies were seen to approach asymptotically those occurring with CO chemisorbed on Ni surfaces. A stable class of such systems is found with the so-called polynuclear metal carbonyls (Chini et al. 1976) which were, for example suggested by Muetterties (1975) as attractive models of metal surfaces for chemisorption and catalysis.

A rather stable candidate from this class is $Rh_6(CO)_{16}$ whose structure (Corey et al. 1963) is reproduced in fig. 5.4. The six Rh atoms form an octahedron with internuclear distances of 2.78 Å which varies only slightly from the nearest-neighbour distance in metallic rhodium (2.69 Å). (The same holds for a whole series of Fe, CO, Ni, Ru, Rh, Os and Pt cluster compounds (Chini et al. 1976).) The eight faces may thus be considered as models for the (111) surface of fcc metallic Rh. Twelve CO molecules are located in terminal positions ($\nu_{C\text{-}O} \approx 2050\ cm^{-1}$) whereas the remaining four are in threefold (face-bridging) coordination ($\nu_{C\text{-}O} = 1800\ cm^{-1}$). The C–O i.r. bands correlate well with those observed after chemisorption of CO on an evaporated Rh film (Harrod et al. 1967, Garland et al. 1965) as well as with those observed after decomposition of $Rh_6(CO)_{16}$ (causing the formation of metallic Rh) and readmission of CO, the bridge-band however now being shifted to about $1850\ cm^{-1}$ (Conrad et al. 1976b). Figure 5.5 shows ultraviolet photoelectron spectra from $Rh_6(CO)_{16}$ as well as from CO adsorbed on a Pd(111) surface (Conrad et al. 1976b). A broad maximum in both cases between 0 and 6 eV is attributed to emission from d-states. The additional two peaks at 8 and 11 eV occur at exactly the same energies and are attributed in the usual way (Conrad et al. 1976, Fuggle et al. 1975, Gustafsson et al. 1975) to states derived from the $(5\sigma + 1\pi)$- and 4σ-

TABLE 5.2
Average bond energies in carbonyls (kcal/mole) (Cartner et al. 1973)

$Cr(CO)_6$	$Mo(CO)_6$	$W(CO)_6$	$Mn_2(CO)_{10}$	$Fe(CO)_5$	$Co_2(CO)_8$	$Ni(CO)_4$
26	36	42	24	29	33	35

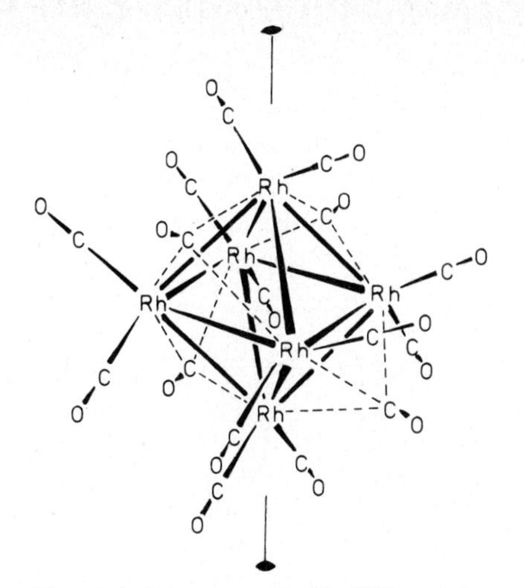

Fig. 5.4. Structure of $Rh_6(CO)_{16}$ (after Corey et al. 1963).

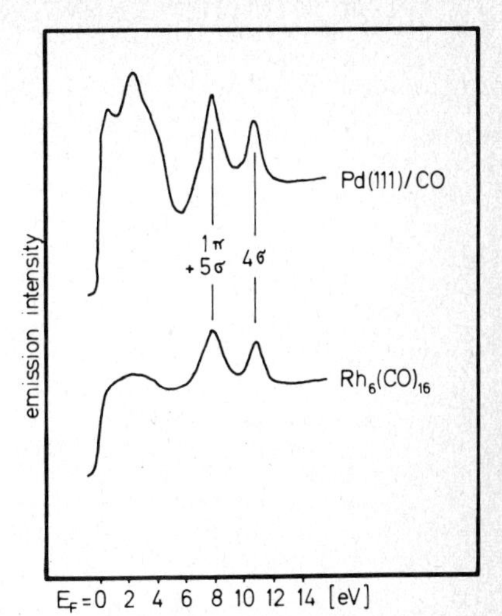

Fig. 5.5. Ultraviolet photoelectron spectra ($h\nu = 40.8$ eV) from $Rh_6(CO)_{16}$ and from CO/Pd(111) (Conrad et al. 1976b).

orbitals of CO, respectively. The similarity between both spectra is striking and indicates also quite similar valence electronic properties in both systems.

$Rh_6(CO)_{16}$ decomposes at 204°C (Chini and Martinengo 1969) which is in exactly the same temperature range where the thermal desorption spectra of CO from the (111) planes of several platinum metals exhibit their maximum (Ertl and Koch 1970, Küppers and Plagge 1976, Comrie and Weinberg 1976, Ertl et al. 1977 and Neumann 1977). From the standard enthalpy of formation of $Rh_6(CO)_{16}$ Brown et al. (1975) concluded the strength of the Rh–CO bond to be about 39 kcal/mole which has to be compared with CO adsorption energies ranging between 30 and 40 kcal/mole on different crystal planes of a series of platinum metals (see tables 5.3 and 5.5).

There are additional analogies between both classes of systems: by means of ^{13}C NMR spectroscopy it was found that, for example, with $Rh_4(CO)_{12}$ the ligands are highly fluxional at room temperature (Cotton et al. 1972) indicating low activation barriers for concerted movement as well as only relatively small energy differences between bridge and terminal positions. The latter conclusion is also indicated by the small

TABLE 5.3

Strength of the substrate–adsorbate bond E_b (kcal/mole) on the most densely packed planes of some transition metals

	N	O	H	CO	NO
Fe(110)	140[a)]		64[a)]		
Ni(111)	135[b)]		63[c)]	27[d)]	25[b)]
Cu(111)			56[e)]	12[f)]	
Ru(001)				29[g)]	
Pd(111)	130[h)]	87[h)]	62[i)]	34[j)]	31[h)]
Ag(111)		80[k)]		6.5[l)]	25[m)]
W(110)	155[n)]			27[p)]	
Ir(111)	127[q)]	93[r)]	63[s)]	34[t)]	20[q)]
Pt(111)	127[u)]		57[v)]	32[w)]	27[u)]

[a)] Bozso, F., G. Ertl and M. Weiss, 1977, J. Catalysis **50**, 519.
[b)] Conrad, H., G. Ertl, J. Küppers and E.E. Latta, 1975, Surface Sci. **50**, 296.
[c)] Christmann, K., O. Schober, G. Ertl and M. Neumann, 1974, J. Chem. Phys. **60**, 4528.
[d)] Christmann, K., O. Schober and G. Ertl, 1974, J. Chem. Phys. **60**, 4719.
[e)] Balooch, M., M.J. Cardillo, D.R. Miller and R.E. Stickney, 1974, Surface Sci. **46**, 358.
[f)] Pritchard, J., 1972, J. Vac. Sci. Techn. **9**, 895.
[g)] Madey, T.E. and D. Menzel, 1974, Jap. J. Appl. Phys. Suppl. 2, Pt. 2, 229.
[h)] Conrad, H., G. Ertl, J. Küppers and E.E. Latta, 1977, Surface Sci. **65**, 235, 245.
[i)] Conrad, H., G. Ertl and E.E. Latta, 1974, Surface Sci. **41**, 435.
[j)] Ertl, G. and J. Koch, 1970, Z. Naturforsch. **25a**, 1906.
[k)] Engelhardt, H.A. and D. Menzel, 1976, Surface Sci. **57**, 591 [(110)-plane].
[l)] McElhiney, G., H. Papp and J. Pritchard, 1976, Surface Sci. **54**, 617.
[m)] Marbrow, R.A. and R.M. Lambert, 1976, Surface Sci. **61**, 317.
[n)] Tamm, P.W. and L.D. Schmidt, 1971, Surface Sci. **26**, 286.
[o)] Tamm, P.W. and L.D. Schmidt, 1971, J. Chem. Phys. **54**, 4775.
[p)] Yates, J.T. and T.E. Madey, 1966, J. Chem. Phys. **45**, 1623.
[q)] Weinberg, W.H., personal communication.
[r)] Ivanov, V.P., G.K. Boreskov, V.I. Savchenko, W.F. Egelhoff and W.H. Weinberg, 1976, Surface Sci. **61**, 207.
[s)] Küppers, J. and A. Plagge, to be published.
[t)] Küppers, J. and A. Plagge, 1976, J. Vac. Sci. Techn. **13**, 259; Comrie, C.M. and W.H. Weinberg, 1976, J. Chem. Phys. **64**, 250.
[u)] Comrie, C.M., W.H. Weinberg and R.M. Lambert, 1976, Surface Sci. **57**, 619.
[v)] Christmann, K., G. Ertl and T. Pignet, 1976, Surface Sci. **54**, 365.
[w)] Ertl, G., M. Neumann and K.M. Streit, 1977, Surface Sci. **64**, 393.

energy difference between bridged and non-bridged isomers of $Co_2(CO)_8$ (Noack 1964), as well as of dinuclear Ru and Fe carbonyls (Noack 1967). The CO:M ratio usually decreases with increasing cluster size (i.e. with a decreasing "surface:volume" ratio) indicating that the maximum number of ligands is determined by steric restrictions rather than by the exhaustion of available valencies. Chini et al. (1976) estimated that for a compound like $Ru_6(CO)_{18}H_2$ on the level of the C atoms about 96% of the available "surface" is occupied. As will be shown in sect. 6, exactly the same features (namely small energy differences between different locations on the surface and the tendency for formation of close-packed layers at high coverages) are found with CO chemisorption systems. In fact large clusters show sometimes the tendency toward close packing of the metal atoms as, for example, clearly evident in $[Rh_{13}(CO)_{24}H_{5-n}]^{n-}$ ($n = 2, 3$) where one central Rh atom is surrounded by 12 Rh atoms at a distance of 2.81 Å. This is presumably the smallest possible unit of a close-packed "metal" which demonstrates that the metal skeleton can be regarded as a "round surface" (Albano et al. 1975, Chini et al. 1976). Also a construction of (idealized) polyhedra described by the oxygen atoms of $Rh_6(CO)_{16}$ (Chini et al. 1976), shows that these may be considered as essentially forming a spherical particle.

Despite still existing differences between cluster compounds (each metal atom in the "surface" has a maximum of 4 or 5 nearest neighbours) and extended transition metal surfaces (up to 9 nearest neighbours) the analogy in the chemical behavior is clearly evident and demonstrates the local character of the chemisorption bond as well as interesting perspectives for bridging the gap between homogeneous and heterogeneous catalysis.

2.2. Chemisorption on alloys – the "ensemble" effect

Since the electronic properties of an isolated metal atom differ considerably from those in the crystalline state, it is not very surprising that the properties of the bonds of a ligand attached to these two systems are at variance too. However, this problem may be tackled in a more sophisticated way by investigating chemisorption on alloys where the chemical properties of the pure constituents differ strongly from each other. Cu/Ni and Ag/Pd alloys were among the most frequently studied systems in this context. Since the surface composition of alloys may deviate considerably from that of the bulk – for thermodynamic reasons – as well as depending on the mode of surface preparation, earlier investigations suffered from a lack of experimental techniques for analyzing the topmost atomic layers of a solid, and only indirect conclusions on this problem could be drawn (van der Planck and Sachtler 1967, 1968). Introduction of Auger electron spectroscopy and more recently of low energy ion back-scattering enabled a direct solution of this problem to be found; however discrepancies may still be found in the literature (Bouwman et al. 1976, Nelson 1976). If no long-range order exists (causing the appearance of superlattice LEED spots as, for example, with AuCu (Potter and Blakeley 1975)) there is still no experimental technique available to determine the short-range order parameters of the surface atoms in an alloy, i.e. to answer the question whether in an A–B alloy the atoms A are preferentially surrounded by A atoms (tendency for cluster formation) or by B atoms (tendency for superlattice formation). This aspect will certainly be of great importance for a complete understanding of chemisorption on alloys and will probably be solved by the use of a surface-sensitive version of extended X-ray absorption fine structure spectroscopy (Lytle et al. 1975, Sayers et al. 1976, Landman and Adams 1976, Gurman and Pendry 1976).

Apart from this difficulty in obtaining a quantitative description of the composition and atom distribution of an alloy surface its chemistry will be complicated over that of the pure components by the following two factors:

(i) The (average) valence electronic properties of atoms A are influenced by atoms B (and vice versa) thus altering the chemisorption bond on A sites. This comprises the so-called "ligand"-effect (Sachtler and van der Plank 1969, Sachtler 1973, van Santen 1975) since it is due to an influence of the metal atoms neighbouring a certain adsorption site, and will be discussed in some more detail in subsect. 3.2. As will be shown with group VIII/Ib alloys this effect may be neglected to a first approximation since the constituents of these alloys more or less retain the properties of the pure components (Ehrenreich and Schwartz 1976).

(ii) If the "ligand" effect can be neglected and if only a single metal atom is involved in the individual chemisorption bond the chemisorptive properties of an alloy surface will simply reflect a superposition (weighted by the corresponding mole fractions) of the behavior of the pure constituents. If more than one surface atom is involved in the bond formation the situation will become more complex since now the distributions and configurations of possible "ensembles" A_xB_y will be the determining factor. This is the so-called "ensemble" effect (Sachtler 1973, van Santen 1975, Dowden 1973, Christmann and Ertl 1972) which in principle can give information on the number and geometric configuration of surface atoms involved in the chemisorption bond.

Although a series of experimental studies on this effect were recently performed (Christmann and Ertl 1972, Sachtler 1971, Takasu and Yamashima 1973, Yu et al. 1976, Mathieu and Primet 1976, Dalmon et al. 1975, Soma-Noto and Sachtler 1974, Burton and Hyman 1975, Burton et al. 1976, Stephan et al. 1976, Araki and Ponec 1976), a complete quantitative understanding of the observed phenomena is still missing. The procedure usually consists of alloying two metals with rather different adsorptive properties (such as Cu–Ni and Ag–Pd in the case of CO or H_2 chemisorption) and studying chemisorption (or catalytic activity) as a function of alloy surface composition. By measuring the amount of H_2 irreversibly adsorbed at room temperature in this sense, van der Plank and Sachtler (1967, 1968) we able to demonstrate conclusively for the first time that surface and bulk compositions may deviate considerably from each other, thus confirming earlier evidence mainly based on work-function data (Sachtler and Dorgelo 1965, Sachtler and Jongepier 1965). The suggestion of Sachtler was that the amount of adsorbed hydrogen is directly proportional to the Ni concentration in the surface, i.e. that only a single Ni atom is responsible for the individual chemisorption bond of a hydrogen atom. Although this idea of "titration" of the surface composition by selective adsorption is rather attractive, recent results (where the surface composition was independently checked by AES) have clearly shown that this simple picture doesn't hold, i.e. that the amount of H_2 adsorbed at room temperature decreases much more rapidly than linearly with increasing Cu content (Yu et al. 1976).

In order to rationalize the observed effect the most primitive model considers that only ensembles of n (or more) neighbouring atoms of kind A are able to form a chemisorption bond which is comparable in strength to that observed with the pure component A. If a random distribution of the constituents in the alloy surface is assumed then the density of "active" adsorption sites is expected to be proportional to $(X_A)^n$, where X_A is the mole fraction of atoms A in the surface layer. A log–log plot of the saturation coverage against X_A should therefore yield a straight line with slope n. This procedure was applied for the first time by Burton and Hyman (1975) to the analysis of data by Sinfelt et al. (1972) on the catalytic activity of Cu–Ni alloys for the hydrogenolysis of ethane to methane, from which they concluded that the C_2H_6 molecule needs *two* neighbouring Ni atoms in the surface in order to react. Figure 5.6 shows a plot of the relative amount of CO adsorbed on a (111) single crystal surface of Ni–Au alloys as a function of the mole fraction X_{Ni} of Ni in the surface (Burton et al. 1976). The straight line has a slope $n = 3$

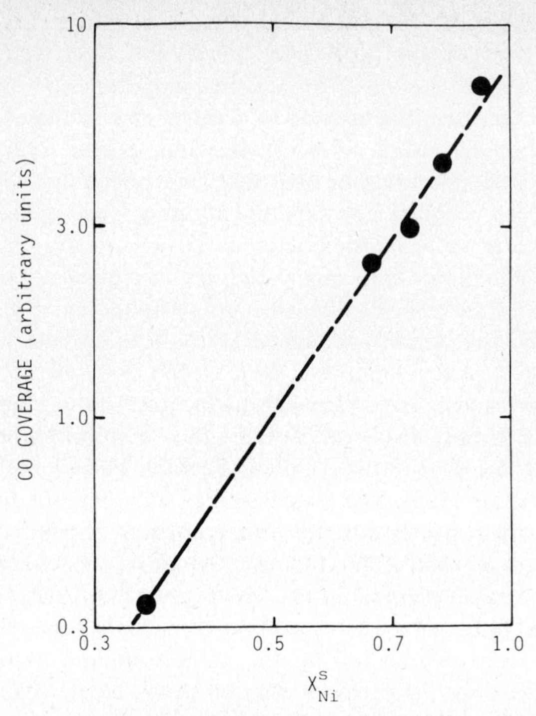

Fig. 5.6. CO coverage as a function of the mole fraction of Ni, x_{Ni}, in the surface of Ni/Au alloys (Burton et al. 1976).

indicating that at least three neighbouring Ni atoms are forming an "ensemble" for (strong) CO adsorption. Analysis of corresponding data for (111) planes of Ag–Pd alloys (Christmann and Ertl 1972) revealed a similar value of $n \approx 3.5$ (Ertl 1976). For the α-state of H_2 on (110) Cu–Ni surfaces – this is the part of hydrogen irreversibly adsorbed at room temperature and therefore corresponding to Sachtler's titration technique – Yu et al. (1976) derived an even stronger decrease of the coverage ($n = 4$), although a quantitative analysis of these data is somewhat complicated. Qualitative hints for a decrease of the adsorbed amount much stronger than proportional to the alloy composition can be found in a series of other papers (Takasu and Yamashima 1973, Mathieu and Primet 1976, Soma-Noto and Sachtler 1974, Stephan et al. 1976, Araki and Ponec 1976) where however an independent determination of the surface composition is lacking.

Some more detailed insight was obtained by looking at the infrared bands of "linearly" and "bridge"-bonded CO. Soma-Noto and Sachtler

(1974) reported that with Ag–Pd alloys shifting of the two bands with alloy composition was a minor effect compared to the drastic variations of their relative intensities, thus demonstrating that with this system the "ensemble" effect is more pronounced than the "ligand" effect. It was found that the concentration of "bridge" (or multi-centered) sites decreased rapidly with increasing Ag content and disappeared completely around 70% Pd, whereas apparently the concentration of the "linear" species at first increased and passed through a maximum at about 70% Pd. The adsorption energy of the linear species was derived to be about 15 kcal/mole smaller than that for the "bridge" species. The conclusion is that with increasing Ag content the number of large Pd clusters rapidly decreases but that less favorable bond formation is possible with smaller ensembles (but evidently not with isolated Pd atoms embedded into a complete Ag atom surrounding since also the "linear" band disappears below about 30% Pd).

If the chemisorption bond involves only a single metal atom the heat of adsorption as a function of coverage should be constant until all A atoms are occupied, and then drop steplike to the value for pure B (provided that the "ligand" effect and interactions between adparticles may be neglected). If on the other hand n metal atoms are contributing to the bond, alloying should lead to the formation of ensembles A_n, $A_{n-1}B$, $A_{n-2}B_2 \ldots B_n$ with different adsorption energies, the latter being further dependent on the specific geometric arrangement. As a result one expects the surface to become energetically heterogeneous with already small concentrations of B producing a spectrum of sites with practically continuously varying adsorption energies. Figure 5.7 shows the variation of adsorption energy of CO as a function of the work function change (which is a measure for the coverage) for the (111) planes of a series of Ag–Pd alloys (Christmann and Ertl 1972) which clearly show this effect: whereas on pure Pd E_{ad} is constant up to high coverage even 5% Ag in the surface (as determined by AES) can cause an appreciable reduction of the adsorbed amount and the onset of a continuous decrease of E_{ad} after about half the saturation coverage is reached. Above about 20% Ag the concentration of pure Pd ensembles obviously becomes negligibly small so that E_{ad} immediately decreases. Above 45% Ag CO adsorption, which was irreversible at room temperature (i.e. $E_{ad} \gtrsim 25$ kcal/mole), could no longer be observed, which is in qualitative agreement with the infrared measurements by Soma-Noto and Sachtler (1974).

The occurrence of "mixed" adsorption sites due to the ensemble effect with adsorption energies between those for the pure constituents becomes even more evident from the inspection of series of thermal

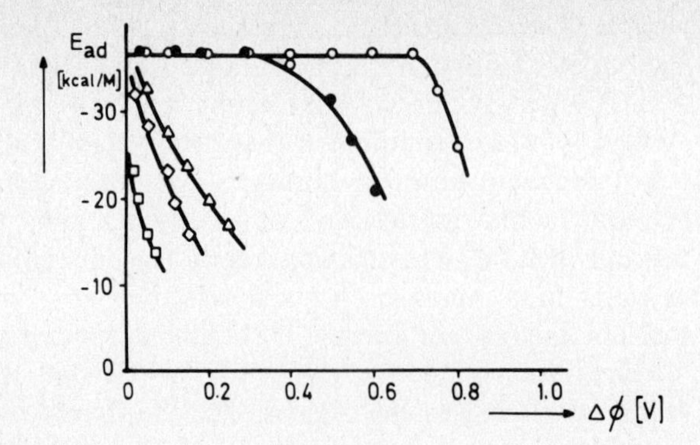

Fig. 5.7. Isosteric heats of CO adsorption as a function of the work function change $\Delta\phi$ on (111) planes of Ag/Pd alloys with different surface composition. ○: 100% Pd; ●: 95% Pd; △: 78% Pd; ◇: 70% Pd; □: 55% Pd (Christmann and Ertl 1972).

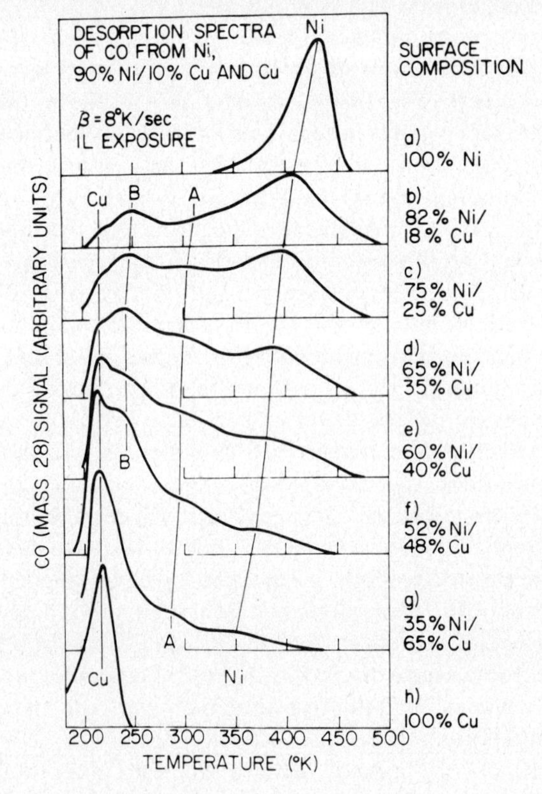

Fig. 5.8. Thermal desorption spectra of CO on (110) planes of Cu/Ni alloys with varying surface composition (Yu et al. 1976).

desorption spectra recorded recently by Yu et al. (1976) for CO adsorbed on Cu/Ni alloys (fig. 5.8). The traces from the alloy surfaces also exhibit appreciable desorption in the temperature range *between* the two single peaks for pure Cu and pure Ni. The authors interpret the apparent shift of the Ni peak with increasing Cu content as a weakening of the bond on pure Ni ensembles due to the "ligand" effect (i.e. change of the valence electronic properties of Ni atoms due to long-range interactions with Cu atoms), but inspection of the curves up to 40% Cu indicates a tail extending to temperatures as high as in the case of pure Ni. The desorption "peaks" with the alloy surfaces would therefore probably arise from the superposition of a whole spectrum of states with different adsorption energies, the energy of pure Ni clusters remaining essentially constant but their concentration decreasing rapidly.

Any quantitative conclusions from the findings described in this section would certainly be premature. The results however strongly suggest, at least with hydrogen and CO, not that single surface atoms form the individual chemisorption bond, but that small ensembles of neighbouring atoms (their number being probably in the order of 3) are causing the essential contributions. This fits well with the structural information from the analysis of LEED data where on pure metals the adsorbates prefer locations at multi-centered sites on the surface (see ch. 4).

A more detailed discussion of the effects connected with chemisorption and catalysis at alloy surfaces may be found in a recent review article by Sachtler and van Santen (1977).

2.3. Surface stoichiometry

To finish this section we may ask what would be the effect on surface composition if chemisorption were a localized two-centre bond between the adparticle and the adjacent surface atom. According to the periodic arrangement of the latter in a single crystal surface at saturation one would necessarily expect an adparticle : surface atom ratio of 1 : 1 and the formation of true 1×1 overlayer structures. As becomes evident from ch. 4, this is in fact sometimes observed. (H/Fe(110) is another example of this kind (Bozso et al. 1977a).) However a 1 : 1 stoichiometry is no proof for the existence of two-centre bonds: this is certainly a good approximation for the bonding of the H-atoms in benzene, but not for the description of the chemical bond in (solid) NiO. On the other hand 1×1 overlayer structures are more the exception than the rule, i.e. in most cases one ends up with a surface stoichiometry differing from a simple 1 : 1 ratio. This may have different reasons:

(i) Steric effects, i.e. the "size" of the adparticles, prevents the formation of a 1×1 structure. This is obviously the case with CO adsorption where the bond energy is relatively small (see sect. 6).
(ii) Surface reconstruction. If the metal–adsorbate bond is very strong (e.g. with atomic nitrogen or oxygen) a rearrangement of the metal atoms is probably energetically more favourable and frequently causes the appearance of coincidence lattices with large unit cells (as for example with the system N/Fe(111) (Bozso et al. 1977b)). These cases may certainly be considered as a special case of solid compound formation whose thickness is obviously restricted to the topmost atomic layers.
(iii) Formation of simple superlattices with small, strongly bound adsorbates (such as O or N). There are many examples of this kind where saturation of the adlayer is characterized by a simple structure and a coverage <1, such as the $c2\times 2$ structures with $\theta = 0.5$ formed by O on Ni(100) (Demuth and Rhodin 1974, Marcus et al. 1975) or by N on Cu(100) (Burkstrand et al. 1976) or Fe(100) (Bozso et al. 1977b). This effect can be interpreted in terms of a "consumption" of the free surface valencies of the atoms surrounding the (for example four-fold) adsorption site so that occupation of neighbouring sites is energetically not favourable. A formally equivalent description consists of the introduction of strong repulsive interactions between particles on neighbouring sites (see sect. 9). In both cases one arrives at the picture that strongly bound adsorbates may be considered as analogues of three-dimensional solid compounds with definite stoichiometries and multi-centered bond formation.

3. The "electronic" factor

3.1. The chemical nature of the substrate and the adsorbate

The chemical natures (as determined by the nuclear charge) of the metal atoms forming the substrate and the adsorbate molecules are of course the most important parameters determining the strength of the chemisorption bond. The theoretical aspects and the experimental evidence from electron spectroscopic techniques have been discussed extensively in the preceding chapters. However, the present state of progress in these fields does not yet allow the derivation or even prediction of adsorption energies with enough accuracy (≤ 0.1 eV), so this branch will certainly be based in the near future on direct determinations from thermodynamic or kinetic measurements.

There were various attempts in the past to correlate energies of chemisorption of a certain adsorbate over a series of different metals with a single (bulk) property of the substrate (Hayward and Trapnell 1964, Bond 1962, Clark 1970, 1974) among which Eley's approach (1954) based on Pauling's approximation of the covalent bond (Pauling 1939) was particularly popular. The heat of sublimation of the metal and its electronegativity were the main parameters but comparison with experiment showed no correlation at all with the strength of the chemisorption bond (Ehrlich 1959, Bond 1962). A better answer was expected by considering the mechanism of chemisorption as being mainly determined by the presence of empty states in the d-bands of the metals, and correlations with the Pauling percentage d-character (Pauling 1949) were attempted, but also failed (Hayward and Trapnell 1964). The only trend which probably may be recognized is a continuous decrease of the adsorption energy in passing across the Periodic Table from left to right (Hayward and Trapnell 1964), but since enough reliable experimental data are still not available this statement has to be considered with some precaution.

Since adsorption energies are sensitively influenced by the presence of impurities, the surface structure, and the coverage, a true comparison of different metals is in principle only possible by looking at the initial energies (i.e. values extrapolated to $\theta = 0$) on clean single crystal surfaces with identical crystallographic orientation. Such data are still rather scarce and only in a few cases available with an accuracy of about ± 1 kcal/mole. Table 5.3 therefore contains also some data which do not completely fulfil these criteria (being obtained e.g. with polycrystalline material or representing an average over a finite range of coverages) but which are believed to represent the behaviour of clean surfaces with enough reliability for reasonable comparisons between different adsorbates and substrates to be made. This list is by no means intended to be complete; some data obtained for different crystal planes of the same substrate will be discussed in sect. 4. Among the various adsorbates oxygen is certainly the most problematic from an experimental point of view since chemisorption is frequently irreversible and not easily distinguishable from the onset of oxide formation. So far no clear data for *molecular* N_2, O_2 or H_2 *chemisorption* on a metal surface is known. The numbers for these adsorbates refer always to strengths of the bond of the metal surface with the *atomic* species, and experimental adsorption energies have therefore to be converted by the inclusion of the dissociation energy of the corresponding diatomic molecule. CO and NO also show some tendency for dissociative chemisorption which may complicate the experimental situation. (For example with the system CO/W the situation is still so complex that no reliable number for the initial adsorption energy is available despite numerous studies with this system (Gomer 1975).)

From a first glance at table 5.3 it becomes evident that the nature of the adsorbate dominates essentially this pattern, i.e. the kind of transition metal is of lesser importance. This is quite reasonable since the electronic properties of the various adsorbates clearly differ much more from each other than do those of the different substrates. With respect to the adsorbate the strength of the bond follows the order

$$N > O > H > CO \gtrsim NO,$$

which one would also expect on the basis of simple chemical arguments. (For example, the dissociation energies of the diatomics N_2, O_2, H_2 follow the same sequence $N > O > H$ which is also seen to be plausible from the LCAO–MO level schemes of these molecules.) The influence of the metal becomes more pronounced with the more weakly bound molecular adsorbates. It is interesting to note that with N and H all bond strengths are equal within $\pm 10\%$ (140 and 62 kcal/mole, respectively), i.e. the differences are less pronounced than with the corresponding diatomic molecules M–H (compare table 5.1).

The strongest influence of the nature of the susbstrate metal is observed with CO adsorption, where the bond energies with the Ib metals are much smaller than those derived with the transition metals where the Fermi level is located within the d-bands. Regarding CO adsorption on the Ib metals Bradshaw and Pritchard (1970) pointed out that the heat of adsorption follows the same sequence as the shift of the C–O stretching vibration towards lower wave numbers, viz. Cu > Au > Ag. With metal carbonyls the latter effect is mainly ascribed to the extent of back-donation of d-electrons into the ligand's $2\pi^*$ orbitals (Brown and Darensbourg 1967). With respect to chemisorption on a metal surface this back-donation effect should be smaller the lower the energy of the d-band edge. This quantity in fact follows the right sequence such that this example probably illustrates a correlation between a single bulk electronic property and the chemisorption energy. It would also qualitatively explain the difference between Ni and Cu, but of course not that between Ni and Pd.

3.2. Chemisorption on alloys – the "ligand" effect

Many earlier attempts to rationalize the "electronic" factor in catalysis were based on studies with VIII–Ib alloys and a collective description of their electronic properties (Schwab 1955, Eley and Couper 1950, Dowden 1950, Hall and Emmett 1959). According to the then adopted rigid band model (Mott and Jones 1958) the constituents of an alloy such as CuNi form a common d-band which according to the Cu concentration is continuously filled with electrons. Filling of all d-states should be completed at 60% Cu, and since empty d-states were believed to play a dominant role in the formation of the surface bond, drastic variations in the catalytic activation were expected to occur at this composition. The experimental data were quite inconsistent (which has to be ascribed to the deviation of the composition of the surface from that of the bulk) and this effect could never be clearly established. Later it became evident from theoretical (Lang and Ehrenreich 1968) as well as experimental (Seib and Spicer 1970, Hüfner et al. 1972) investigations that the rigid band model was for this type of alloys a rather poor description. Instead it turned out that to a first approximation no charge transfer between the constituents should take place and they therefore retain essentially the properties of the pure components. Together with the localized picture of the chemisorption bond this concept leads to the "ensemble" effect in chemisorption on alloys as discussed in subsect. 2.2, where only the nature of the metal atoms directly involved in the bond formation is of importance.

For Cu/Ni alloys a theoretical treatment within the coherent potential approximation (CPA) confirmed the idea of negligible charge transfer between the constituents (Stocks et al. 1971). This is not completely preserved with Ag–Pd alloys (Stocks et al. 1973) where there is also some experimental evidence of transfer of s charge onto Pd (Snodgrass 1971). Calculations for Ag–Au alloys revealed that the situation may be even more complicated: conduction electrons are transferred onto Au sites and are partly compensated for by a smaller d-electron flow from Au to Ag sites (Gelatt and Ehrenreich 1974). Experimental evidence for the mutual influence of the valence electronic properties of the constituents in an alloy may be obtained by core-level spectroscopic techniques (XPS, APS) which reflect variations of the *local* potentials. Such effects were for example observed with Ni–Cu, Ni–Pd and Ni–Fe alloys (Wandelt and Ertl 1975, 1976).

Whenever such effects occur the local environment of an adsorption site will influence the local electronic properties of the atoms forming this ensemble and thereby in an indirect manner also its chemisorptive properties. This (long-range) phenomenon was – by analogy to the through-bond interactions observed in complex chemistry – called the "ligand" effect (Sachtler 1973) and is related to the indirect interactions between adsorbed particles as discussed in subsect. 9.3. Experimental evidence for the operation of this effect is relatively scarce, and obviously the "ensemble" effect dominates in all examples so far investigated.

In an infrared spectroscopic study on the adsorption of CO on Pd–Ag alloys Soma-Noto and Sachtler (1974) observed mainly drastic variations of the relative concentrations of "linear" and "bridge" species (the ensemble effect), but also minor shifts of the i.r. bands which were interpreted by the ligand effect. The observations that, with increasing Ag content the adsorption energy for pure Pd is no longer reached even at the smallest coverage, point in the same direction (Christmann and Ertl 1972).

Yu et al. (1976) interpreted the apparent shift of the high-temperature peak in their CO thermal desorption spectra towards lower temperatures with increasing Cu content (see fig. 5.8) in terms of the operation of the

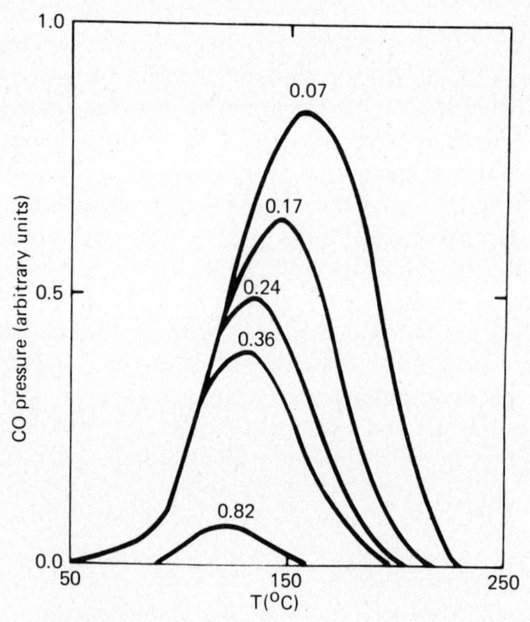

Fig. 5.9. Thermal desorption spectra for CO on (111) planes of Au/Ni alloys with different surface composition (Burton et al. 1976).

ligand effect leading to a continuous lowering of the adsorption energy on Ni adsorption sites by about 5 kcal/mole.

A series of CO desorption spectra from the work by Burton et al. (1976) is reproduced in fig. 5.9 and shows again such an effect, namely a shift of the peak towards lower temperatures with increasing Au content from which a variation of the adsorption energy on Ni-sites by about 3 kcal/mole may be estimated.

4. The crystallographic orientation of the surface

It is quite natural to assume that the geometric arrangement of the surface atoms has some influence on the chemisorption bond, particularly if we believe that the latter in general involves an ensemble of more than a single metal atom. Quite early H.S. Taylor (1926) introduced the concept of "active centres" which were regarded to be regions characterized by particular configurations of the surface atoms. Their actual geometry was discussed particularly by Balandin (1929a,b) in the framework of his so-called "multiplet" theory of catalysis. With "real" surfaces this effect may be studied by using small particles with varying size, thus exposing varying proportions of the different crystal planes (Boudart 1969).

According to a naive picture one would assume that the strength of the chemisorption bond increases as the number of "unsaturated" valencies of the surface atoms increases, i.e. if their coordination number decreases. That would mean a lower adsorption energy at the most densely packed planes than on the planes with higher Miller indices. A look to molecular chemistry however shows that this assumption will not generally hold: the energy for dissociation of an H atom from H_2O is 119 kcal/mole, but only 102 kcal/mole from OH (CRC Handbook of Physics and Chemistry 1967)!

From the model of a surface molecule variations of the adsorption energy are to be expected if the surface geometry markedly influences either the energies (and/or occupancies) of those metal orbitals which are involved in the bond formation, or the overlap between the adatom orbitals and the corresponding group orbitals of the surface atoms. Presumably this effect will be more pronounced with solids exhibiting strongly directed bonds in the bulk which probably also persist at the surface as "dangling" bonds such as with the elemental semiconductors. The existence of electronic surface states represents a further complication of this model.

In the following a series of examples for chemisorption on metal single crystal surfaces with different orientation will be presented. The general conclusion will be that the influence of the surface orientation on the bond energy is smaller than one would probably have expected. However there may be marked effects with respect to the *kinetics* of adsorption – a field which is still very far from being theoretically explored. As far as data are available initial heats of adsorption as derived from isosteric (equilibrium) measurements are listed. In some cases also values as determined from thermal desorption spectra were taken where appreciable variation of the adsorption energy in the studied range of coverages may be excluded. The accuracy is even with the most carefully studied examples not better than ±1 kcal/mole.

Probably the most extensive study with different single crystal planes was performed by Domke et al. (1974) with the system H_2/W by means of field emission microscopy. Their data are listed in table 5.4 together with a value for W(110) as derived for a macroscopic W(110) single crystal plane (Tamm and Schmidt 1969, 1971). In the latter case, however, it has to be noted that Polizzotti and Ehrlich (1975) concluded that a perfect W(110) plane would be completely unable to react directly with molecular hydrogen striking from the gas phase so that dissociation is probably assisted by the presence of a low concentration of surface defects. The data by Domke et al. (1974) are generally in good agreement with others available for a series of W planes as reviewed by Schmidt (1974). They show a variation between 33 and 40 kcal/mole, that means that the strength of the M–H bond varies only between 68 and 71.5 kcal/mole, i.e. by 5%. A similar behaviour is found with nickel (Christmann et al. 1974b) ((111): E_{ad} = 23 kcal/mole; (100): 23; (110): 21.5) and with palladium as well (Conrad et al. 1974a) ((111): 21; (110): 24.5).

Equal or nearly equal adsorption energies were also found for the interaction of N_2 with different single crystal planes of W (Delchar and Ehrlich 1965) ((100) and (110): 78 kcal/mole), Mo, (Mahnig and Schmidt 1972) ((110) and (100): 87 kcal/mole) and Fe (Bozso et al. 1977b,c) ((110): 49; (100): 53; (111): 51 kcal/mole). The latter case is particularly remarkable since with Fe(110) and (111) there is strong evidence that the surface reconstructs under the influence of interaction with nitrogen so

TABLE 5.4
Initial heats of H_2 adsorption on different W planes

Plane	(110)	(100)	(211)	(111)	(013)	(122)	(123)	(114)
E_{ad}(kcal/mole)	33	35	40	36.5	33	36.5	39	34

that adsorption and desorption involve a complex sequence of elementary steps.

Table 5.5 lists a series of data for CO adsorption on different single crystal planes of Ni, Cu and Pd. There is a general tendency for an increase of the adsorption energy with decreasing coordination number of the surface atoms among the three most densely packed planes, viz. (111) < (100) < (110). The more open planes however exhibit no further increase but in the case of Pd have even values comparable to those determined with the (111) plane. The total variation between different planes of the same metal is less than 20%, which is also smaller as one probably might have expected. The last row of table 5.5 contains some theoretical values for palladium which were evaluated by a semi-empirical method (a modified Anderson formalism (Doyen and Ertl 1974)) by coupling metallic d-orbitals to the $2\pi^*$-levels of CO and adjusting the parameters in such a way that agreement with the experimental value for the (110) plane was achieved. A later extension of the calculations to include also coupling of the CO 5σ-state to the metallic

TABLE 5.5
Initial heats of CO adsorption (kcal/mole) on different planes of Ni, Cu and Pd

Plane	(111)	(100)	(110)	(210)	(211)	(311)
Ni	26.5[a]	30[b]	30[c]			
Cu	12[d]	13.5[e]			14.5[f]	14.5[f]
Pd	34[g]	36.5[h]	40[i]	35[i]		35.5[i]
Pd (theory[j])	34	36	40	35		35

[a] Christmann, K., O. Schober and G. Ertl, 1974, J. Chem. Phys. **60**, 4719.
[b] Tracy, J.C., 1972, J. Chem. Phys. **56**, 2736.
[c] Taylor, T.N. and P.J. Estrup, 1973, J. Vac. Sci. Techn. **10**, 26; Madden, H.H., J. Küppers and G. Ertl, 1973, J. Chem. Phys. **58**, 3401.
[d] Conrad, H., G. Ertl, J. Küppers and E.E. Latta, 1975, Solid State Comm. **17**, 613.
[e] Tracy, J.C., 1972, J. Chem. Phys. **56**, 2748;
Chesters, M.A., J. Pritchard and M.L. Sims, 1972, in: Adsorption–desorption phenomena (F. Ricca, ed.) (Academic Press, New York) p. 277.
(Tracy observed below $\theta = 0.1$ an increase of E_{ad} to about 16.5 kcal/mole which is not believed to be representative for the perfect Cu(100) plane).
[f] Papp, H. and J. Pritchard, 1975, Surface Sci. **53**, 371.
[g] Ertl, G. and J. Koch, 1970, Z. Naturforsch. **25a**, 1906.
[h] Tracy, J.C. and P.W. Palmberg, 1969, J. Chem. Phys. **51**, 4852.
[i] Conrad, H., G. Ertl, J. Koch and E.E. Latta, 1974, Surface Sci. **43**, 462.
[j] Doyen, G. and G. Ertl, 1974, Surface Sci. **43**, 197. In this semi-empirical calculation the value for Pd(110) was adjusted with the experimental data

sp-bands did not alter the general agreement between the trends shown by experiment and theory (Doyen and Ertl 1977). Using the same method, variations of the same order of magnitude ($\sim$15%) of the adsorption energies of hydrogen, oxygen and nitric oxide between different crystal planes were derived (Doyen and Ertl 1978).

Experimental data for other adsorbates are scarce so far. It will only be mentioned that Comrie et al. (1976) derived adsorption energies for NO of 27.5 kcal/mole on Pt(111) and of 28 kcal/mole on Pt(110) which fit into the general picture.

Looking at the variation of the adsorption energy with coverage, the differences between different crystal planes turn out to be at least as large as those between the initial heats of adsorption. This indicates that the interactions between adsorbed particles and the energetic differences between different adsorption sites are of about the same order of magnitude as the variations of the single-adsorbate bond energies between different crystal planes. Since thermal desorption spectra are sensitively reflecting energetic variations (1 kcal/mole corresponds roughly to a peak shift of 20 K) it becomes plausible as to why their shape is sometimes markedly influenced by the surface orientation as well as by the coverage (Schmidt 1974).

Marked influences of the surface orientation are sometimes also observed with respect to the adsorption kinetics, particularly with the initial sticking probabilities s_0. Adams and Germer (1971) performed a thorough investigation of this effect with the system N_2/W. They observed relatively large sticking coefficients (>0.2) at 300 K with a series of planes, whereas others appeared to be essentially inactive, i.e. s_0 being at least two orders of magnitude smaller. A stereographic plot of the normals to various *W* planes as reproduced in fig. 5.10 revealed that active and inactive planes could be categorized on the basis of their surface atom configuration, which is reflected by their location in the stereographic triangle. It was found that the [$\bar{1}$11]-axis divides the active from the inactive planes, but on the other hand it also divides the stereographic unit triangle into regions containing planes with surface atoms with coordination number 4 in the configuration as found on the (100) plane, and planes with coordination number 4 in the configuration found on the (111) plane. It was thus concluded that dissociative adsorption of N_2 on tungsten at 300 K takes place only with a measurable rate at (100) sites. Field emission measurements by Polizzotti and Ehrlich (1975) revealed that a perfect W(110) plane is even unable to dissociatively adsorb hydrogen at room temperature. A similar result was obtained for N_2 adsorption on the basal (0001) plane of Re (Liu and Ehrlich 1976).

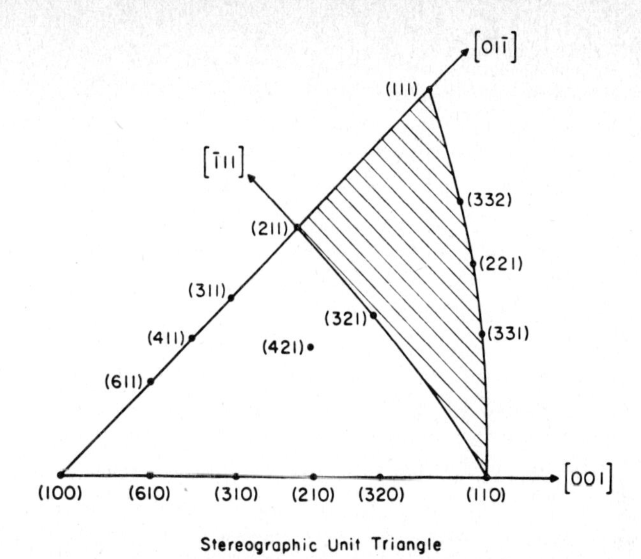

Fig. 5.10. Stereographic triangle of the normals to various planes of tungsten. Planes located in the shaded area are predicted to be unreactive for dissociative N_2 chemisorption at 300 K (Adams and Germer 1971).

The latter finding is of particular interest with respect to O_2 and H_2 adsorption on Pt(111). Reports on the sticking coefficients of these systems were very controversial and it was even claimed that these gases do not adsorb at all at room temperature at clean and perfect Pt(111) surfaces (Bernasek and Somorjai 1975, Lang et al. 1972, Weinberg et al. 1973). More recent measurements however revealed unequivocally that both systems behave rather normally, i.e. exhibit sticking probabilities of the order of about 0.1 at 300 K. The reason for the failure to detect adsorption is caused in the case of H_2 by the relatively small heat of adsorption leading to no irreversible adsorption at 300 K (Lu and Rye 1974, Christmann et al. 1976), and in the case of O_2 by the high reactivity of O_{ad} with spurious amounts of CO or H_2 being present in the residual gas atmosphere (Bonzel and Ku 1973, Joebstl 1975). The extended discussion of these systems in the literature demonstrates that at the present stage of sophisticated surface research even such a simple question as to whether a gas adsorbs or not may still be controversial.

The "true" cases of small reactivity are frequently connected with an activation energy for dissociative adsorption. In a careful molecular beam study Barlooch et al. (1974) derived activation energies for H_2 adsorption on Cu of 3 kcal/mole for the (110) plane and of 5 kcal/mole for the (100) and (310) planes, which will be discussed in more detail in sect. 8.

An activation barrier was also derived to exist for dissociative N_2 adsorption on Fe (Bozso et al. 1977b,c) ((110): 8 kcal/mole, (100): 5 kcal/mole, (111): 0) giving rise for example to a factor of 60 difference for the initial rates of adsorption on Fe(110) and Fe(111) at 680 K. However, even if the activation energy is taken into account the sticking coefficients with these systems are still too small by a factor of 10^6 compared with those expected on the basis of a simple collision model. Probably a very particular orientation of the impinging N_2 molecules is needed for a successful collision.

5. The role of surface steps

Taylor's original concept of "active" sites (1926) was particularly concerned with structural imperfections of low-index planes which are present in a great variety and concentration on "real" catalyst surfaces. In the framework of their so-called "Adlineationstheorie", Schwab and Pietsch (1928) later suggested that linear defects on the surface such as present for example along a step exhibit an enhanced activity due to the lower coordination numbers of the respective surface atoms. It was recognized that periodic arrays of steps on single crystals are sometimes easily formed and rather stable, so such systems are suitable models for studying these aspects under well defined conditions. The (average) orientation, height, and periodicity may conveniently be monitored by LEED (Ellis and Schwoebel 1968, Henzler 1970, Perdereau and Rhead 1971, Lang et al. 1972).

Extended studies with surfaces of this type, mainly with respect to the reactivity with hydrocarbons, were performed by Somorjai and his co-workers (Somorjai 1975, 1977) who also proposed a nomenclature which is illustrated by fig. 5.11. Sometimes dramatic differences between flat and stepped surfaces were reported, for example even with respect to the adsorption of H_2 on Pt(111) (Bernasek and Somorjai 1975, Lang et al. 1972).

Systematic studies, however, revealed that the influence of the steps on the energetics of adsorption was of about the same order of magnitude as the difference between different crystal planes, i.e. at the most by a few kcal/mole.

Measurements of the isosteric heat of CO adsorption on Pd(111)- and Pd(S)-[9(111) × (111)] surfaces revealed in both cases practically identical results (Conrad et al. 1974b). Similar conclusions were reached by Hagen et al. (1976) with Ir(111)- and Ir(S)-[6(111) × (100)] surfaces where

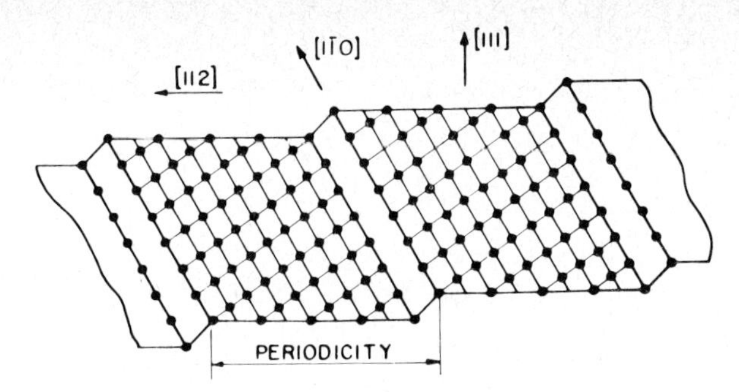

Fig. 5.11. Schematic representation of a Pt(S)-[9(111) × (100)] surface. 9(111) designates a terrace of (111) orientation 9 atomic rows in width, and (100) indicates a step with (100) orientation 1 atomic layer high (Somorjai 1977).

no significant differences in thermal desorption spectra of CO were observed. On the other hand a marked effect was found by Collins and Spicer (1977) with CO adsorption on Pt(111) who concluded from thermal desorption experiments that the adsorption energy associated with atoms along steps is higher by about 6 kcal/mole than sites on the "flat" parts of the surface. Iwasawa et al. (1977) concluded that steps on Pt(111) also provide an enhanced probability for CO dissociation at room temperature.

Pronounced effects were also observed by Hagen et al. (1976) in the case of O_2 adsorption on Ir(111)- and Ir(S)-[6(111) × (100)] surfaces. In the latter case besides a TDS maximum around 800°C (which was also observed with the low-index plane) an additional high-temperature state desorbing around 1200°C was observed. However it cannot be excluded that oxide formation (which is a process caused by nucleation and therefore might be favoured by the presence of steps) complicates a straightforward interpretation of these results.

Thermal desorption spectra of H_2 adsorbed on flat and stepped Ir(111) surfaces revealed nearly equal peak temperatures. The spectra from the stepped surfaces, however, exhibited a tail which was more extended to higher temperatures suggesting that part of the hydrogen atoms is more strongly bound at the steps (Nieuwenhuys et al. 1976). Quantitative evidence for this conclusion was reached with isosteric heat of adsorption measurements for H_2 adsorbed on Pd(111)- and Pd(S)-[9(111) × (111)] surfaces (Conrad et al. 1974a). As shown in fig. 5.12 with Pd(111) the adsorption energy remains constant up to medium coverages whereas with the stepped surface it increases with decreasing coverage so that the initial heats of adsorption differ by 3 kcal/mole which is

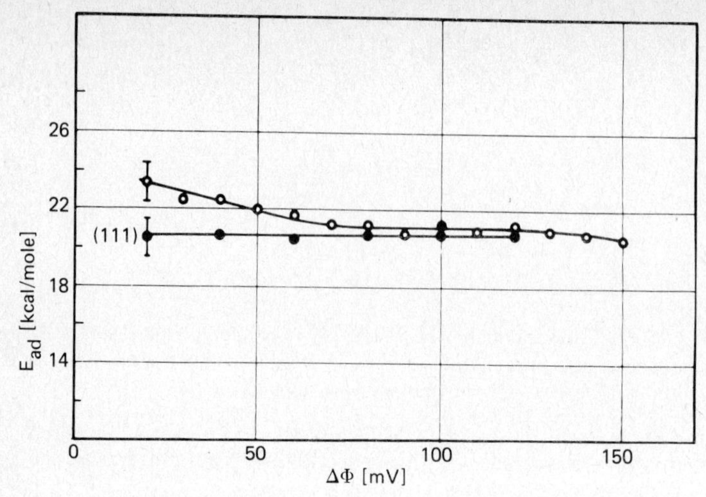

Fig. 5.12. Isosteric heat of hydrogen adsorption on Pd(111) (full circles) and on a stepped Pd(S)-[9(111) × (111)] surface (open circles) as a function of the work function variation $\Delta\phi$ (Conrad et al. 1974a).

considered to be caused by a corresponding difference in bond energy between step sites and atoms located on terraces.

A quite similar result (i.e. again a difference by 3 kcal/mole in the initial adsorption energies) was derived for H_2 adsorption on Pt(111)- and Pt(S)-[9(111) × (111)] surfaces (Christmann and Ertl 1976). In this case, it was possible to characterize the adsorption sites associated with the steps in even more detail: as can be seen from fig. 5.13 with Pt(111) the work function decreases continuously with increasing coverage, whereas with the stepped surface this quantity at first increases, exhibits a first break at $\theta \approx 0.12$ and a second break at $\theta \approx 0.25$, from where on the dipole moments of the additionally adsorbed particles change their sign. Keeping in mind that with the latter system about 11% of the surface atoms are located at steps and that under the chosen experimental conditions the saturation coverage is $\theta_{sat} \approx 1$ (i.e. a 1 : 1 ratio of surface and adsorbate atoms) the interpretation appears to be straightforward. As illustrated schematically by fig. 5.14 obviously *two* H atoms (with slightly differing positive dipole moments) have to be associated with each surface atom at the steps whereas the dipole moment of the H-atoms on the terraces has a reverse sign. Since the metal–H bond strength on the flat portions of the surface is slightly smaller (56.3 kcal/mole compared to 57.6 kcal/mole at the steps) the step sites are occupied first. A more detailed inspection reveals that at $T > 0$ K

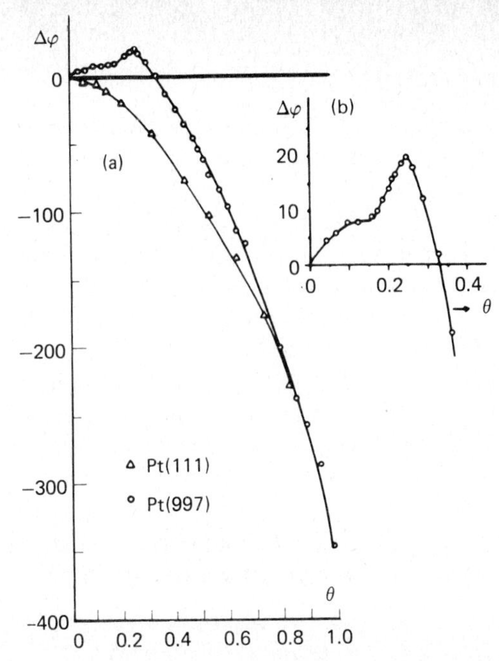

Fig. 5.13. Variation of the work function as a function of hydrogen coverage on a "flat" Pt(111) surface (triangles) and on a stepped Pt(S)-[9(111) × (111)] plane (circles) (Christmann and Ertl 1976).

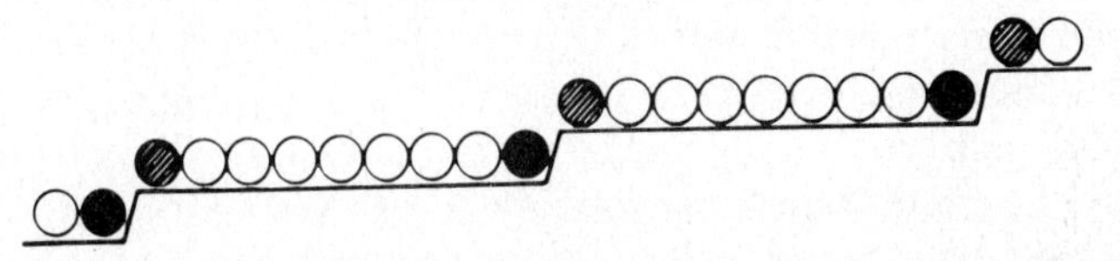

Fig. 5.14. Model for the Pt(S)-[9(111) × (111)] at monolayer coverage with hydrogen, illustrating the existence of three different types of adsorbed H atoms (Christmann and Ertl 1976).

even at very low coverages according to Boltzmann statistics part of the terrace sites should be occupied. This conclusion is consistent with the observation that the magnitude of the work function maximum decreases with increasing temperature.

As in the case of the influence of the crystallographic orientation of low-index planes again the kinetics of adsorption is much more affected than their energetics, which becomes particularly evident from Somorjai's (1975, 1977) studies on hydrocarbon catalysis reactions. In the case of H_2 adsorption on platinum Christmann and Ertl (1976) reported an initial sticking coefficient of 0.34 on the Pt(S)-[9(111) × (111)] surface which is larger by a factor 4 than that derived for a "flat" Pt(111) surface

($s_0 = 0.08$). This effect may still be explained by a local picture, namely by assuming a sticking probability near unity near the steps and the "normal" value on the terrace sites.

Recently Hopster et al. (1977) performed a study on the interaction of O_2 with Pt(111) surfaces with varying step concentrations and concluded that the initial sticking coefficient increases *exponentially* with increasing step concentration, an effect which was already observed earlier with the oxygen adsorption on cleaved Si surfaces (Ibach et al. 1973). Evidently such a behaviour can hardly be explained with a local model and it was therefore suggested that the steps lower the activation barrier for dissociative adsorption through the correlation to their lowering of the work function (Ibach 1975). Although this model looks intriguing and can also account for the exponential decrease of the sticking coefficient with coverage, it would certainly be very important to determine the activation energy for O_2 adsorption *directly* (by varying the gas and the surface temperature). Another possibility for checking this hypothesis would consist in preadsorbing a small amount of an alkaline metal which should equally cause a lowering of the work function.

6. Energy profiles: variation of the adsorption energy across the surface

Figure 5.2 illustrates schematically how the adsorption energy varies along a certain crystallographic orientation with the position of the adparticle. A complete description needs the knowledge of a two-dimensional energy profile across the unit cell of the substrate whereby the vertical distance always corresponds to the minimum of the potential energy with respect to movement perpendicular to the surface. This energy profile determines the mobility of adsorbed particles into different directions on the surface (the energy maxima correspond to the activation barriers); the positions with minimum energy will be the adsorption sites at low temperature and coverage (as long as interactions between adsorbed particles may be neglected) and therefore determine the (macroscopic) initial adsorption energy; the energy difference between various "sites" will determine the relative populations at non-zero temperature; the overall variations will (apart from intermolecular interactions) be responsible for the formation of incoherent structures or lattice–gas type arrangements of the adparticles at high coverages.

A rough insight into the energetic variations across a metal surface may be obtained by looking at experimental data on the activation energy for surface diffusion: most information has been obtained by observing the spreading of non-uniform adsorbate layers on small tips by means of field emission microscopy (Gomer et al. 1957, Gomer 1975).

In this way activation energies E^* for atomic hydrogen diffusion on tungsten between 6 and 16 kcal/mole (depending on coverage and surface orientation) were observed (Gomer et al. 1957). For H/Pt a value of about 5 kcal/mole was derived (Lewis and Gomer 1969). Keeping in mind that the effective strength of the metal–hydrogen bond E_b is about 60 kcal/mole these data mean that E^*/E_b attains values between 0.1 and 0.2, a result which appears to be quite general (with respect to the order of magnitude) for transition metal/gas systems. The field emission method, however, is unable to yield very accurate and detailed results, since transport over adjacent regions also influences diffusion on a given plane. Analysis of current fluctuations from small regions would therefore provide a more unequivocal way (Kleint 1971, Gomer 1973), but so far no experimental results from applications of this technique have been reported.

Direct observation of individual adsorbate atoms by means of the field ion microscope provides a much more straightforward means for determining diffusion parameters (Ehrlich and Hudda 1966). Unfortunately this technique is restricted to metallic adsorbates, but is for example very powerful for studying the surface self-diffusion on metals with a high degree of accuracy (Ehrlich and Hudda 1966, Bassett and Parsley 1970). For the self-diffusion on different rhodium planes Ayrault and Ehrlich (1974) determined the following values for the activation energies: (111): 3.6; (311): 12.4; (110): 13.9; (331): 14.8; (100): 20.2 kcal/mole. Although these numbers vary strongly with the crystallographic orientation they are still considerably smaller than the binding energy. Interestingly, with the (311), (110) and (331) planes atom movement was observed to be strictly one-dimensional along close-packed [110] rows, indicating strong anisotropies for movements in different directions. Similar results may be obtained by using macroscopic techniques for the study of surface diffusion of metals (Bonzel 1973). Different coefficients for self-diffusion on Ni(110) along the [100]- and [110]-directions were for example recently determined by Bonzel and Latta (1977).

In summary, experimental data on activation energies for surface diffusion are still rather scarce and provide only rather limited insight into the problem under discussion. It may best be concluded that the barriers are usually much smaller than the energies for removing the adsorbed particles completely into the gas phase, and that pronounced anisotropies may exist with respect to variation of the surface orientation as well as concerning the movement into different directions on a given single-crystal plane.

If the variation of the adsorption energy across the surface is $\Delta E_b < kT$, the adsorbed particles are delocalized, i.e. they are mobile like a two-dimensional gas. While such situations may occur in cases of physisorption the observation of ordered chemisorbed layers (as a rule) demands that the adparticles are localized, i.e. $\Delta E_b > kT$. Nevertheless there may be different locations within the unit cell of the substrate lattice where the adsorption energies are nearly equal. As a consequence even at rather low coverage both types of sites should be occupied, their ratio of occupation being determined by the energy difference ΔE and kT through Boltzmann statistics. An example of this type has recently been found with the system CO/Pt(111) (Ertl et al. 1977). From an analysis of LEED and work function data it was concluded that for $\theta \leqslant \frac{1}{3}$ the system may be regarded as consisting of two types of adsorption sites (presumably threefold and twofold coordinated) exhibiting different adsorbate dipole moments. From the variation of the work function change with temperature at fixed coverage ($\theta = \frac{1}{3}$) the difference in adsorption energy between both sites was derived to be only 0.5 kcal/mole. Investigation of the vibrational properties by high resolution electron loss spectroscopy confirmed these conclusions, since at room temperature the existence of two different types of adsorbates with comparable concentrations was detected (Froitzheim et al. 1977). Similar observations were made with the system CO/Ni(100) (Andersson 1977) and will be discussed in the next section.

The minima of adsorption energy are always located at sites with high symmetry, a fact which becomes particularly evident by looking at the adsorbate structures so far analyzed by LEED (see ch. 4). However a series of systems involving adsorption of CO was observed to form incoherent layers at high coverages, i.e. the adsorbed molecules exhibit a tendency for the formation of close-packed layers where they are no longer fixed exclusively to high-symmetry sites. The first quantitative investigation of this effect was performed by Tracy and Palmberg (1969) with CO/Pd(100). These authors observed that at $\theta = \frac{1}{2}$ an out-of-registry transformation of the arrangement of the adsorbed particles occurs in a manner as illustrated by fig. 5.15 which is accompanied by a steplike drop of the adsorption energy by about 7 kcal/mole. This number has to be regarded as characterizing the degree of variation of the CO adsorption energy over the Pd(100) surface. (The influence of mutual repulsion comes into play only at higher coverages as discussed in subsect. 8.2). Similar effects were observed to occur with CO adsorbed on Pd(111) (Ertl and Koch 1970) and on Ni(111) (Christmann et al. 1974a) where at $\theta = \frac{1}{3}$ the overlayers become out of registry, this being

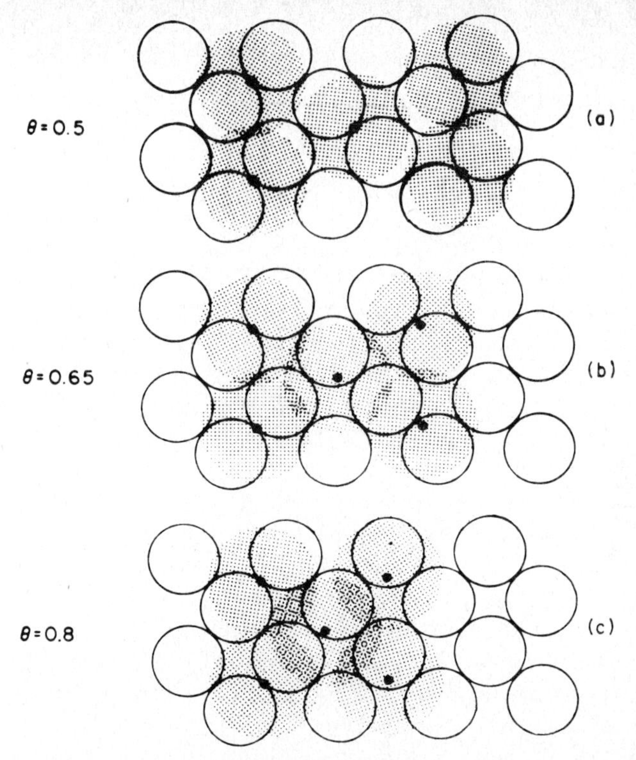

Fig. 5.15. Structure models for CO adsorbed on Pd(100) at different coverages (Tracy and Palmberg 1969).

associated with a steplike decrease of the adsorption energies by 2 and 3 kcal/mole, respectively. These latter results suggest that the (111) plane is with regard to CO adsorption energetically "smoother" than the (100) surface.

Several theoretical attempts were performed to evaluate data on the variation of the adsorption energy across a single crystal surface:

Gunnarsson et al. (1976) reported first results of a self-consistent calculation for H chemisorption on the (100)Al surface. The "bridge" position was found to be energetically favoured over the "on-top" position by about 3 kcal/mole, this energy difference also being the activation energy for surface diffusion. It will be very interesting to see further results from this rigorous treatment.

Fassaert and van der Avoird (1976) treated chemisorption of hydrogen on different nickel surfaces by a tight-binding extended Hückel method

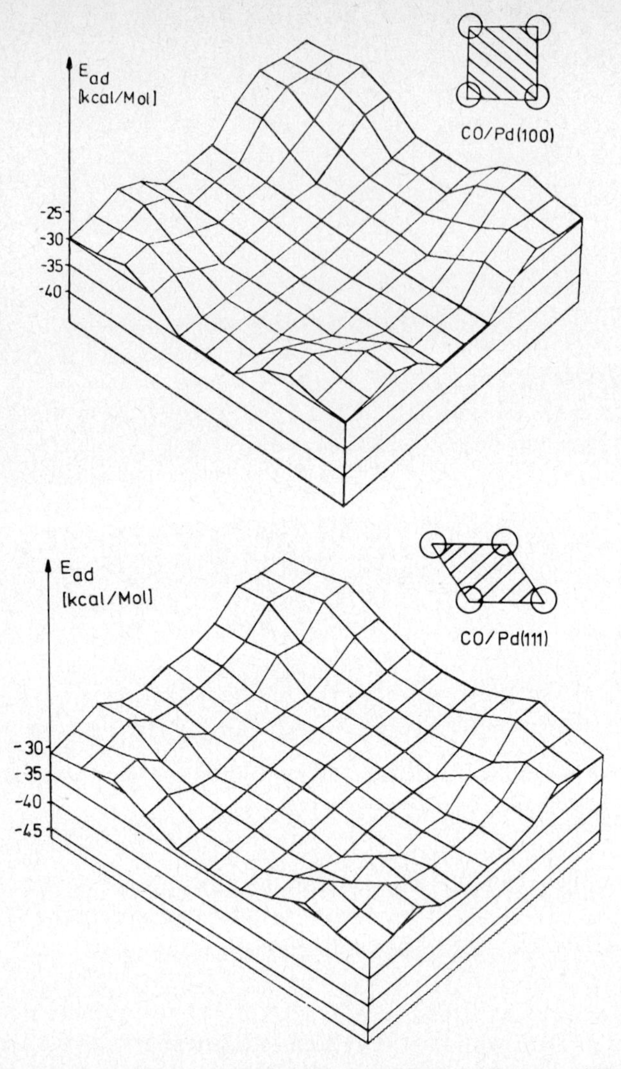

Fig. 5.16. Energy profiles for CO/Pd(100) and CO/Pd(111) (Doyen and Ertl 1977).

and concluded that the strength of the adsorption bond decreases in all cases in the order "on top" > "bridge" > "centered".

Qualitative data by applying the CNDO approximation to CO chemisorption on nickel clusters were obtained by Blyholder (1975). As a general result it was found that the most favourable sites are those with the highest coordination of the adsorbate by surface atoms, in general agreement with the experience from the analysis of LEED data (see ch. 4).

These findings are confirmed by the results of extended calculations of energy profiles by means of a semi-empirical method (Doyen and Ertl 1978). The results for CO/Pd(100) and CO/Pd(111) are reproduced in fig. 5.16. In the former case "bridge" and fourfold coordinated sites are predicted to be practically equivalent. Since at $\theta = 0.5$ the CO–CO distance for particles in bridge-position forming a $c4 \times 2$ structure (see fig. 5.15) is slightly larger than with fourfold coordinated adsorbates forming a $c2 \times 2$ structure this configuration is predicted to be more favourable, and is in agreement with the experimentally proposed structure model of fig. 5.15 (Park and Madden 1968, Tracy and Palmberg 1969). The overall variation of the adsorption energy across the surface is furthermore in good agreement with the steplike decrease at $\theta = 0.5$ connected with the onset of formation of out-of-registry configurations as discussed above.

The energetic variation across the (111) plane is smaller, which also agrees with experiment. Bridge and threefold sites are nearly equivalent, in agreement with the above mentioned conclusions for the system CO/Pt(111).

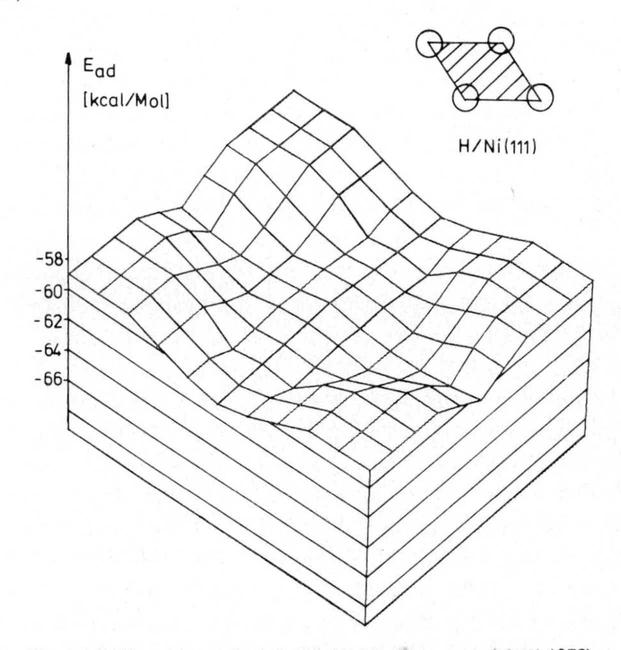

Fig. 5.17. Energy profile for H/Ni(111) (Doyen and Ertl 1978).

Figure 5.17 shows the calculated energy profile for the system H/Ni(111) (Doyen and Ertl 1978) where the threefold coordinated sites should be preferred, which is in agreement with the results of a recent analysis of LEED data (van Hove et al. 1978). Interestingly, the energy profile for O/Pd(110) predicts the locations between two surface atoms along the close-packed direction to be energetically favoured instead of the sites in the centre of the unit cell, which again is confirmed by a LEED analysis for the O/Ni(110) system (Demuth 1977).

The general conclusion of these calculations is that the adsorbed particles mostly prefer to stay in highly coordinated and symmetric sites on the surface and that the activation barrier for surface diffusion is always roughly one order of magnitude smaller than the strength of the chemisorption bond itself. Both findings are in general agreement with experimental data.

7. Vibrations

Infrared spectroscopy is widely used for identifying adsorbates on small particles (Little 1966, Hair 1967), but may however also be applied as a reflection technique for single crystal surfaces (Greenler 1969, Pritchard 1972, Yates et al. 1973). With a few exceptions (Blyholder 1964b), however, this method probes with sufficient sensitivity mainly vibrational modes within the adsorbed molecules and not the metal–adsorbate vibration. The latter is on the other hand interesting in the present context since it reflects the variation of the potential energy of the chemisorption bond normal to the surface, i.e. the shape of the potential curve of fig. 5.1 near the ground state. Information of this type is obtained by high resolution electron energy loss spectroscopy (EELS) as first applied by Propst and Piper (1967) and later developed mainly by Ibach (1972). Similar to the case of infrared spectroscopy with this technique only vibrational modes with a component of the dipole moment perpendicular to the surface will yield a signal. The local geometry of the adsorption site defines its point group symmetry and thereby also the maximum number of different observable vibrations. Furthermore a simple picture shows that some information on the geometry of the adsorption site may also be obtained from the energy of vibration (Froitzheim et al. 1976a): if different sites (e.g. on-top, bridge and fourfold sites on a (100) plane) are considered with equal bond strength then the force constant effective for vibration normal to the surface will be highest for the single atom position, since in the other cases the bond is shared between two and four (or five) atoms, respectively, whereas for vibrations normal to the surface the bond forces are only partially effective. This "spring" model therefore predicts $\omega_{top} > \omega_{bridge} > \omega_{fourfold}$ and is justified by some experimental evidence as discussed below.

Hydrogen adsorbed on tungsten is so far the most extensively studied system in this respect: with H/W(100) Propst and Piper (1967) observed an energy loss of about 140 meV. Later this result was theoretically confirmed by calculations by Smith et al. (1974). By applying the so-called density functional formalism they derived from the second derivative of the hydrogen–metal interaction potential a vibrational energy of 200 meV. More recently more sophisticated measurements were performed by Froitzheim et al. (1976b) who found the appearance of two bands at 155 and 135 meV (their intensity ratio depending on coverage) which were ascribed to adsorption at "on-top" and "bridge" sites, according to the above-mentioned arguments.

This result was later confirmed by Backx et al. (1977) who performed similar measurements also with W(110) and W(111) surfaces: interestingly in all three cases a vibration with identical energy of 157 ± 3 meV was found, suggesting that adsorption on single sites is associated with essentially identical force constants irrespective of the local geometry of the surrounding surface atoms. It is remembered that according to table 5.4 the energy of the metal–hydrogen bond varies only by about 2.5% between these three different planes. With H/W(110) Backx et al. (1977) observed an additional band at 95 meV which they attributed to bonding in centered sites, whereas with W(111) no additional losses were observed.

The situation is considerably more complex with the system O/W(100) which is connected with its complicated structural properties (Froitzheim et al. 1976a): at low coverages ($\theta < 0.2$) a single loss at 75 meV was observed which was ascribed to adsorption in fourfold sites. The occupation of different sites as well as the formation of double rows and surface reconstruction were made responsible for the complex nature of the spectra at higher coverages.

The structural properties are much simpler for the system O/Ni(100), where a p2 × 2 structure is formed at $\theta = 0.25$ and a c2 × 2 structure at $\theta = 0.5$. Different independent LEED studies (Demuth et al. 1973, van Hove and Tong 1975) came to the conclusion that in both cases the O atoms occupy fourfold sites 0.9 Å above the plane through the centres of the topmost Ni atoms. Interestingly Andersson (1976) found quite different vibration energies for both configurations, namely 53 meV for the p2 × 2 and 39.5 meV for the c2 × 2 structure. The simplest explanation would consist in again assuming the occupation of different adsorption sites which, however, would be in disagreement with the LEED results. If the geometry of the chemisorption bond remains constant the only possibility for the change of the vibration energy is to assume a considerable variation of the shape of the potential curve upon occupation of nearest-neighbour sites associated with the transition p2 × 2 → c2 × 2. An interpretation along these lines was proposed by Andersson who tried to connect the change of the potential curve with the relatively low activation barrier for bulk diffusion. Unfortunately no data on the variation of the oxygen adsorption energy with coverage are available which would certainly shed more light on this problem.

Very interesting results were found by Andersson (1977) for CO adsorbed on Ni(100). This system forms a c2 × 2 layer at $\theta = 0.5$ (Tracy 1972) for which vibrational losses at 59.5 and 256.5 meV were observed (fig. 5.18a). The latter is the well-known C–O stretching mode whereas the former corresponds to the vibration of the CO molecule against the surface. Its energy is quite similar to those of the corresponding modes in $Ni(CO)_4$ (46 and 52.5 meV, Jones et al. 1968), whereas that of the C–O stretching vibration is practically identical to that in $Ni(CO)_4$ (Jones et al. 1969). From these findings it was concluded that under the chosen conditions all CO molecules are located in identical "on-top" positions.

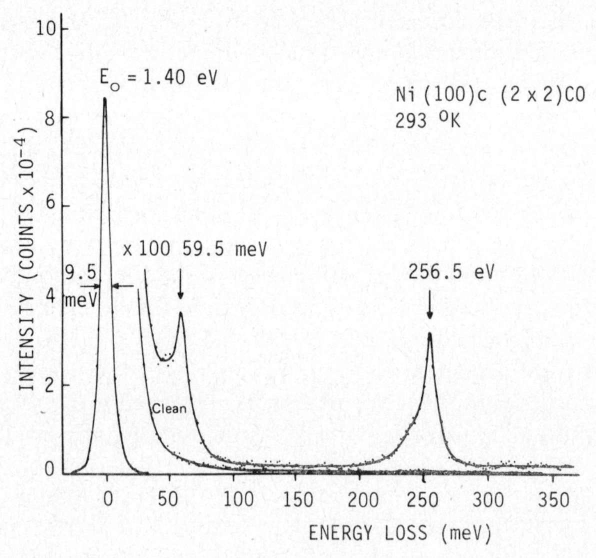

Fig. 5.18a. High resolution electron energy loss spectra for CO adsorbed on Ni(100), $\theta = 0.5$ (Andersson 1977).

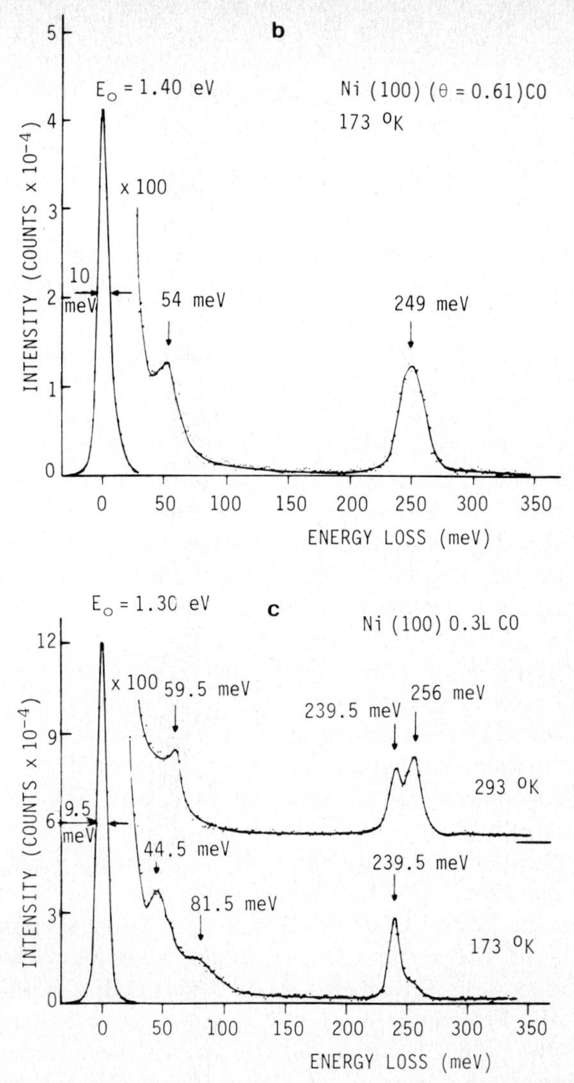

Fig. 5.18b,c. High resolution electron energy loss spectra for CO adsorbed on Ni(100), (b) $\theta \approx 0.6$, (c) $\theta < 0.5$, T = 173 K and 293 K (Andersson 1977).

This result is further remarkable in view of the close resemblance between complex chemistry and chemisorption which in this context also manifests itself obviously with respect to the shapes of the potential curves, apart from the binding energies themselves which were discussed in subsect. 2.1.

Saturation of the adsorbate layer at $\theta \approx 0.6$ is characterized by the formation of an incoherent hexagonal surface structure where the CO molecules are no longer in identical adsorption sites. This effect reflects itself in a broadening and shift of both loss peaks (fig. 5.18b) which now have to be considered as representing envelopes from continuously distributed contributions due to the occupation of a whole spectrum of adsorption sites.

Another interesting feature was observed at very low coverages prior to the formation of an ordered overlayer structure (fig. 5.18c). At 173K the C–O vibration now appears at 239.5 meV (instead of 256 meV at $\theta = 0.5$) and the excitation in the energy region of the M–C vibrations is split into a doublet at 44.5 and 81.5 meV. Both observations suggest that under these conditions the CO molecules are not located in "on-top" positions but rather in twofold (or probably even fourfold) coordinated sites. The splitting is ascribed to excitations of the symmetric and asymmetric metal–CO stretching vibration modes. This result indicates in fact not that the single-atom adsorption site is energetically most favourable (as indicated by the results at $\theta = 0.5$), but that interactions between adsorbed CO molecules are responsible for the occupation of "on-top" sites in the $c2 \times 2$ structure. That in fact the difference in adsorption energy ΔE between single and multiple coordinated sites is rather small becomes evident from experiments at 293 K (upper curve of fig. 5.18c): peaks at 239.5 and 256 meV of about equal intensity indicate that now comparable amounts of CO molecules are adsorbed (with the same total coverage as with the experiment at 173 K) in single and "bridge" (or fourfold) sites. This effect has to be ascribed to operation of Boltzmann statistics (through $\exp(-\Delta E/kT)$) and ΔE is estimated to be roughly of the order of 1 kcal/mole – an effect quite similar to that discussed for the system CO/Pt(111) in the preceding section.

Excitations of the metal–CO vibrations were also observed with W(100) ($\hbar\omega = 45$ meV, Froitzheim et al. 1976c) and with Pt(111) ($\hbar\omega =$ 58 and 45 meV, Froitzheim et al. 1977), indicating close similarities of the shape of the potential curves for CO adsorption on different metals, the force constants being mainly influenced by the local geometry of the chemisorption bond rather than by the crystallographic orientation or the chemical nature of the substrate metal.

8. Activation barriers

Dissociative chemisorption as illustrated schematically by the potential diagram of fig. 5.1 occurs very frequently with a high reaction probability and without any noticeable activation energy, i.e. with negligible temperature dependence. However in a few cases there is some experimental evidence for the existence of the sketched activation barrier.

Molecularly adsorbed nitrogen is held on iron surfaces with an adsorption energy of about 5–10 kcal/mole (Bozso et al. 1977), and Kishi and Roberts (1977) concluded, on the basis of XPS measurements, the existence of two different species, the molecular axis being either parallel or perpendicular to the surface. With Fe(100) and Fe(110) surfaces the formation of atomically adsorbed nitrogen proceeds with apparent activation energies of about 5 and 8 kcal/mole respectively, the metal–N bond energy being about 140 kcal/mole. These numbers give some indication about the energy quantities involved in fig. 5.1.

The molecular species may, of course, in principle be held even stronger as for example in the case of NO and CO. Then eventually desorption of the molecularly adsorbed species becomes faster than dissociation. With the system NO/Ni(111) the activation energy for dissociation was observed to be dependent on the oxygen coverage of the surface (Conrad et al. 1975). At low coverages the activation energy for dissociation is rather small so that adsorbed NO ($E_{ad} = 25$ kcal/mole) already dissociates below room temperature as followed by UPS. The energies of the metal–O and metal–N bonds are estimated to be in the order of about 100 and 135 kcal/mole. With NO/Pd(111) on the other hand E^* is considerably higher so that dissociation only takes place above about 200°C. The M–O and M–N bonds are now somewhat weaker (about 90 and 125 kcal/mole), whereas the adsorption energy for the molecular species is quite similar. Figure 5.19 illustrates how the observed increase of E^* with decreasing bond strength of the atomic

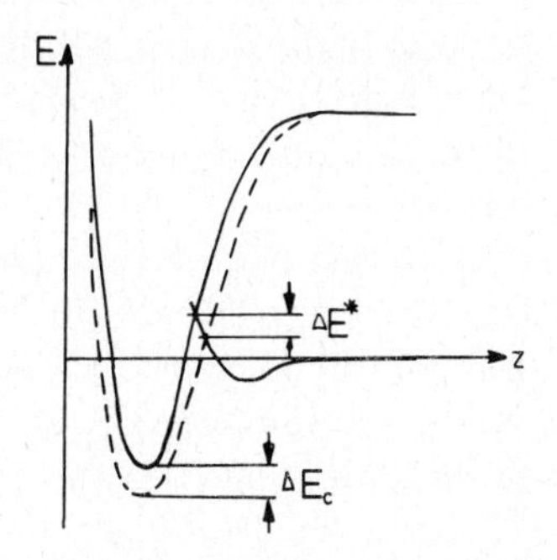

Fig. 5.19. Connection between the change of chemisorption energy, ΔE_c, and the variation of the activation energy for dissociation, ΔE^*.

species can qualitatively be understood. Such a behaviour resembles the linear dependence of activation energy on reaction energy, as for example found for proton transfer processes in acid–base catalysis (Brønsted relation) whose interpretation is based on quite similar arguments.

Perhaps the most detailed investigation on the mechanism of activated dissociative chemisorption was performed by Balooch et al. (1974) with hydrogen on different copper surfaces. By using a supersonic molecular beam the mean energy of the impinging H_2 molecules could be varied between 1.6 and 10.7 kcal/mole. The surfaces were simultaneously exposed to a highly dissociated beam of deuterium. HD, formed through recombination of $H_{ad} + D_{ad}$, and desorbing from the surface was detected mass spectroscopically and could thus be used as a measure for the rate of dissociative H_2 adsorption. Figure 5.20 shows the dissociative adsorption probabilities as a function of the component of the translational energy of the impinging H_2 molecules normal to the surface $E_{\perp}$ for three different Cu planes. In all cases this quantity starts to increase at about $E_{\perp} = 3$–4 kcal/mole. Qualitative agreement of these data could be achieved by a simple one-dimensional potential energy diagram as illustrated by fig. 5.21. A more realistic description has to use a two-dimensional distribution of potential curves, i.e. the activation barrier depends on the point within the substrate unit cell where the impinging molecule strikes the surface. The orientation of the molecular axis with the plane of surface atoms is presumably a further complicating factor unless its lifetime in the physisorbed (molecular) state is not large compared to the characteristic time constant for rotational reorientation. An even more severe simplification in the present model arises from the fact that the one-dimensional potential diagram refers to two adsorbed hydrogen atoms on identical sites. However the interatomic distance of H_2 (0.75 Å) is much smaller than that of the surface atoms (2.5 Å) so that this condition will only be fulfilled for rather special configurations of the incident H_2 molecule with respect to the geometry of the surface atoms.

Another interesting result in this context was the observation that the angular distribution of HD molecules desorbing from a Cu(310) surface exhibited a maximum normal to the (310) plane rather than to (100) terraces. Similarly the angular distribution of scattered H_2 was peaked at the specular angle measured with respect to the ⟨310⟩ cut of the crystal as illustrated by fig. 5.22. These results suggested that the equipotential lines for the repulsive portion of the potential are approximately parallel to the (310) plane at distances farther from the surface than the crossing

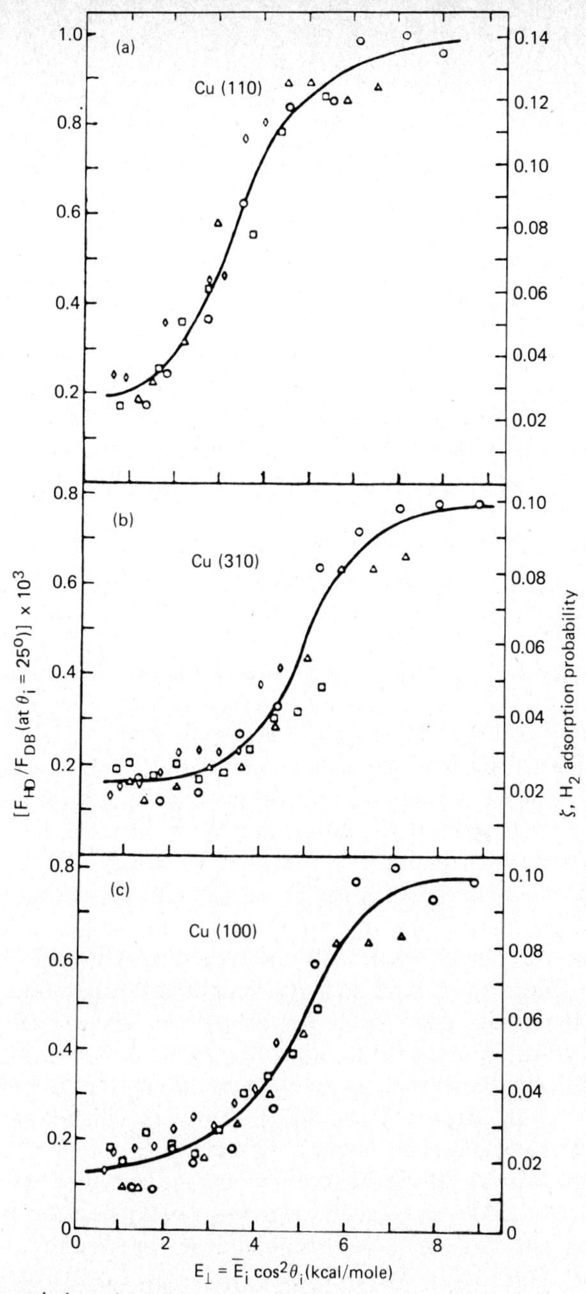

Fig. 5.20. Dissociative adsorption probabilities for H_2, ζ_{H_2}, (right ordinate) for Cu(110), (100), and (310) planes as a function of the translational energy normal to the surface $E_\perp$ of the impinging hydrogen molecules (Balooch et al. 1974).

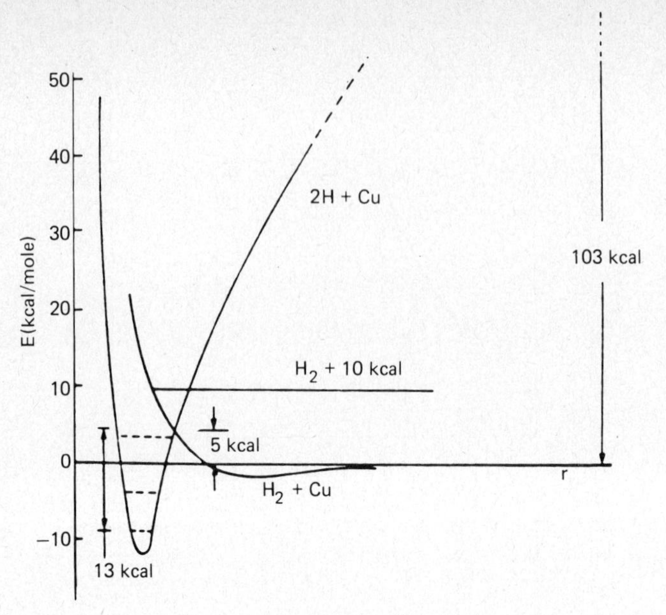

Fig. 5.21. Schematic one-dimensional potentials for the interaction of hydrogen with copper (Balooch et al. 1974).

points of the atomic and molecular potentials as drawn schematically in fig. 5.22. This idea is supported by previous studies with Cu(110) where the angular distribution of desorbing H_2 molecules did not depend significantly on the azimuthal orientation (Balooch and Stickney 1974).

The adsorption probability for D_2 on Cu(110) was found to be higher by about a factor of 1.4 $[=(m_{D_2}/m_{H_2})^{1/2}]$ compared to that for H_2 (Balooch et al. 1974). A similar result was obtained for hydrogen adsorption on W(100) by Tamm and Schmidt (1970). This could simply be a result of the fact that the time spent by the hydrogen molecule near the surface where dissociation is possible is proportional to the square root of its mass. Also the density of available vibrational states varies with the square root of the mass and would probably account for this effect.

Much more pronounced isotope effects were observed with the dissociative adsorption of methane on tungsten. In the temperature range between about 1000 and 2000°C Winters (1976) derived the following ratios of the adsorption probabilities (= sticking coefficients): $s(CD_3H)/s(CD_4) = 1.7$, $s(CD_2H_2)/s(CD_4) = 2.7$, $s(CDH_3)/s(CD_4) = 3.5$, and $s(CH_4)/s(CD_4) = 4.5$. These ratios were found to exhibit an unexpectedly small temperature dependence, whereas the apparent activa-

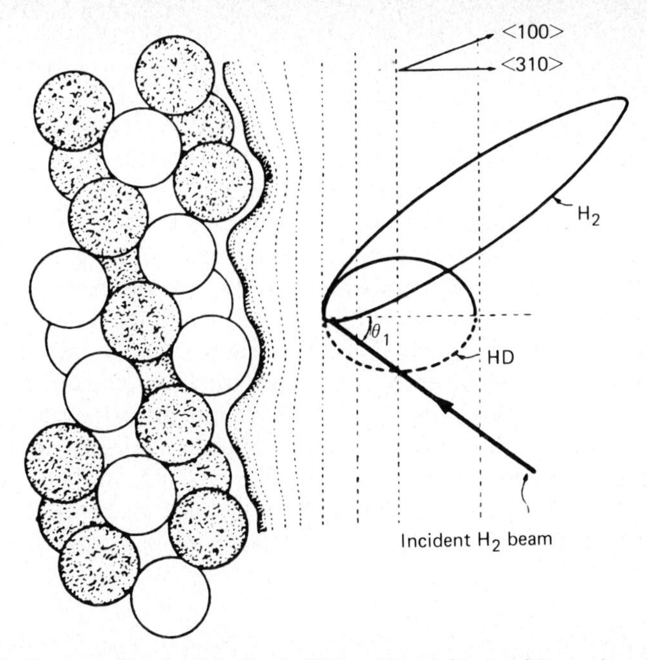

Fig. 5.22. Cross section through the Cu(310) surface with possible equipotentials for the H_2 interaction potential. The angular distributions of reflected and desorbing hydrogen molecules are indicated in polar coordinates (Balooch et al. 1974).

tion energies were all 9.3 ± 1 kcal/mole. In order to interpret this unexpected behaviour a model was proposed where besides vibrational excitation of the methane molecule tunnelling of a hydrogen atom through the potential barrier plays an important role. Model calculations then revealed that the kinetic isotope effect can essentially be accounted for by the tunnelling process. It will be very interesting to see whether further evidence for such a mechanism will be found. On the other hand, however, Stewart and Ehrlich (1975) observed a similar isotope effect with the dissociative adsorption of CH_4 and CD_4 on rhodium at $T < 700$ K and explained their results within the framework of Slater's theory of unimolecular reactions without assuming the operation of a tunnel effect. This model, however, does not account for Winters' results, since it predicts a strong temperature dependence for the ratio $s(CH_4)/s(CD_4)$ which was not observed.

9. Interactions between adsorbed particles

The discussion so far was concerned with the properties of the single-particle chemisorption bond and therefore strictly holds only for the limit of zero coverage. At finite coverages additional effects come into play, the most obvious of which is that a certain point on the surface can not be occupied by more than one chemisorbed particle. As outlined in the introductory section the most convenient (and also from the experimental point of view justified) means of treatment consists in splitting the multi-particle system into properties inherent in the single-particle chemisorption bond and into effects caused by interactions between adsorbates which then may be regarded as a perturbation of the potential as sketched by fig. 5.3. This section discusses the physical origin of these interactions and some of the available information, whereas the remainder of this chapter will be devoted to their consequences on structural and thermodynamic properties of chemisorption systems.

9.1. Dipole–dipole interactions

Since chemisorption always involves different species (namely the surface atoms and the adsorbate) a perfectly covalent bond is never to be expected. Only in exceptional cases will the centres of positive and negative charge be located in the same plane parallel to the surface so that usually a dipole layer will be built up. The effective dipole moment μ of an individual adsorbate complex can easily be derived from the measurable change of the work function $\Delta\phi$ through the relation $\Delta\phi = 4\pi n_s \mu$, where n_s is the coverage in particles/cm^2, $\Delta\phi$ and μ being given in electrostatic units. The dipole moment μ is in turn related with the dipole length d and the charge separation q through $\mu = qd$. Estimates for these quantities are given in chapter 4, but it has to be kept in mind that such an analysis of dipole moments is very questionable since usually the centres of charge distributions will not be identical with the locations of the atomic nuclei.

As a consequence of the build-up of a dipole layer, pairwise repulsive interaction between adsorbed particles will always occur, which is given by $U = \mu^2/r^3$. The summation $W = \frac{1}{2}\Sigma\, \mu^2/r^3$ over all interacting dipoles per unit area within a uniform overlayer was performed by Topping (1927) leading to $W = \frac{1}{2} n_s^{5/2} \mu^2 \kappa$, where $\kappa \approx 9$ is a factor which varies only very slightly with the geometric arrangement of the adsorbates. The repulsive interaction per adsorbed particle is thus given by $w \approx 4.5 n_s^{3/2} \mu^2$

[erg]. This effect should for example cause a continuous decrease of the (effective) differential adsorption energy with increasing coverage. Estimates for typical systems however demonstrate that this effect is expected to be rather small: CO adsorption on Pd(111) causes a change of the work function of $\Delta\phi = 1$ eV at $n_s = 10^{15}$ cm^{-2} (Ertl and Koch 1970) leading to $w = 0.14$ kcal/mole. The dipole moments of non-metallic adsorbates are usually even smaller than with this example so that the influence of dipole–dipole interactions on the energetics is practically negligible.

The situation is somewhat different in the case of alkali metal adsorption: Gerlach and Rhodin (1970) for example determined $\Delta\phi = -2.55$ V at $n_s = 1.5 \times 10^{15}$ cm^{-2} with the system Na/Ni(100) from which $w = 0.75$ kcal/mole is derived. This value is still very small and would probably be within the limits of error for measurements of the adsorption energy, but it is already in the order of kT (at room temperature) and comparable to typical values for the indirect interaction energies between adsorbed particles. As a consequence the dipole–dipole interactions in these systems may influence the geometric arrangement of the adsorbed particles on the surface as discussed for example by Carroll and May (1972).

A further consequence of dipole–dipole interactions between adsorbates is the mutual depolarization of the dipole moments leading to a non-linear variation of the work function change with coverage. This phenomenon may be described again by a classical electrostatic model (Topping 1927) leading to $\Delta\phi = (4\pi\mu_0 n_s/(1 + q\alpha n_s^{3/2}))$, where μ_0 is the dipole moment at zero coverage and α the polarizability. This formula probably describes the experimental situation rather well in cases of physisorption such as Xe/Pd(100) (Palmberg 1971) but is not satisfactory for chemisorption systems where the degree of charge transfer is of vital importance (Schmidt and Gomer 1965, 1966). Microscopic theories such as developed by Lang (1971) for alkali adsorption on jellium are sometimes able to explain at least the qualitative features. It has to be mentioned that in some cases [H/W(110), Barford and Rye (1974); H/Pt(111), Christmann et al. (1976)] the average dipole moment even increases with coverage. A physical explanation of this effect is still missing.

9.2. Orbital overlap

One might at first assume that the chemisorption complexes themselves still contain some degree of "unsaturated" valence, that means that they are able to interact with other particles to form chemical bonds by means of orbital overlap. This process is certainly the basis of heterogeneous catalysis by which, however, the chemical nature of the adsorbate and its chemisorption bond is completely altered and whose treatment would be outside the scope of this chapter. Apart from this effect, however, "normal" adsorbate complexes may be regarded to be saturated, i.e. there are no strong attractive forces between neighbouring particles over short distances nor any evidence for the formation of strongly held second layers on top of the monolayers. The direct interactions between adsorbates via orbital overlap then depends in detail of course on the orbital geometry and population but exhibits presumably characteristic features similar to those describing the interactions between free stable molecules (such as CO or N_2) or noble gas atoms. These interactions are strongly repulsive on a short-range scale and will ultimately limit the closest possible approach between neighbouring adsorbates and therefore their maximum coverage, unless this effect is not dominated by indirect interactions (as is certainly the case with strongly held atomic adsorbates).

The problem of deriving the interaction potential was discussed in the literature to some extent for physisorbed systems whose general features can presumably also be adopted for chemisorbed layers. The most convenient starting point consists in the assumption of a Lennard-Jones potential of the well-known form $U(r) = 4\varepsilon[-(r_0/r)^6 + (r_0/r)^{12}]$ where the parameters ε and r_0 are taken from corresponding gas-phase data. It was found that ε is probably smaller (Sams et al. 1962) and r_0 slightly larger (Barker and Everett 1962) than in the gas phase. More refined treatments (Sinanoglu and Pitzer 1960, McLachlan 1964) revealed that probably another repulsive interaction ($\sim r^{-3}$) has to be included for short distances, but Everett (1965) demonstrated that a rigorous application of these refined theories should cause much stronger perturbations than those deduced from experiment, so that a (modified) Lennard-Jones potential will probably still reveal the best description.

In the case of chemisorption there is strong evidence that this type of interaction dominates the adsorption of CO on transition metals at high coverages: as outlined in sect. 6 with this adsorbate the surfaces appear to be energetically rather smooth enabling the formation of incoherent structures with saturation densities of about 1×10^{15} molecules/cm^2, irrespective of the chemical nature and the crystallographic orientation of the surface. This indicates that adsorbed CO molecules may to a good approximation be considered as particles occupying circular areas with about 3 Å diameter on the surface and recalls observation with polynuclear carbonyl complexes where the ligands tend to form a densely packed layer around the central cluster (see subsect. 2.1).

Tracy and Palmberg (1969) were the first to derive a pairwise interaction potential as a function of distance for CO adsorbed on Pd(100) from the experimentally evaluated variation of the heat of adsorption and the (from LEED observations) known mutual configurations. They found fairly good agreement with the interaction potential between gaseous CO molecules. Their results are shown in fig. 5.23 and indicate that the adsorbed CO molecule is slightly "bigger" than in the gaseous state. Similar conclusions were reached by Ertl and Koch (1970) with the system CO/Pd(111).

So far CO is the only chemisorbed molecule where the operation of direct orbital-overlap interactions has been established. The chemisorp-

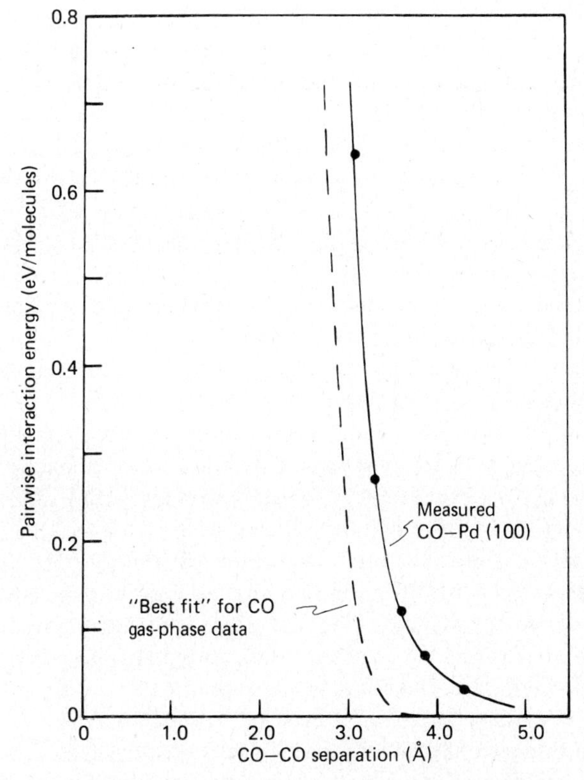

Fig. 5.23. Pairwise interaction potential between CO molecules adsorbed on Pd(100) as a function of their mutual distance, and a comparison with gas-phase data (Tracy and Palmberg 1969).

tion bond is relatively weak and this is probably the reason why there is no indication of a "saturation" of free surface valencies with increasing coverage. With all other adsorbates the situation is more complex, i.e. it has to be assumed that indirect interactions play a dominant role in determining the maximum coverage as well as the variation of the adsorption energy with surface concentration.

9.3. Indirect interactions

If chemisorption of a molecule involves coupling not only to a single metal atom (as usually will be the case) then the binding properties of the neighbouring sites (with identical geometry) will be altered, i.e. neighbouring adsorbed particles will interact with each other in an indirect manner through their bond with the metal surface. This concept of through-bond interaction has also been successfully applied in molecular chemistry for several years and can conveniently be described there in terms of a LCAO–MO picture (Hoffmann 1971). The situation found with chemisorption (of equal particles) on metal surfaces was treated theoretically in different ways; these theoretical aspects are outlined in ch. 1. The general results from model calculations are as follows:

(1) The interactions exhibit an oscillatory character, i.e. they may be attractive as well as repulsive.

(2) Their values are about one order of magnitude smaller than the strength of the chemisorption bond itself.

(3) They decay within distances of very few (2–3) lattice constants to values below kT.

Successful attempts for a direct observation of the operation of these types of interactions were performed by field ion microscopy: Graham and Ehrlich (1973) observed for W atoms adsorbed in alternate [111]-rows of a W(211) surface (mutual distance $d = 4.48$ Å) a correlated motion of pairs and a strong reduction of the barrier for surface diffusion compared with the situation for isolated atoms, leading to the conclusion of the existence of relatively strong attractive forces. The same authors (Graham and Ehrlich 1974) found that for distances >7 Å the interaction energies became smaller than $\frac{1}{2}kT$, i.e. they are rapidly decaying.

The formation of ordered adsorbed layers as observed with LEED is frequently a consequence of the operation of indirect interactions and in turn the analysis of LEED data provides a means for deriving the interaction parameters, as will be outlined in sect. 10. So far very little quantitative work has been performed in this direction. As an example fig. 5.24 shows (pairwise) interaction potentials in two different crystallographic directions between oxygen atoms on W(211) as derived from a simulation of experimental LEED data (Ertl and Schillinger 1977). Although the shapes of the potential curves are only estimates (merely the interaction energies between distinct lattice points were derived) the general features of indirect interactions as outlined above become clearly evident.

Of course this type of interaction is not restricted to equal particles, but exhibits very interesting aspects in the case of different particles: co-adsorption of a second species may either increase or decrease the effective adsorption energy, unless no phase separation occurs on the surface. For example admission of CO to a Ni(111) surface precovered with hydrogen causes either a lowering or an increase of the desorption energy of H_2 (by a few kcal/mole), depending on coverage (Conrad et al. 1977a), indicating the complex nature of these phenomena.

Indirect interactions between different species may also cause variations of their electronic properties. If CO is adsorbed on a Ni(111) surface precovered with oxygen the ionization energy of the CO-4σ level is increased by about 0.3 eV if compared with the data for CO adsorption on a clean surface (Conrad et al. 1976a). Interestingly this orbital is essentially not involved in the chemisorption bond at all, but mainly localized at the O-atom which is directed away from the surface. The simplest explanation for this effect is illustrated by fig. 5.25, where electronic charge is partially transferred from the 4σ-orbital through the CO-molecule and the substrate to a neighbouring adsorbed oxygen atom

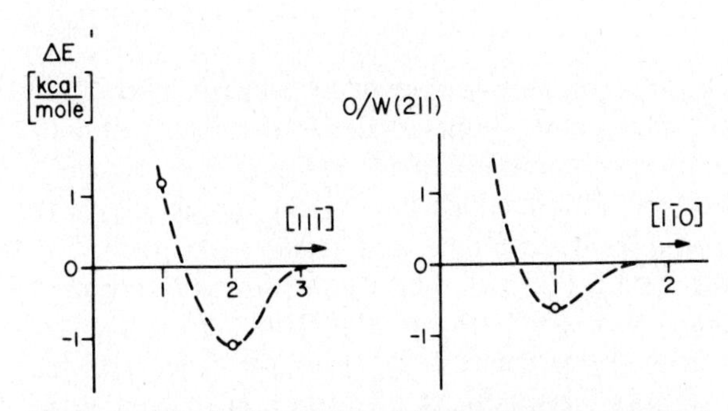

Fig. 5.24. Pairwise interaction potentials between adsorbed oxygen atoms in two directions on a W(211) surface as a function of the number of lattice spacings (Ertl and Schillinger 1977).

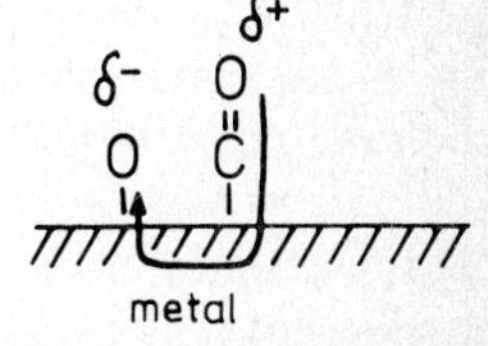

Fig. 5.25. Model for partial charge transfer for $O_{ad} + CO_{ad}$ co-adsorbed on Ni(111) (Ertl 1977).

with its higher electronegativity. The desorption energy of CO is lowered by about 3 kcal/mole by the presence of neighbouring adsorbed oxygen atoms.

Effects of this type are not singular and are also quite common in molecular chemistry where they are often described in terms of Lewis acid–base or electron pair donor–acceptor interactions. Empirical principles correlating these effects with the degree of charge transfer, the variation of bond lengths etc. have for example been developed by Gutman (1975) and will in principle certainly also hold for the chemistry on metal surfaces.

10. Ordered adsorbed phases and thermodynamic properties

The model underlying the deviation of the Langmuir adsorption isotherm assumes that the adsorbed particles occupy the lattice points of the two-dimensional substrate with equal probability, with hard-wall potentials between adsorbed particles preventing double (or multiple) occupancy of any particular site. As a result saturation would be characterized by $\theta = 1$ and the formation of a true 1×1 adsorbate layer. No superstructure should be formed at $\theta < 1$ but instead always a random distribution of the adsorbed particles over the substrate lattice should occur. The observation of additional LEED spots, however, demonstrates that the formation of ordered adsorbed layers (with periodicities deviating from those of the substrate lattice) are more the rule than the exception. Obviously this effect is a direct consequence of the existence of interaction potentials of the type shown in fig. 5.23 (apart from the possibility that the surface periodicity may be changed by reconstruction).

Figure 5.3 illustrates how the superposition of a pairwise interaction potential (a) with the periodic a priori single-particle potentials (full lines in b and c) determine the effective potential (dashed lines) of a particle B in the vicinity of particle A which is located in the origin of these diagrams. Obviously two different situations may result:
(i) The nearest-neighbour distance is governed by the interaction potential and not by the lattice periodicity (fig. 5.3c). This will cause the formation of incoherent adlayer structures as for example observed with adsorbed CO at high coverages (see sects. 8, 9) or with alkali metals on Ni surfaces (Gerlach and Rhodin 1968, 1969).
(ii) The potential minima and therefore the locations of the adsorbed particles are still determined by the surface periodicity. If the substrate

unit cell contains only one preferred adsorption site the adsorbed phase may be described by a "lattice gas" model. This model and its consequences will now be discussed in some more detail and may be defined as follows:
(a) The lattice of adsorption sites with equal a priori binding energies ε_i is identical with the substrate lattice, i.e. is described by $\boldsymbol{r}_i = n\boldsymbol{a} + m\boldsymbol{b}$ where n, m = integers and $\boldsymbol{a}$, $\boldsymbol{b}$ the vectors defining the unit cell.
(b) The effective binding energy of an adsorbed particle is modified by pairwise interactions described by an interaction potential $\varphi(\boldsymbol{r}_i - \boldsymbol{r}_j)$. The occupation of an adsorption site i may be defined by an occupation number b_i: $b_i = 1$ if the site is occupied, and $b_i = 0$ if the site is empty. The total coverage θ is then given by $\theta = \langle b_i \rangle$.
The total energy of this system may then be described by

$$E = \sum_i \varepsilon_i b_i + \sum_{i \neq j} b_i b_j \varphi(\boldsymbol{r}_i - \boldsymbol{r}_j).$$

Contributions from vibrational and translational excitations are irrelevant in the present context and are therefore neglected.

Evidently the energy of the system now depends on the mutual configuration of the adsorbed particles and therefore no longer the configuration with maximum entropy (= random distribution) will characterize the equilibrium. As a result long-range order may occur, depending on the (microscopic) interaction potential $\varphi(\boldsymbol{r}_i - \boldsymbol{r}_j)$ and the thermal energy kT, which may be treated with the methods of statistical thermodynamics.

If $\varphi(\boldsymbol{r}_i - \boldsymbol{r}_j)$ is known, the equilibrium configuration at given temperature and coverage may be computer simulated by means of the Monte Carlo technique (Ertl and Küppers 1970). The simplest system is represented by a square lattice and where only repulsive interactions of magnitude ε between particles on neighbouring sites are operating. At $\theta = 0.5$ and $T = 0$ K the adsorbed particle would obviously form a perfect c2 × 2 structure in equilibrium. At finite temperatures the statistical properties of this system may be described by the two-dimensional Ising model leading to an order–disorder transition at a critical temperature T_c (Doyen et al. 1975). This breakdown of long-range order manifests itself by a decrease of the intensity of the half-order LEED spots whose "configurational" part is given (within the kinematic approximation) by

$$I_{\text{conf}}(\tfrac{1}{2}\tfrac{1}{2}) - c \sum_i \sum_j \langle b_i b_j \rangle \exp[\mathrm{i}\, \Delta \boldsymbol{k}(\boldsymbol{r}_i - \boldsymbol{r}_j)].$$

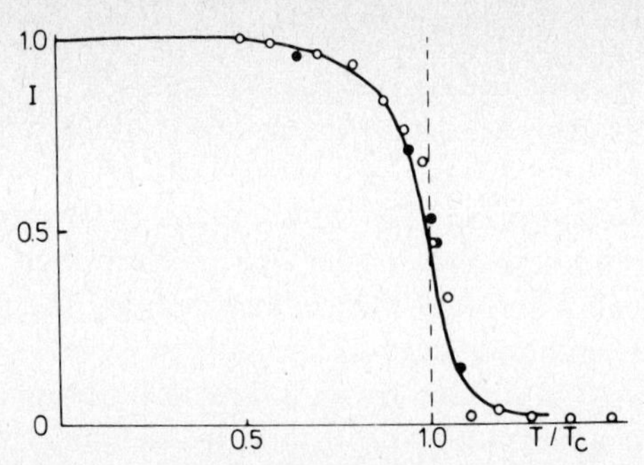

Fig. 5.26. Order–disorder transition for H/W (100) at $\theta = 0.5$. Intensity of the half-order LEED spots from the c2 × 2 structure as a function of temperature ($T_c = 500$ K). Full circles: experimental data (Estrup 1969); open circles: results from Monte Carlo calculations; full line: analytical solution (Doyen et al. 1975).

$\boldsymbol{\Delta k}$ is the difference between the wave vectors of the scattered and primary electron beams. The summation may be performed from a knowledge of the occupation numbers b_i (e.g. resulting from the computer simulation of the equilibrium configuration at a given ε/kT or from the analytic solution of the Ising problem). The finite lattice size and coherence width may additionally be taken into account. The resulting variation of $I(\frac{1}{2}\frac{1}{2})$ with temperature is reproduced in fig. 5.26 as a function of T/T_c. An adjustment was made to $\varepsilon = 1.76\,kT_c$ so as to attain agreement with experimental results by Estrup (1969) for the c2 × 2 structure formed by hydrogen adsorption on W(100). From $T_c = 500$ K a repulsive interaction energy $\varepsilon = 1.75$ kcal/mole results which is about 5% of the adsorption energy.

A more general treatment of adsorbate phase transitions was recently performed by Binder and Landau (1976). Figure 5.27a illustrates the phase diagram for the just discussed case of a square lattice with nearest-neighbour repulsions. Evidently the transition temperature is highest for $\theta = 0.5$ whereas for $\theta \leqslant 0.33$ and $\theta \geqslant 0.66$ no long-range order at all will be possible.

If besides these nearest neighbour (nn) interactions additional next nearest neighbour (nnn) interactions are included the situation may become much more complicated. Figure 5.27b shows the phase diagram resulting for $\varepsilon_{nnn}/\varepsilon_{nn} = 0.25$. Besides the 2 × 2 structure (I) two additional p2 × 2 structures (II) and (III) may be formed, and the phase diagram contains additional triple and tricritical points. If the fact is taken into account that besides "lattice gas" overlayers in principle also incoherent structures may be formed it becomes quite evident that realistic systems

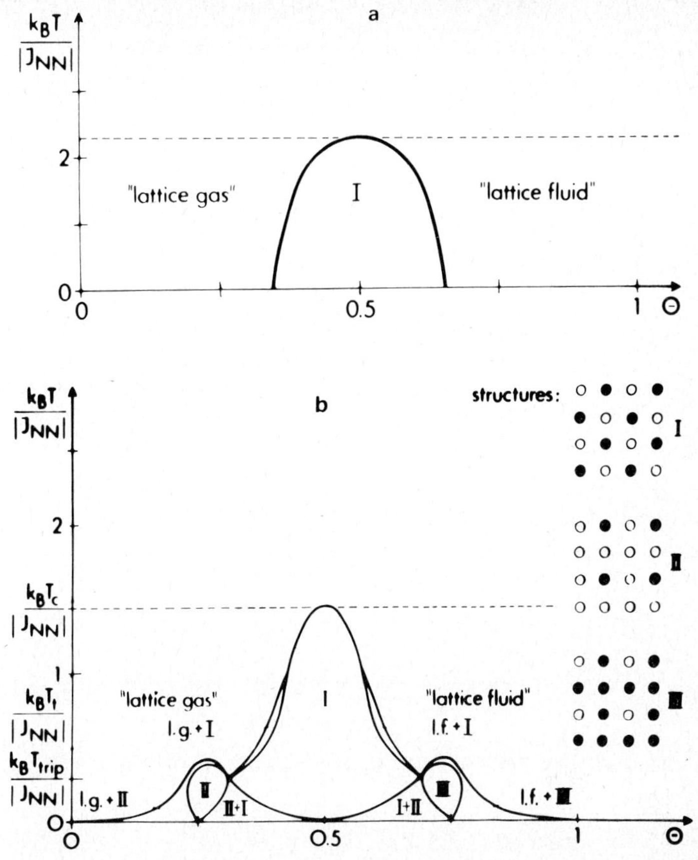

Fig. 5.27. Theoretical phase diagrams for adsorbed layers on a square lattice. (a) $\varepsilon_{nn} > 0$ (only repulsive interactions between nearest neighbours). (b) $\varepsilon_{nnn}/\varepsilon_{nn} = 0.25$ (repulsive interactions between nearest and next-nearest neighbours) (Binder and Landau 1976).

may exhibit very complicated phase diagrams and a great variety of ordered and disordered structures.

Experimental determinations of phase diagrams for adsorbed layers are so far rather scarce. Figure 5.28 shows a diagram derived by Tracy (1972) for the system CO/Ni(100) from careful LEED observations. The boundaries of the two-phase region (c2 × 2 and hexagonal (incoherent) structure) and the possible occurrence of a triple point are uncertain but nevertheless this example provides a nice insight into the two-dimensional structural properties of an adsorbate system.

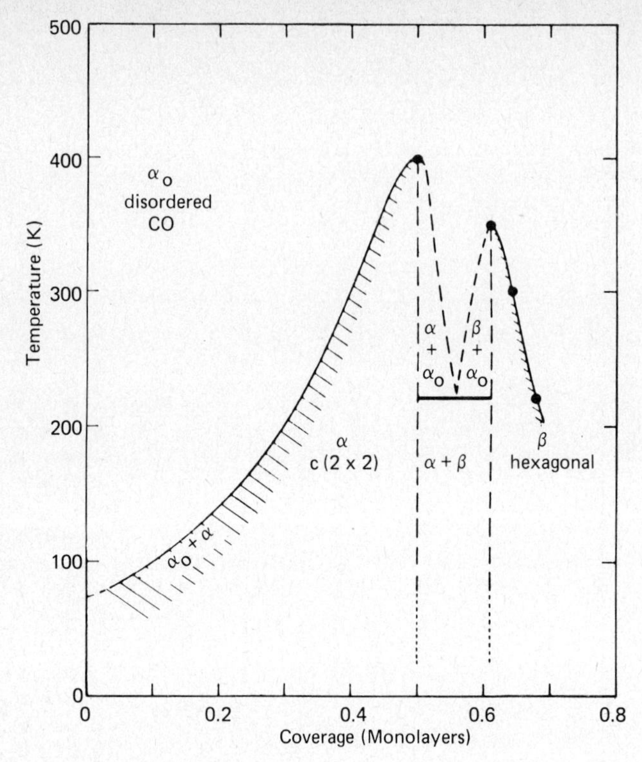

Fig. 5.28. Experimental phase diagram for CO/Ni(100) (Tracy 1972).

Figure 5.26 illustrates the simplest case of how quantitative information on the interaction energy between neighbouring adsorbates may be derived from an analysis of LEED intensities at varying temperature. In more general cases at least the signs (and eventually relative magnitudes) of the interactions in different surface directions and even beyond nearest neighbours may often be determined from an inspection of the adsorbate unit cells. Quantitative data may then be obtained by a trial-and-error procedure: by assuming certain values for the ε_{ij} the equilibrium configuration is computer simulated by the Monte Carlo technique and the resulting diffraction intensity and profile (i.e. beam width) compared with corresponding LEED data until agreement between theory and experiment is achieved (Ertl and Plancher 1975). The interaction potentials reproduced in fig. 5.23 were obtained in this way. However it has to be kept in mind that the results usually will not be unique but rather provide a means for describing the experimental findings with a minimum set of adjustable parameters in terms of a microscopic model.

Additional features come into play if the adsorbed layer consists of more than a single component. If two different species A and B are present the interaction energies ε_{AA}, ε_{BB} and ε_{AB} have to be taken into consideration and the sign and magnitude of $\Delta\varepsilon = \varepsilon_{AA} + \varepsilon_{BB} - 2\varepsilon_{AB}$ determines, as in the thermodynamics of "ordinary" mixed phases, the shape of the phase diagram. The following limiting cases may be distinguished (Ertl 1969):

(a) If $\Delta\varepsilon \gg kT$ the formation of a regular mixed phase will take place (cooperative adsorption) as frequently characterized by the appearance of new surface periodicities and therefore new LEED patterns. As a consequence also the effective adsorption energies of A and B may be altered, as well as their dipole moments or orbital ionization energies. The system O + CO/Ni(111) offers an example of this kind where all these features were observed (Conrad et al. 1976).

(b) If $\Delta\varepsilon \ll kT$ both species will not be miscible but rather coexist on the surface in separate domains (competitive adsorption). If the mean domain size exceeds the coherence width of the electrons used with the LEED experiments the diffraction pattern exhibits a superposition of spots arising from both ordered phases. Obviously only the adparticles located at the boundaries of the islands may now be influenced by the presence of a second species.

The creation of an induced heterogeneity by the operation of interactions between adsorbed particles has of course also consequences for the thermodynamic properties of adsorbed layers on single crystal surfaces: the differential adsorption energy is no longer independent of coverage and adsorption isotherms do not follow the simple Langmuir equation.

In the case of the lattice gas model (i.e. all particles occupy sites with identical symmetry of the substrate atoms) the variation of the adsorption energy with coverage will be given by

$$E_{ad} = E_{ad}^0 - \sum_n \langle b_n \rangle \varepsilon_n,$$

where the index n refers to the different pairs of possible interactions ($n = 1$: nearest neighbours, $n = 2$: next nearest neighbours etc.) $\langle b_n \rangle$ is the probability that neighbour n is occupied and ε_n is the corresponding interaction energy. In the case of repulsive interactions ($\varepsilon_n > 0$) evidently E_{ad} will continuously decrease with increasing coverage, which is also observed very often experimentally. The detailed shape of the

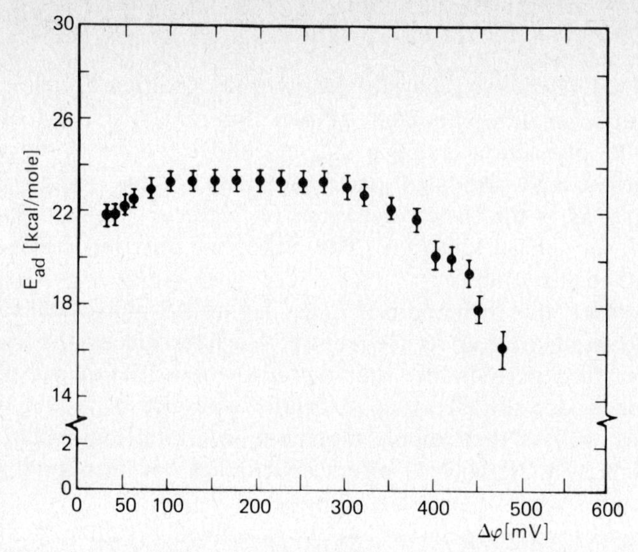

Fig. 5.29. Isosteric heat of adsorption for H/Ni(110) as a function of the work function change $\Delta\varphi$ (Christmann et al. 1974b).

curve $E_{ad}(\theta)$ however needs a knowledge of ε_n as well as of $\langle b_n \rangle$, the latter quantity in turn being determined by the statistical properties of the system. A square lattice with only repulsive interactions ε between nearest neighbours is again the simplest possible case and was treated by Wang (1937). As a result E_{ad} vs. θ should exhibit a sigmoid form, the detailed shape being dependent on ε/kT.

If attractive interactions are dominating, E_{ad} should increase with θ at low coverages and the adsorbed particles should condense to islands on the surface. Such an effect was observed with the system H/Ni(110): fig. 5.29 shows a plot of E_{ad} as a function of the work function change $\Delta\phi$ which served as a relative measure of the coverage.

Although theoretical models for adsorption isotherms under the influence of adsorbate interactions were extensively discussed in the literature (see e.g. Clark 1970, Honig 1961, Steele 1974), so far nearly no experimental data for chemisorbed systems on well-defined single-crystal surfaces are available. An interesting situation again occurs in the case of attractive interactions between nearest neighbours for which case a proper description is given by the Fowler (or Frumkin) isotherm (Fowler and Guggenheim 1956)

$$p = p_0 \frac{\theta}{1-\theta} \exp(z\varepsilon\theta/kT),$$

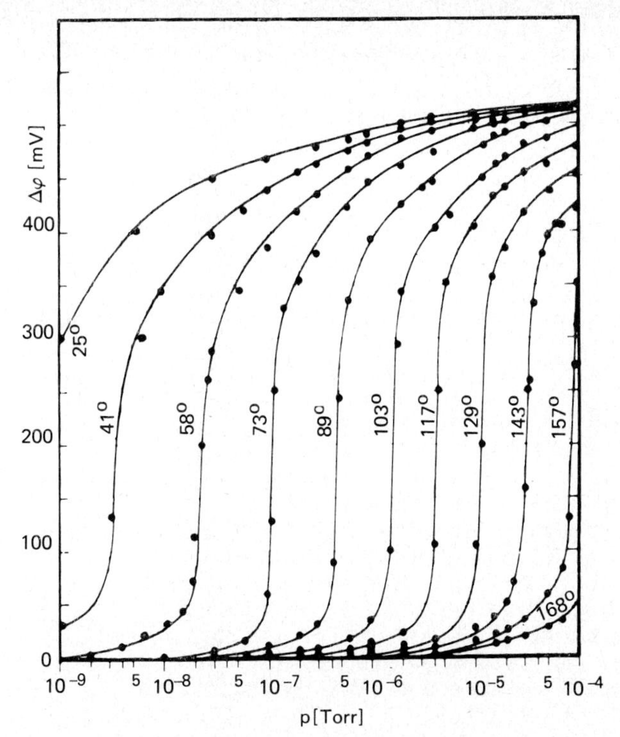

Fig. 5.30. Adsorption isotherms ($\Delta\varphi$ vs. H_2 pressure) for H/Ni(110) (Christmann et al. 1974b).

where z is the number of nearest-neighbour sites surrounding each adsorption site. When $z\varepsilon < 4kT$, i.e. the pairwise attraction is of the order of 1 kcal/mole or more, this equation predicts discontinuities in the coverage due to two-dimensional phase transitions by increasing the pressure beyond critical values. The physical adsorption of Xe on (0001) graphite as followed by Auger spectroscopy is a nice example for this type of behaviour (Suzanne et al. 1973). The only example for a chemisorption system of which I am so far aware is presented by the system H/Ni(110) where the initial increase of the adsorption energy with coverage by about 2 kcal/mole (fig. 5.29) suggests a pairwise attractive interaction energy of the necessary order of magnitude. A set of corresponding adsorption isotherms is reproduced in fig. 5.30 and indicates a two-dimensional "gas–solid" transition, the latter being characterized by the appearance of a 1×2 LEED pattern (Christmann et al. 1974b).

Finally it should be noticed that many characteristic features of ordered adsorbed layers and their thermodynamic properties as they become evident by LEED were already recognized by Lander (1964) in the earlier period of modern surface physics.

References

Adams, D.M., 1967, Metal ligand and related vibrations (Arnold) p. 84.
Adams, D.L. and L.H. Germer, 1971, Surface Sci. **27**, 21.
Albano, V.G., A. Ceriotti, P. Chini, G. Ciami, S. Martinengo and M. Auber, 1975, JCS Chem. Comm. 859.
Andersson, S., 1976, Solid State Comm. **20**, 229.
Andersson, S., 1977, Solid State Comm. **21**, 75.
Araki, M. and V. Ponec, 1976, J. Catal. **44**, 439.
Ayrault, G. and G. Ehrlich, 1974, J. Chem. Phys. **60**, 281.
Backx, C., B. Feuerbacher, B. Fitton and R.F. Willis, 1977, Surface Sci. **63**, 193.
Balandin, A.A., 1929a, Z. Phys. Chem. **B2**, 289.
Balandin, A.A., 1929b, Z. Phys. Chem. **B3**, 167.
Balooch, M. and R.E. Stickney, 1974, Surface Sci. **44**, 310.
Balooch, M., M.J. Cardillo, D.R. Miller and R.E. Stickney, 1974, Surface Sci. **46**, 358.
Barford, B.D. and R.R. Rye, 1974, J. Chem. Phys. **60**, 1046.
Barker, J.A. and D.H. Everett, 1962, Trans. Faraday Soc. **58**, 1608.
Bassett, D.W. and M.J. Parsley, 1970, J. Phys. **D3**, 707.
Benton, A.F. and T.A. White, 1930, J. Am. Chem. Soc. **52**, 2325.
Bernasek, S.L. and G.A. Somorjai, 1975, J. Chem. Phys. **62**, 3149.
Binder, K. and D.P. Landau, 1976, Surface Sci. **61**, 577.
Blyholder, G., 1964a, J. Phys. Chem. **68**, 2772.
Blyholder, G., 1964b, Proc. 3rd Int. Cong. on Catalysis (North-Holland, Amsterdam) p. 657.
Blyholder, G., 1974, J. Vac. Sci. Techn. **11**, 865.
Bond, G.C., 1962, Catalysis by metals (Academic Press, New York).
Bonzel, H.P., 1973, Structure and properties of metal surfaces (Maruzen Company, Tokyo) p. 247.
Bonzel, H.P. and R. Ku, 1973, Surface Sci. **40**, 85.
Bonzel, H.P. and E.E. Latta, 1977, Phys. Rev. Lett. **38**, 839.
Boudart, M., 1969, Advances in Catalysis, **20**, 153.
Bouwman, R., L.H. Toneman, M.A.M. Boersma and R.A. van Santen, 1976, Surface Sci. **59**, 72.
Bozso, F., G. Ertl, M. Grunze and M. Weiss, 1977a, Appl. Surface Sci. **1**, 103.
Bozso, F., G. Ertl, M. Grunze and M. Weiss, 1977b, J. Catalysis **49**, 18.
Bozso, F., G. Ertl and M. Weiss, 1977c, J. Catalysis **50**, 519.
Bradshaw, A.M. and J. Pritchard, 1970, Proc. Roy. Soc. **A316**, 169.
Brown, D.L.S., J.A. Connor and H.A. Skinner, 1975, JCS Faraday I, **71**, 699.
Brown, T.L. and D.J. Darensbourg, 1967, Inorg. Chem. **6**, 971.
Burkstrand, J.M., G.G. Kleiman, G.G. Tibbetts and J.C. Tracy, 1976, J. Vac. Sci. Techn. **13**, 291.
Burton, J.J. and E. Hyman, 1975, J. Catal. **37**, 114.

Burton, J.J., C.R. Helms and R.S. Polizzotti, 1976, J. Chem. Phys. **65**, 1089.
Carroll, C.E. and J.W. May, 1972, Surface Sci. **29**, 60.
Cartner, A., B. Robinson and P.J. Gardner, 1973, JCS Chem. Comm., 317.
Chini, P. and S. Martinengo, 1969, Inorg. Chim. Acta **3**, 315.
Chini, P., G. Longoni and V.G. Albano, 1976, Adv. Organomet. Chem. **14**, 285.
Christmann, K. and G. Ertl, 1972, Surface Sci. **33**, 254.
Christmann, K. and G. Ertl, 1976, Surface Sci. **60**, 365.
Christmann, K., O. Schober and G. Ertl, 1974a, J. Chem. Phys. **60**, 4719.
Christmann, K., O. Schober, G. Ertl and M. Neumann, 1974b, J. Chem. Phys. **60**, 4528.
Christmann, K., G. Ertl and T. Pignet, 1976, Surface Sci. **54**, 365.
Clark, A., 1970, The theory of adsorption and catalysis (Academic Press, New York).
Clark, A., 1974, The chemisorptive bond: basic concepts (Academic Press, New York).
Collins, D.M. and W.E. Spicer, 1977, Surface Sci. **69**, 85.
Comrie, C.M. and W.H. Weinberg, 1976, J. Chem. Phys. **64**, 250.
Comrie, C.M., W.H. Weinberg and R.M. Lambert, 1976, Surface Sci. **57**, 619.
Conrad, H., G. Ertl and E.E. Latta, 1974a, Surface Sci. **41**, 435.
Conrad, H., G. Ertl, J. Koch and E.E. Latta, 1974b, Surface Sci. **43**, 462.
Conrad, H., G. Ertl, J. Küppers and E.E. Latta, 1975, Surface Sci. **50**, 296.
Conrad, H., G. Ertl, J. Küppers and E.E. Latta, 1976a, Surface Sci. **57**, 475.
Conrad, H., G. Ertl, H. Knözinger, J. Küppers and E.E. Latta, 1976b, Chem. Phys. Lett. **42**, 115.
Conrad, H., G. Ertl, J. Küppers and E.E. Latta, 1977a, Proc. 6th Int. congr. on Catalysis Chemical Society, London, 427.
Conrad, H., G. Ertl, J. Küppers and E.E. Latta, 1977b, Surface Sci. **65**, 235, 245.
Corey, E.R., L.F. Dahl and W. Beck, 1963, J. Amer. Chem. Soc. **85**, 1202.
Cotton, F.A., A.K. Fischer and G. Wilkinson, 1959, J. Amer. Chem. Soc. **81**, 800.
Cotton, F.A., L. Kruczynski, B.L. Shapiro and L.F. Johnson, 1972, J. Amer. Chem. Soc. **94**, 6191.
CRC Handbook of physics and chemistry, 1967, F 149, F 154.
Dalmon, J.A., M. Primet, G.A. Martin and B. Imelik, 1975, Surface Sci. **59**, 95.
Darling, J.H. and J.S. Ogden, 1972, Inorg. Chem. **11**, 666.
De Boer, J.H., 1953, The dynamic character of adsorption (Oxford University Press, Oxford).
Delchar, T.A. and G. Ehrlich, 1965, J. Chem. Phys. **42**, 2686.
Demuth, J.E., 1977, J. Colloid and Interface Sci. **58**, 184.
Demuth, J.E. and T.N. Rhodin, 1974, Surface Sci. **42**, 261.
Demuth, J.E., D.W. Jepsen and P.M. Marcus, 1973, Phys. Rev. Lett. **31**, 540.
Domke, M., G. Jahring and M. Drechsler, 1974, Surface Sci. **42**, 389.
Dowden, D.A., 1950, J. Chem. Soc., 242.
Dowden, D.A., 1973, Proc. 5th Int. Cong. on Catalysis (Hightower, J.T., ed.) (North-Holland, Amsterdam) Pt. 1, p. 621.
Doyen, G. and G. Ertl, 1974, Surface Sci. **43**, 197.
Doyen, G. and G. Ertl, 1977, Surface Sci. **69**, 157.
Doyen, G. and G. Ertl, 1978, submitted to J. Chem. Phys.
Doyen, G., G. Ertl and M. Plancher, 1975, J. Chem. Phys. **62**, 2957.
Ehrenreich, H. and L.M. Schwartz, 1976, Solid State Phys. **31**, 149.
Ehrlich, G., 1959, J. Chem. Phys. **31**, 1111.
Ehrlich, G. and F.G. Hudda, 1966, J. Chem. Phys. **44**, 1039.

Eischens, R.P. and W.A. Pliskin, 1958, Advances in Catalysis **10**, 1.
Eley, D.D., 1954, Catalysis and chemical bond (Univ. of Notre Dame Press, Indiana).
Eley, D.D. and A. Couper, 1950, Disc. Faraday Soc. **8**, 172.
Ellis, W.P. and R.L. Schwoebel, 1968, Surface Sci. **11**, 82.
Ertl, G., 1969, in: Molecular processes on solid surfaces (Drauglis, Gretz, Jaffee, eds.) (McGraw Hill, New York) p. 147.
Ertl, G., 1976, unpublished.
Ertl, G., 1977, J. Vac. Sci. Techn. **14**, 435.
Ertl, G. and J. Koch, 1970, Z. Naturforsch. **25a**, 1906.
Ertl, G. and J. Küppers, 1970, Surface Sci. **21**, 61.
Ertl, G. and M. Plancher, 1975, Surface Sci. **48**, 364.
Ertl, G. and D. Schillinger, 1977, J. Chem. Phys. **66**, 2569.
Ertl, G., M. Neumann and K.M. Streit, 1977, Surface Sci. **64**, 393.
Estrup, P.J., 1969, in: The structure and chemistry of solid surfaces (Somorjai, G.A., ed.) (Wiley, New York) 19–1.
Everett, D.H., 1965, Disc. Faraday Soc. **40**, 177.
Fassaert, D.J.M. and A. van der Avoird, 1976, Surface Sci. **55**, 291.
Fowler, R. and E.A. Guggenheim, 1956, Statistical mechanics (Cambridge University Press, London).
Froitzheim, H., H. Ibach and S. Lehwald, 1976a, Phys. Rev. **B14**, 1362.
Froitzheim, H., H. Ibach and S. Lehwald, 1976b, Phys. Rev. Lett. **36**, 1549.
Froitzheim, H., H. Ibach and S. Lehwald, 1976c, Proc. Int. Symp. on Photoemission (Nordwijk) (R.F. Willis et al., eds.) p. 277.
Froitzheim, H., H. Hopster, H. Ibach and S. Lehwald, 1977, Appl. Phys. **13**, 147.
Fuggle, J.C., M. Steinkilberg and D. Menzel, 1975, Chem. Phys. **11**, 307.
Garland, C.W., R.C. Lord and P.F. Triano, 1965, J. Phys. Chem. **69**, 1188.
Gelatt, C.D. and H. Ehrenreich, 1974, Phys. Rev. **B10**, 398.
Gerlach, R.L. and T.N. Rhodin, 1968, Surface Sci. **10**, 446.
Gerlach, R.L. and T.N. Rhodin, 1969, Surface Sci. **17**, 32.
Gerlach, R.L. and T.N. Rhodin, 1970, Surface Sci. **19**, 403.
Goddard, W.H., 1977, J. Vac. Sci. Techn. **14**, 416.
Gomer, R., 1973, Surface Sci. **38**, 373.
Gomer, R., 1975, Solid State Phys. **30**, 94.
Gomer, R., R. Wortmann and R. Lundy, 1957, J. Chem. Phys. **26**, 1147.
Graham, W.R. and G. Ehrlich, 1973, Phys. Rev. Lett. **31**, 1407.
Graham, W.R. and G. Ehrlich, 1974, Phys. Rev. Lett. **32**, 1309.
Greenler, R.G., 1969, J. Chem. Phys. **50**, 1963.
Gunnarsson, O., H. Hjelmberg and B.I. Lundqvist, 1976, Phys. Rev. Lett. **37**, 292.
Gurman, S.J. and J.B. Pendry, 1976, Solid State Comm. **20**, 287.
Gustafsson, T., E.W. Plummer, D.E. Eastman and J.L. Freeouf, 1975, Solid State Comm. **17**, 341.
Gutmann, V., 1975, Coord. Chem. Rev. **15**, 207.
Hagen, D.I., B.E. Nieuwenhuys, G. Rovida and G.A. Somorjai, 1976, Surface Sci. **57**, 632.
Hair, M.L., 1967, Infrared spectroscopy in surface chemistry (Marcel Dekker, New York).
Hall, W.K. and P.H. Emmett, 1959, J. Phys. Chem. **63**, 1102.
Harrod, J.H., R.W. Roberts and E.F. Rissmann, 1967, J. Phys. Chem. **71**, 343.
Hayward, D.O. and B.M.W. Trapnell, 1964, Chemisorption (Butterworths, London).
Henzler, M., 1970, Surface Sci. **19**, 159.
Hoffmann, R., 1971, Acc. Chem. Res. **4**, 1.

Honig, J.M., 1961, Adv. Chem. **33**, 239.
Hopster, H., H. Ibach and G. Comsa, 1977, J. Catalysis **46**, 37.
Huber, H., P. Kündig, M. Moskovits and G.A. Ozin, 1972, Nature Phys. Sci. **235**, 98.
Hüfner, S., G.K. Wertheim, R.L. Cohen and J.H. Wernick, 1972, Phys. Rev. Lett. **28**, 488.
Hulse, J.E. and M. Moskovits, 1976, Surface Sci. **57**, 125.
Ibach, H., 1972, J. Vac. Sci. Techn. **9**, 713.
Ibach, H., 1975, Surface Sci. **53**, 444.
Ibach, H., K. Horn, H. Lüth and R. Dorn, 1973, Surface Sci. **38**, 433.
Iwasawa, Y., R. Mason, M. Textor and G.A. Somorjai, 1977, to be published.
Joebstl, J.A., 1975, J. Vac. Sci. Techn. **12**, 347.
Jones, L.H., R.S. McDowell and M. Goldblatt, 1968, J. Chem. Phys. **48**, 2663.
Jones, L.H., R.S. McDowell and M. Goldblatt, 1969, Inorg. Chem. **8**, 2349.
Kishi, K. and M.W. Roberts, 1977, Surface Sci. **62**, 252.
Kleint, C., 1971, Surface Sci. **25**, 394.
Küppers, J. and A. Plagge, 1976, J. Vac. Sci. Techn. **13**, 259.
Lander, J.J., 1964, Surface Sci. **1**, 125.
Landman, U. and D.L. Adams, 1976, Proc. Nat. Acad. Sci. (USA) **73**, 2550.
Lang, B., R.W. Joyner and G.A. Somorjai, 1972, Surface Sci. **30**, 440, 454.
Lang, N.D., 1971, Phys. Rev. **B4**, 4234.
Lang, N.D. and H. Ehrenreich, 1968, Phys. Rev. **168**, 605.
Langmuir, I., 1918, J. Amer. Soc. **40**, 1361.
Lennard-Jones, J.E., 1932, Trans. Faraday Soc. **28**, 28.
Lewis, R. and R. Gomer, 1969, Surface Sci. **17**, 333.
Little, L.H., 1966, Infrared spectra of adsorbed species (Academic Press, New York).
Liu, R. and G. Ehrlich, 1976, J. Vac. Sci. Techn. **13**, 310.
Lu, K.E. and R.R. Rye, 1974, Surface Sci. **45**, 677.
Lytle, F.W., D.E. Sayers and E.A. Stern, 1975, Phys. Rev. **B11**, 4825.
Mahnig, M. and L.D. Schmidt, 1972, Z. Phys. Chem. N.F. **80**, 71.
Marcus, P.M., J.E. Demuth and D.W. Jepsen, 1975, Surface Sci. **53**, 501.
Mathieu, M.V. and M. Primet, 1976, Surface Sci. **58**, 511.
McLachlan, A.D., 1964, Mol. Phys. **1**, 381.
Mott, N.F. and H. Jones, 1958, The theory of the properties of metals and alloys (Dover, New York).
Muetterthies, E.L., 1975, Bull. Soc. Chim. Belg. **84**, 959.
Nelson, G.C., 1976, Surface Sci. **59**, 310.
Neumann, M., 1977, personal communication.
Nieuwenhuys, B.E., D.I. Hagen, G. Rovida and G.A. Somorjai, 1976, Surface Sci. **59**, 155.
Noack, K., 1964, Chem. Ber. **47**, 1064.
Noack, K., 1967, J. Organomet. Chem. **7**, 151.
Palmberg, P.W., 1971, Surface Sci. **25**, 598.
Park, R.L. and H.H. Madden, 1968, Surface Sci. **11**, 158.
Pauling, L., 1939, The nature of the chemical bond (Cornell Univ. Press, Ithaca).
Pauling, L., 1949, Proc. Roy. Soc. **A196**, 343.
Perdereau, J. and G.E. Rhead, 1971, Surface Sci. **24**, 555.
Polizzotti, R.S. and G. Ehrlich, 1975, Bull. Am. Phys. Soc. **20**, 857.
Potter, H.C. and J.M. Blakely, 1975, J. Vac. Sci. Techn. **12**, 635.
Pritchard, J., 1972, J. Vac. Sci. Techn. **9**, 895.
Propst, F.M. and T.C. Piper, 1967, J. Vac. Sci. Techn. **4**, 53.
Sachtler, W.M.H., 1971, J. Vac. Sci. Techn. **9**, 828.

Sachtler, W.M.H., 1973, Le Vide **164**, 67.
Sachtler, W.M.H. and G.J.H. Dorgelo, 1965, J. Catal. **4**, 654.
Sachtler, W.M.H. and R. Jongepier, 1965, J. Catal. **4**, 665.
Sachtler, W.M.H. and P. van der Plank, 1969, Surface Sci. **18**, 62.
Sachtler, W.M.H. and R.A. van Santen, 1977, Adv. Catalysis **26**, 69.
Sams, J.R., G. Constabaris and G.D. Halsey, 1962, J. Chem. Phys. **46**, 1334.
Sayers, D.E., E.A. Stern and J.R. Herrich, 1976, J. Chem. Phys. **64**, 427.
Schmidt, L.D., 1974, Catal. Rev. Sci. Eng. **9**, 115.
Schmidt, L.D. and R. Gomer, 1965, J. Chem. Phys. **42**, 4573.
Schmidt, L.D. and R. Gomer, 1966, J. Chem. Phys. **45**, 1605.
Schwab, G.M., 1955, Angew. Chemie **67**, 433.
Schwab, G.M. and E. Pietsch, 1928, Z. Phys. Chem. **B1**, 385.
Seib, D.H. and W.E. Spicer, 1970, Phys. Rev. **B2**, 1694.
Sinanoglu, O. and K.S. Pitzer, 1960, J. Chem. Phys. **32**, 1279.
Sinfelt, J.H., J.L. Carter and D.J.C. Yates, 1972, J. Catal. **24**, 283.
Smith, J.R., S.C. Ying and W. Kohn, 1974, Solid State Comm. **15**, 1491.
Snodgrass, R.J., 1971, Phys. Rev. **B1**, 3738.
Soma-Noto, Y. and W.M.H. Sachtler, 1974, J. Catal. **32**, 315.
Somorjai, G.A., 1975, The physical basis for heterogeneous catalysis (Drauglis, E. and R.I. Jaffee, eds.) (Plenum Press, New York) p. 395.
Somorjai, G.A., 1977, Adv. Catalysis **26**, 2.
Steele, W.A., 1974, The interaction of gases with solid surfaces (Pergamon Press, Oxford).
Stephan, J.J., P. L. Franke and V. Ponec, 1976, J. Catal. **44**, 359.
Stewart, C.N. and G. Ehrlich, 1975, J. Chem. Phys. **62**, 4672.
Stocks, G.M., R.W. Williams and J.S. Faulkner, 1971, Phys. Rev. **B4**, 4390.
Stocks, G.M., R.W. Williams and J.S. Faulkner, 1973, J. Phys., **F3**, 1688.
Suzanne, R., J.P. Coulomb and M. Bienfait, 1973, Surface Sci. **40**, 414.
Takaishi, T., 1958, Z. Phys. Chem. N.F. (Frankfurt) **14**, 164.
Takasu, T. and J. Yamashima, 1973, J. Catal. **28**, 171.
Tamm, P.W. and L.D. Schmidt, 1969, J. Chem. Phys., **51**, 5352.
Tamm, P.W. and L.D. Schmidt, 1970, J. Chem. Phys. **52**, 1150.
Tamm, P.W. and L.D. Schmidt, 1971, J. Chem. Phys. **54**, 4775.
Taylor, H.S., 1926, J. Phys. Chem. **30**, 145.
Taylor, H.S., 1931, J. Amer. Chem. Soc. **53**, 578.
Topping, J., 1927, Proc. Roy. Soc. **A114**, 67.
Tracy, J.C., 1972, J. Chem. Phys. **56**, 2736.
Tracy, J.C. and P.W. Palmberg, 1969, J. Chem. Phys. **51**, 4852.
Van der Plank, P. and W.M.H. Sachtler, 1967, J. Catal. **7**, 300.
Van der Plank, P. and W.M.H. Sachtler, 1968, J. Catal. **12**, 25.
Van Hove, M., G. Ertl, W.H. Weinberg, J. Behm and K. Christmann, 1978, in preparation.
Van Hove, M. and S.Y. Tong, 1975, J. Vac. Sci. Techn. **12**, 230.
Van Santen, R.A., 1975 Surface Sci. **53**, 35.
Wandelt, K. and G. Ertl, 1975, Ber. Bunsenges. **79**, 1101.
Wandelt, K. and G. Ertl, 1976, Z. Naturforsch. **31a**, 205.
Wang, J.S., 1937, Proc. Roy. Soc. **A161**, 127.
Weinberg, W.H., R.M. Lambert, C.M. Comrie and J.W. Linnett, 1973, Proc. 5th int. cong. on Catalysis (J.W. Hightower, ed.) (North-Holland, Amsterdam p. 519.
Winters, H.F., 1976, J. Chem. Phys. **64**, 3495.
Yates, J.T., R.G. Greenler, I. Ratajczykowa and D.A. King, 1973, Surface Sci. **36**, 739.
Yu, K.Y., D.T. Ling and W.E. Spicer, 1976, J. Catal. **44**, 373.

SUBJECT INDEX

*table

**table*

**table*

*table

*table

*table

*table

**table*

*_table_

*table

*table

AUTHOR INDEX*

* *Numbers in brackets indicate quotation in reference lists or tables.*